AF509985

POPULATION BALANCE OF PARTICLES IN FLOWS

The population balance methodology provides a powerful framework for studying polydisperse entities such as aerosols, crystals and bubbles. This self-contained and accessible book explains how this theoretical framework can be employed across a wide range of scientific, engineering and environmental problems. The methodology is explained step by step, showing readers how to use these techniques by formulating the population balance problem, choosing models and implementing appropriate solution methods. Particular focus is given to the coupling of the population balance with fluid mechanics and computational fluid dynamics (CFD), in both laminar and turbulent flows. Applications of the population balance methodology are explored in case studies including nanoparticle synthesis, soot formation and crystallisation, and sample open-source code is provided. This book will be valuable to researchers across a range of disciplines including chemical and mechanical engineering, physics and environmental science and can be used as a resource for advanced undergraduate and graduate courses.

STELIOS RIGOPOULOS is Reader in Thermofluids at Imperial College London. He has conducted research on the population balance methodology and its applications for over 20 years, receiving a Royal Society University Research Fellowship for his work. His group has pioneered methods for solving the population balance and coupling it with fluid dynamics, including turbulent flow, with applications that included aerosols, crystallisation, soot formation and nanoparticle synthesis.

POPULATION BALANCE OF PARTICLES IN FLOWS

From Aerosols to Crystallisation

STELIOS RIGOPOULOS
Imperial College London

CAMBRIDGE
UNIVERSITY PRESS

CAMBRIDGE
UNIVERSITY PRESS

Shaftesbury Road, Cambridge CB2 8EA, United Kingdom

One Liberty Plaza, 20th Floor, New York, NY 10006, USA

477 Williamstown Road, Port Melbourne, VIC 3207, Australia

314–321, 3rd Floor, Plot 3, Splendor Forum, Jasola District Centre, New Delhi – 110025, India

103 Penang Road, #05–06/07, Visioncrest Commercial, Singapore 238467

Cambridge University Press is part of Cambridge University Press & Assessment, a department of the University of Cambridge.

We share the University's mission to contribute to society through the pursuit of education, learning and research at the highest international levels of excellence.

www.cambridge.org
Information on this title: www.cambridge.org/9781316512579

DOI: 10.1017/9781009067829

© Stelios Rigopoulos 2024

This publication is in copyright. Subject to statutory exception and to the provisions of relevant collective licensing agreements, no reproduction of any part may take place without the written permission of Cambridge University Press & Assessment.

When citing this work, please include a reference to the DOI 10.1017/9781009067829

First published 2024

A catalogue record for this publication is available from the British Library

Library of Congress Cataloging-in-Publication Data
Names: Rigopoulos, Stelios, 1974– author.
Title: Population balance of particles in flows : from aerosols to crystallisation / Stelios Rigopoulos.
Description: Cambridge ; New York, NY : Cambridge University Press, 2024. | Includes bibliographical references and index.
Identifiers: LCCN 2024006572 | ISBN 9781316512579 (hardback) | ISBN 9781009067829 (ebook)
Subjects: LCSH: Granular flow – Mathematical models. | Particle dynamics. | Computational fluid dynamics. | Discretization (Mathematics) | Stochastic processes – Mathematical models.
Classification: LCC QA911 .R54 2024 | DDC 532/.052–dc23/eng/20240830
LC record available at https://lccn.loc.gov/2024006572

ISBN 978-1-316-51257-9 Hardback

Additional resources for this publication at www.cambridge.org/9781316512579

Cambridge University Press & Assessment has no responsibility for the persistence or accuracy of URLs for external or third-party internet websites referred to in this publication and does not guarantee that any content on such websites is, or will remain, accurate or appropriate.

Dedicated to my parents, Vasileios and
Ermioni, and to my wife, Hsing-fen, with gratitude.

Contents

Preface

Aims and Content of the Book

The population balance is a methodology applicable to a wide range of problems of scientific, engineering and environmental interest. These include atmospheric aerosols, nanoparticle synthesis, crystallisation, cloud formation, sprays, bubble flow, granulation, biochemical engineering, polymerisation and astrophysics. In all of these problems, it is desired to predict the distribution of properties (such as particle size) of a population of entities such as aerosol particles, colloids and droplets. The interest in these properties may arise from a need to tailor them to an application (as in the case of functional nanoparticles), to mitigate their environmental impact (as for a harmful aerosol such as soot) or to build models for a process for which these properties are of importance (such as a surface reaction involving the particle surface area). The evolution of the properties is shaped by a number of processes such as aggregation, growth and transport, and the population balance is an approach for building models that link the evolution of the desired distribution to these processes.

The objective of this book is to present the population balance as a general methodology, to explain the details of its deployment and to show how it can be applied via a number of case studies. In particular, the book discusses the formulation of population balance models, the physical and chemical models that act as constitutive equations, the solution methods available for the population balance equation (PBE) and, finally, the coupling of PBE with fluid dynamics, which includes the effect of turbulence and the implementation of the PBE in computational fluid dynamics (CFD) codes. The methods, which are presented in the first five chapters of this book, are demonstrated in the final chapter via a number of case studies drawn from the research of the author's group, in problems of nanoparticle synthesis, soot formation and crystallisation. All of these feature a spatially dependent population balance and show the potential of the coupled fluid dynamics–population balance approach.

The methodologies employed in the various fields where the population balance is applied are often not described under the term 'population balance'. For example, the term General Dynamic Equation is usually employed in aerosol science, while in other fields, the name 'Smoluchowski equation' is used in honour of one of the pioneering figures in this field. One of the main objectives in the present book is to describe these methodologies under a unified perspective. The term 'population balance' has been chosen both because of its increasing use across several fields and of its descriptive nature.

One of the main developments in the field of population balance modelling during the last two decades has been the coupling of the PBE with fluid dynamics. The developments in CFD have enabled the emergence of coupled CFD-PBE simulations with a variety of methods. One of the aims of the present book is to communicate these developments, with special attention paid to turbulent flow, which raises unique modelling challenges.

While the formulation is presented from a unified perspective, the particular models will be motivated by the case studies in the last chapter. This is necessitated by the very wide range of problems that can be approached by the population balance methodology, each of which requires its own set of physical, chemical and sometimes biological models to be incorporated. It is hoped that workers in other fields will find relevant material in the discussion of the formulations and solution methods, which are quite general.

The book's approach is sufficiently general to render it relevant to workers from several disciplines including chemical engineering, mechanical engineering, environmental engineering, physics and chemistry, while some parts cover in more detail focussed subject areas such as aerosol science and crystallisation. The book has been written to serve a dual purpose, as a textbook for advanced undergraduate and graduate students, as well as a research monograph.

Structure of the Book

The content of the book is organised in the following way.

In Chapter 1, the population balance is presented as a general methodology, along with its range of applications. Several of the main concepts and terminology that will be used throughout the book are introduced. A brief discussion of preliminary concepts, such as distributions and their properties, are also included.

Chapter 2 presents a unifying framework for a hierarchy of population balance formulations. These include spatially homogeneous and inhomogeneous, one-dimensional and multidimensional or coupled (multiple) PBEs, as well as simplified monodisperse models. The coupling of PBE with CFD is also discussed.

The physical and chemical models that act as constitutive equations are discussed in Chapter 3. Emphasis is, as mentioned, on the models used in the case studies in this book. The inverse problem, that is, the determination of kinetic parameters from experiments, is also discussed.

Chapter 4 discusses the solution methods that have been developed for the various forms of the PBE. At first, the methods are described in the context of the spatially homogeneous PBE. The coupling of the PBE with fluid dynamics and CFD is addressed in the later sections.

Chapter 5 discusses certain unique theoretical and computational problems that are introduced into population balance modelling by the presence of turbulence.

Finally, in Chapter 6, the theory and methodology presented in the previous chapters is applied to three case studies, chosen from the research of the author's group. The topics are silica nanoparticle synthesis in a laminar flame, soot formation in laminar and turbulent flames and precipitation in turbulent flow. They all involve coupling of PBE and fluid dynamics, thus bringing together the book's main themes.

Use of the Book for Courses

Parts of the book are intended to function as material for courses on population balance. Such a course could be offered at an advanced undergraduate or graduate level. The knowledge required is a typical undergraduate science or engineering background, including mathematics (elements of calculus and differential equations) and elementary fluid dynamics. All methods shown in the book are developed from scratch. Overlap is avoided in topics that have been covered in detail elsewhere, but several appendices are provided in some cases.

The material that could form the basis for a course consists mainly of Chapters 1–3 and parts of Chapter 4. Chapters 1 sets the stage, while Chapter 2 describes the various formulations of the methodology. The instructor may focus on parts of Chapter 3, depending on the focus of the course. The range and coverage of solution methods in Chapter 4 is probably too detailed for a course, but an overview and a few elementary ones (such as the solution of the Smoluchowski equation) would be needed to give the student a complete idea of how to formulate and solve a problem with the population balance methodology.

Accompanying Code

A suite of codes implementing some of the methods discussed in this book can be found at a link provided in the online resources for the book:

https://www.cambridge.org/9781316512579#resources

The codes are in Fortran 90 and are open source.

Acknowledgements

A number of people have contributed to the research that led to this book, to whom I wish to express my heartfelt thanks. These include my former and current PhD students Giovanni di Veroli, Petros Akridis, Fabian Sewerin, Anxiong Liu, Binxuan Sun, Hin Yan Tang, Malamas Tsagkaridis and Daniel O'Sullivan, as well as my PhD supervisor, Alan Jones, and my collaborators Bill Jones, Salvador Navarro-Martinez and George Papadakis. I am also indebted to several people who proofread all or parts of the manuscript and provided valuable feedback. These include Fabian Sewerin, Bill Jones, Salvador Navarro-Martinez, Li Shen, Yannis Drossinos, Luc Vervisch, Bert Diemer, Spyros Samaras, Pavlos Aleiferis, Sarkis Shadarevian and George Efstathiou. I also received useful comments and suggestions from Nondas Mastorakos, Sergey Utyuzhnikov, Maxime Lombart and George Papadakis. Finally, I am grateful to Li Shen for his excellent work on the cover image and index for the book.

The University Research Fellowship that I received from the Royal Society in 2005 was instrumental for the early development of the ideas that have ultimately found their way into this book. Subsequently, my group's research in this field has been supported by grants from the Engineering and Physical Sciences Research Council (EPSRC) and the Leverhulme Trust, as well as by scholarships by Imperial College London and the China Scholarship Council (CSC), to all of which I wish to express my sincere gratitude. I also wish to thank the staff from Cambridge University Press for their valuable assistance in bringing the book to completion.

Both the book itself and the research that led to it would not have been made possible without the support of my parents, Vasilios and Ermioni, my sister, Maria, and my wife, Hsing-fen, particularly in the difficult COVID-19 years during which much of the book was written.

Notation

The following is a brief description of the mathematical notations that will be employed in the present book. All symbols are defined when first encountered.

Cartesian Tensor Notation

- Employed for equations written in Cartesian coordinates.
- Indices (1,2,3) refer to the three Cartesian coordinates (x, y, z), respectively.
- A free index (i.e. one that is not repeated) indicates the set of all components of a vector or a matrix, or a system of equations for all values of the index, for example:

$$u_i = \begin{bmatrix} u_1 \\ u_2 \\ u_3 \end{bmatrix}, \quad u'_i u'_j = \begin{bmatrix} u'_1 u'_1 & u'_1 u'_2 & u'_1 u'_3 \\ u'_2 u'_1 & u'_2 u'_2 & u'_2 u'_3 \\ u'_3 u'_1 & u'_3 u'_2 & u'_3 u'_3 \end{bmatrix}.$$

- An index repeated twice within the same term indicates summation over all values of that index, for example:

$$\frac{\partial u_i}{\partial x_i} = \frac{\partial u_1}{\partial x_1} + \frac{\partial u_2}{\partial x_2} + \frac{\partial u_3}{\partial x_3}.$$

  Each index does not appear more than twice within the same term.

Boldface Notation

- Used when we need to refer to an entire vector with a single symbol, usually inside a function, for example:

$$u_i(\mathbf{x}, t) = u_i(x_1, x_2, x_3, t).$$

Component Notation

- Used when we wish to distinguish between the coordinates and emphasise their physical meaning, for example:

$$\frac{dp}{dz} = -\rho g.$$

  Also used when employing non-Cartesian coordinates.
- The Cartesian components of the velocity vector are indicated by (u, v, w) for the (x, y, z) coordinates, respectively.

1

Introductory Concepts

1.1 Overview of the Population Balance Methodology

1.1.1 Introduction

A number of physical problems are described by a population of entities with a distribution of one or more properties. Some examples are the distributions of sizes and shapes in crystals, molecular weights in polymers and ages in microbial cells. The evolution of these properties may be shaped by processes such as particle formation and aggregation, as well as by transport within a flow field.

There are several reasons why we may be interested in this distribution. It may determine the properties of a product, as in the case of the size distribution of purpose-made nanoparticles. By controlling it, we can tailor the product to particular applications. In other cases, it may determine the impact of the entities on the environment or human health, as in the case of the size distribution of soot and other aerosols, where the smaller particles can penetrate deeper into the body when inhaled. Finally, it may be an important process variable and thus essential for developing a model of the process; for instance, the surface area distribution of crystals in a reaction crystallisation process determines the rate of a surface reaction.

The population balance is a general methodology for describing systems with distributed properties. We will often use the term 'particle' in the present book with the understanding that it may refer to a solid particle, droplet or bubble. The objective of a population balance model is to predict the distribution of the properties of interest by linking it to the physical and chemical processes that shape it. These processes include interaction of particles between themselves, such as aggregation into larger particles or fragmentation into smaller ones, interaction of particles with their environment, such as particle formation from a precursor or disintegration into it, and transport of particles by processes such as convection by a carrier fluid and Brownian motion.

In problems involving transport in physical space, the population balance is intricately connected with fluid dynamics. This is because fluid dynamics, apart from controlling the transport of particles, also 'set the stage' for the physical and chemical processes that appear in the population balance by, for example, determining the local concentration of a chemical species that acts as a precursor for particle formation.

The main purpose of this introductory chapter is to define the scope of the population balance methodology and to introduce certain basic concepts and terminology that will be used in the rest of the book. Before commencing, we first take a brief look at some of the problems to which the population balance can be applied.

1.1.2 Applications of the Population Balance

The processes involved in the population balance, such as the merging of two units into one or the breaking of a unit into fragments, appear in many different (and sometimes seemingly unrelated) problems. While the laws governing these processes may be problem specific, they share enough common features to allow a unified description under the population balance framework. For example, depending on their size range, the aggregation of small aerosol particles may be governed by a collision model based on the kinetic theory of gases, while the coalescence of larger droplets may be due to a mechanism determined by the flow field. As we will see in Chapter 3, both phenomena can be described with the same population balance equation, with the different mechanistic models entering as constitutive equations. A number of representative applications are briefly described below. While only a few of these will feature in the present book, each of them has an extensive literature of its own, and selected references are mentioned to provide an entry point to that literature.

Atmospheric Aerosols

Aerosols are small solid particles or droplets suspended in a gas, with diameters typically in the range of 1 nm–100 μm (although this is not strict). Atmospheric aerosols arise from both natural sources and human activities. They can exhibit a variety of sizes, morphologies and chemical compositions which determine their physical and chemical properties and, consequently, their environmental impact, which includes effects on both human health and climate change. Kreyling et al. (2006) discuss the effect of particle size on particle–lung interactions, while Stettler et al. (2013a,b) and Zhang et al. (2019) focus on aviation black carbon emissions and the importance of particle size.

A number of significant questions about aerosols can be answered by the population balance – for example, how do the distributions of their properties evolve, how are they dispersed in the atmosphere and what is their state when they reach the ground and affect populated areas. The size distribution plays also a role on their action in activating cloud condensation (Calderón et al., 2022). Population balance models of aerosols constitute a major part of atmospheric dispersion simulations, where they are coupled with models for atmospheric flow. Reviews of aerosol science with emphasis on distributions and their modelling can be found in Hidy and Brock (1970), Williams and Loyalka (1991), Friedlander (2000) and Seinfeld and Pandis (2016).

Cloud and Rain Formation

Cloud and rain formation occurs via collision, coalescence and break-up of droplets. The evolution of the drop size distribution in the atmosphere was thus one of the first applications of population balance modelling, as exemplified by early works such as Warshaw (1967). An account of research in this area can be found in Chapter 15 of Pruppacher and Klett (1996). The interaction of aerosols and clouds is also important, as mentioned in the previous paragraph.

Volcanic Ash

The ash resulting from volcanic eruptions poses a major hazard for aeroplanes, and therefore the prediction of the evolution of ash clouds is a major task for meteorologists. Apart from the transport and dispersion of the ash plume in the atmosphere, models have to account for aggregation, which gives rise to bigger particles with different transport properties. The combination of aggregation with transport of volcanic ash is a typical problem that involves coupled population balance and fluid dynamics. Reviews of models for volcanic ash dispersion and aggregation can be found in Brown et al. (2012) and Beckett et al. (2020), while Suman et al. (2019) deal with the impact of ash (alongside that of other particles) on aircraft engines. Fig. 1.1 shows images of volcanic ash particles, where the effect of aggregation is evident.

Soot and Carbonaceous Nanoparticles

Soot is particulate material consisting mainly of carbon (with a small amount of H and other compounds present in the fuel). It is formed in combustion processes such as engines, gas turbines and furnaces. Soot is a special case of aerosol particles and has severe impacts on human health that depend to a large extent on particle size. Smaller particles penetrate into the lungs with potentially adverse effects (Kreyling et al., 2006). For this reason, the minimisation of soot formation is a major objective of designers of internal combustion engines and gas turbines. At the same time, carbon black (another form of particulate carbon) is manufactured for use in tires, inks, batteries and solar cells, while new purpose-made

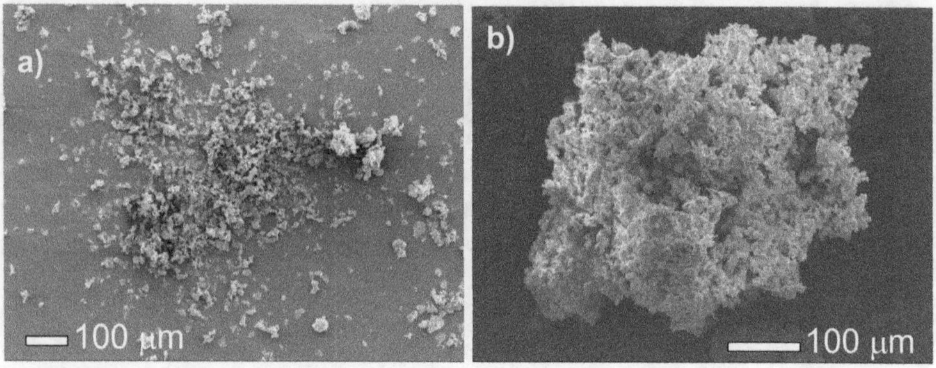

Figure 1.1 SEM images of ash aggregates: (a) a broken ash cluster and (b) an ash cluster. Reprinted from Bonadonna et al. (2011).

carbonaceous nanoparticles can have exceptionally high value; these include carbon quantum dots and carbon-coated nanoparticles for use as magnetic biofluids (Kelesidis et al., 2017). The size and morphology of such particles determine their suitability for particular applications.

The formation of soot and carbonaceous nanoparticles is very complex and its prediction requires a combination of fluid dynamics, chemistry and population balance modelling. Earlier work relied on simplified models and has been reviewed by Kennedy (1997). More recently, detailed population balance models have started to be incorporated into turbulent combustion models and reviews can be found in Raman and Fox (2016) and Rigopoulos (2019).

Fig. 1.2 shows images of soot particles obtained from a laminar flame. It is evident that the particles exhibit a range of size and morphologies. The picture on the right shows a close-up of an aggregate with a fractal structure, a feature that will be discussed in Section 3.2.3. The simulation of sooting flames is the objective of the case study in Section 6.2.

Nanoparticle Synthesis

Engineered nanoparticles such as silica and titania have multiple applications such as in pigments and optical fibers, while carbonaceous nanoparticles were discussed in the previous paragraph. The value of such products depends on particle size and morphology, and therefore the ability to control these properties yields the potential for manufacturing tailor-made nanoparticles for specialised uses. Population balance modelling can provide a predictive approach of the outcome of aerosol synthesis based on the process and equipment design. Reviews of nanoparticle synthesis can be found in Pratsinis (1998), Kruis et al. (1998), Kodas and Hampden-Smith (1999) and Pratsinis (2010), while Raman and Fox

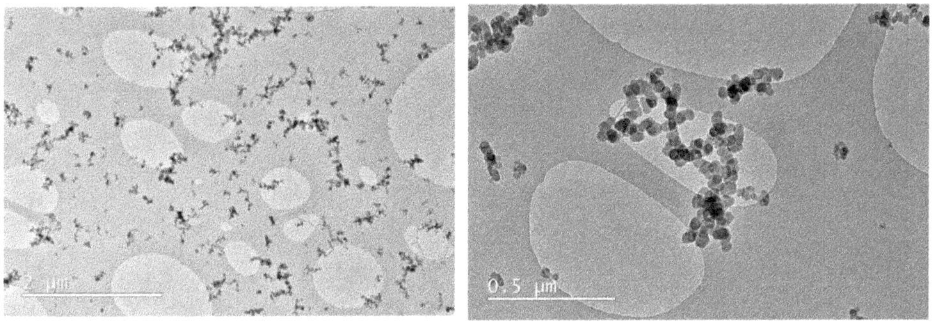

Figure 1.2 Micrographs of soot particles obtained via transmission electron microscopy (TEM); large-scale view (left) and close-up on an aggregate (right). Note that the big 'holes' are from the carbon film. Courtesy of Garcia Gonzalez (2018).

(2016) discuss the modelling of the process based on population balance and fluid dynamics.

The formation of silica particles via flame synthesis is the objective of the case study in Section 6.1.

Metal Particles as Energy Carriers

Metal particles such as aluminium have many applications, including their use as recyclable energy carriers and thus carbon-free alternatives to fossil fuels (Bergthorson, 2018). The size distribution of the oxide smoke is of primary importance, both as a process variable and for its role in the design of the subsequent separation processes. A comprehensive population balance model of this process can be found in Finke and Sewerin (2023).

Crystallisation

Crystallisation and precipitation processes are widely employed in the chemical and pharmaceutical industries for the formation of crystalline products from solutions. The size distribution and morphology of the crystals produced determine their properties and suitability for particular applications, as well as their behaviour during separation processes. Fig. 1.3 shows images of $CaCO_3$ crystals obtained from a precipitation process; the presence of both single crystals and agglomerates can be noted.

The modelling of crystallisation is one of the oldest applications of population balance modelling, having received its first detailed exposition in Randolph and Larson (1971) (see Randolph and Larson, 1988 for the latest edition). More recent reviews can be found in Mersmann (2001), Mullin (2001), Lewis et al. (2015) and Myerson et al. (2019). While the aforementioned literature refers to crystallisation

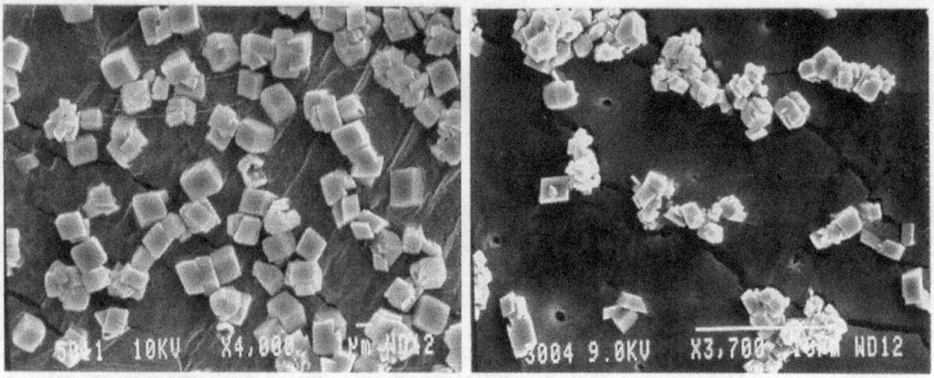

Figure 1.3 Scanning electron microscopy (SEM) images of $CaCO_3$ crystals produced via gas–liquid precipitation at an early (left) and a late (right) stage of the process. Reprinted from Rigopoulos and Jones (2003b) with permission from American Chemical Society.

and precipitation from solution, there are also processes involving precipitation of grains from supersaturated solutions, as discussed in the classic work of Lifshitz and Slyozov (1961).

The application of population balance to crystallisation will be shown in detail in the context of the case study in Section 6.3.

Spray Dynamics

Sprays are encountered in applications such as fuel injection and inhalation of medicines. In combustion, the size of the droplets determines the surface area and, therefore, the rate of evaporation, on which combustion depends. In medicinal sprays, the droplet size distribution determines whether the spray will penetrate and reach the targeted deposition sites in the lungs. The prediction of this distribution requires population balance models of spray break-up and evaporation, coupled with fluid dynamics that describe droplet dispersion. A comprehensive discussion of sprays can be found in Sirignano (2010).

Disease Transmission via Aerosol Droplets

The cloud of droplets resulting from a cough or sneeze is akin to a spray and can be analysed with similar experimental and modelling tools. The droplet size distribution determines their transport properties and therefore how far they reach, which is very important in the transmission of diseases such as COVID-19 (Bourouiba, 2020). For a modelling perspective on the link between droplet physics and disease transmission, one may consult Stilianakis and Drossinos (2010), Robinson et al. (2012), Drossinos and Stilianakis (2020), De Oliveira et al. (2021) and Drossinos et al. (2022).

Colloid Dynamics and Flocculation

The flocculation of suspended matter in waterways and in water treatment plants is of great importance for the destabilisation and treatment of particulate matter. Flocculation is induced by particle collisions and the ensuing size distribution determines the settling properties of the flocs. As such, it is a classical application of population balance modelling and its coupling with hydrodynamics. Reviews can be found in Thomas et al. (1999) and Partheniades (2009).

Asphaltene Fouling

Asphaltenes are carbonaceous compounds of moderately high molecular weight that are present in crude oil. As these compounds aggregate into colloidal particles, they form deposits that result in fouling of oil pipelines and result in heavy costs to the oil industry. An analysis of the formation of asphaltenes with population balance modelling can be found in Vilas Bôas Fávero et al. (2017).

Bubble Flows

In bubble flows, the population of bubble sizes determines the rates of interfacial processes such as mass transfer, as well as the momentum exchange. Reviews of population balance modelling of bubble column reactors can be found in Jakobsen et al. (2005) and in Chapter 8 of Jakobsen (2014).

Nuclear Engineering

Nuclear accidents result in the emission of aerosols with significant amounts of radioactivity. The prediction of such emissions and their impact is required for assessing nuclear safety. Models for such predictions combine population balance modelling with nuclear reactor thermal hydraulics. For an account of such models, the reader may consult Chapter 8 of Williams and Loyalka (1991).

Granulation

Granulation is the process of producing a granular material with desired properties, which depend on the distribution of size and possibly other variables, such as porosity or composition. Reviews of population balance modelling of granulation can be found in Reynolds et al. (2005) and Abberger (2007).

Biology and Biochemical Engineering

Many biological and biochemical problems are described by a population balance. Cell populations have distributed properties such as mass and age that determine various important process parameters. The application of population balance modelling to biology and biochemical engineering has a long history, and one may consult Ramkrishna (2000), Hjortsø (2004) or Ramkrishna and Singh (2014) for more details.

Polymerisation

Polymerisation can be described as a population balance of molecules with a distribution of molecular weights, starting from the monomers. Accounts of the application of population balance modelling to polymerisation can be found in Ziff (1980) and Wulkow (1996).

Another important problem is the design of polymerisation processes. Heterogeneous processes, in particular, such as emulsion polymerisation (Rawlings and Ray, 1988), involve a dispersed phase with a distribution of one or more properties that are important for the process, and their prediction can be approached with population balance modelling. For more information, one may consult the review of Kiparissides (2006).

Astrophysics

Certain astrophysical problems, such as clustering of planets, stars and galaxies have been described with a population balance equation that accounts for the dynamics of coagulation and fragmentation. Examples of such works can be found in Lee (2000) and Lombart and Laibe (2021).

1.1.3 Scope and Methodology of Population Balance

Owing to its generality and wide range of applications, the population balance has appeared in the literature under different guises, and one of the objectives of this book is to present a unifying framework for them. In the context of aerosol science, the population balance has more often been called the General Dynamic Equation (GDE). In dispersed multiphase flow, multi-fluid models (cf. Section 2.7) are forms of population balance for dispersed entities with a distribution of size and velocity. In many fields, ad hoc models have been proposed for integral properties of distributions, such as total number and volume of particles, and these are also forms of population balance.

The population balance approach is built around the formulation of a population balance equation (PBE), which links the dynamics of the distribution of the property or properties under investigation to the physical and chemical models that determine it. The PBE is a conservation equation akin to the equations of conservation of mass, momentum and energy. It includes source terms that account for processes such as growth and aggregation, arising from interaction of particles with their environment or with other particles, as well as terms that depict transport in physical space, such as convection by a fluid flow.

The PBE must be complemented by expressions for the rates of the physical and chemical processes considered, which play the role of constitutive relations. For some processes and problems, these expressions are well established. In other cases, considerable uncertainty may be present and experiments or simulations at the microscopic level (where the dynamics of individual particles are simulated, as opposed to the statistics of their population) may need to be combined with a population balance study. The combination of microscopic methods, population balance modelling and fluid dynamics yields a truly multiscale modelling approach.

The application of the population balance methodology can be summarised in the following four steps:

Step 1: Formulation of the population balance model. This step involves the choice of the appropriate form of the PBE and the identification of the physical and chemical processes to be included.

Step 2: Selection of kinetic models. Once the processes involved have been identified, models and kinetic data for them must be selected. If such data are not available in the literature, their determination may require further experiments or microscopic simulations.

Step 3: Coupling with flow, species and energy transport. In some cases, this coupling can take the form of an ideal reactor model, while in others, the equations of fluid dynamics and transport phenomena have to be coupled with the PBE. Turbulence, if present, may require additional modelling elements.

Step 4: Application of a solution method. In most cases, this will be a numerical method, as analytical solutions are available only for a few special cases. If the population balance is coupled with fluid dynamics, the solution will involve combining the PBE solution method with computational fluid dynamics (CFD).

The execution of these steps will be explained in Chapters 2–5. In Chapter 6, a number of case studies are presented where the procedure above is demonstrated.

1.2 Distributions and Their Properties

1.2.1 Discrete and Continuous Distributions

In the present section, we briefly review the basic features of distributions, with focus on the issues relevant to the material in the present book. For more details, one may consult a book on probability such as Papoulis (1991) or Grimmett and Stirzaker (2001).

Both discrete and continuous distributions are used in the population balance framework, the choice depending on the nature of the problem. Discrete distributions are suited to populations of particles comprising an integer number of units, such as a population arising from the coagulation of a monodisperse colloid. Continuous distributions are required for problems where a smallest building block cannot be defined (e.g. a population of droplets or bubbles whose size can change continuously via evaporation or dissolution respectively). Populations described by discrete distributions can also be described by continuous distribution functions, and this is often preferable for reasons that will be further explained in Section 2.2.4.

A discrete distribution of a single property is defined by a set of variables as $(n_1, \ldots, n_n)$, where n_i is the number of particles comprising i units of that property. A typical example is a population arising from coagulation of monodisperse particles, where n_1 is the number of particles of volume v_0 (the volume of the smallest particle), n_2 is the number of particles of volume $2v_0$ and so on.

For a distribution of a continuous variable, such as particle volume, v, we define the number density function, $n(v)$ (we will refer to it as simply *number density*), as a continuous function such that the number of particles with volume within an infinitesimally small range between v and $v + dv$ is $n(v)dv$, as shown in Fig. 1.4. The unit of the number density, therefore, is the inverse of particle volume (or whichever independent variable is employed). If the distribution is evolving with time, such as the one shown in Fig. 1.4, then the number density is time dependent.

A continuous distribution can be *discretised* to yield a set of particle numbers $(n_1, \ldots, n_n)$, each denoting the number of particles within intervals $(dv_1, \ldots, dv_n)$, not necessarily uniform. Such a discretisation is often employed in the measurement of distributions and in numerical methods for solution of the PBE.

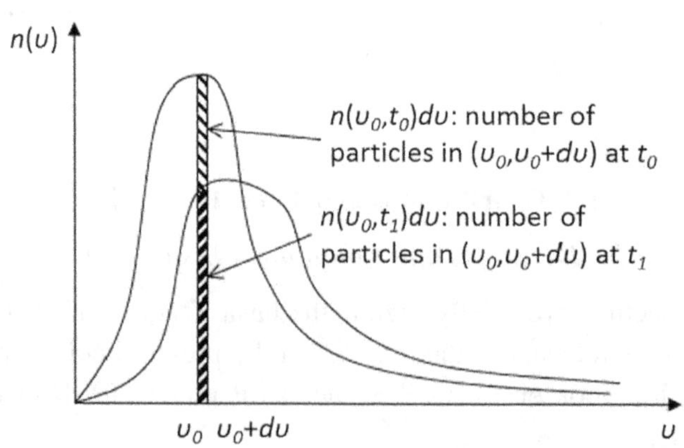

Figure 1.4 The continuous particle size distribution and its time evolution. Reproduced from Rigopoulos (2019) under CC-BY 4.0 license.

Both discrete and continuous distributions can be extended to distributions of several variables (multivariate distributions) in a straightforward manner. For example, if α is the surface area of a particle, then $n(v, \alpha)$ is the number of particles with volume between v and $v + dv$ and surface area between α and $\alpha + d\alpha$.

1.2.2 Number Density and Probability Density Function

An associated concept is the probability density function (PDF). The PDF describing a distribution over particle volume, for example, is defined so that $f(v)dv$ is the probability of finding a particle with volume between v and $v + dv$. The number density function and PDF are closely related, the only difference being that the integral of the number density is the total number of particles (also called the zeroth moment of the distribution, m_0):

$$\int_0^\infty n(v)dv = m_0. \tag{1.1}$$

By contrast, the PDF depicts a probability and therefore has the normalisation property:

$$\int_0^\infty f(v)dv = 1. \tag{1.2}$$

Equations for the evolution of a probability distribution are used in numerous areas of science. The *Boltzmann equation* is an equation for the PDF of velocities and momenta of gas molecules. The *Fokker–Planck equation* is an equation for the PDF of a random variable undergoing drift and diffusion processes. *PDF transport equations* are used in turbulent flow and particularly in reacting flow as closure models for the unclosed correlations involving velocities and scalars (cf. Section 5.5). The number density function and the probability density function are related as follows:

$$n(v) = m_0 f(v). \tag{1.3}$$

In this book, an equation is termed a PBE if formulated in terms of a number density and a PDF transport equation if written in terms of a probability density function. It follows that a PBE can be converted to a PDF transport equation via normalisation with the zeroth moment of the number density function. PDF transport equations are employed when the number of particles is not relevant or has no meaning. For example, in the description of turbulence, we may consider the PDF of velocity as the probability of finding a fluid element with velocity within a range $(\mathbf{u}, \mathbf{u} + d\mathbf{u})$ at a certain location and time. However, no meaning can be ascribed to the number of fluid elements. An approach consisting of a combination of PDF and PBE will be described in Section 5.5.2 for population balance modelling in turbulent flow.

1.2.3 Properties of Distributions

We will now summarise some important properties of distributions, with emphasis on quantities relevant to population balance modelling.

The moments of a distribution are defined as follows. For a discrete distribution of one variable, the k-th moment is:

$$m_k = \sum_{i=1}^{\infty} i^k n_i. \tag{1.4}$$

For the distribution of a continuous variable, such as the particle volume, moments are defined via an integral over the whole range of values of this variable, which is $(0,\infty)$ in this case:

$$m_k = \int_0^{\infty} v^k n(v) dv. \tag{1.5}$$

The moments are defined in the same way for number and probability density functions, while the zeroth moment of the latter is equal to one. Note also that the moments, as defined above, are about the value zero. They may also be defined about an arbitrary value; for example, the moments of the aforementioned distribution about a value v_0 are:

$$m_k = \int_0^{\infty} (v - v_0)^k n(v) dv. \tag{1.6}$$

However, only moments about zero will be used in the present book, and the term 'moment' will be always interpreted in this sense.

Physical meaning can be ascribed to certain low order moments of number density functions. As mentioned in Section 1.2.2, the zeroth moment represents the total number of particles. Other moments may have a physical meaning that depends on the choice of independent variable. In the case of a distribution of particle volume, for example, the first moment represents the total volume of the particles, as can be seen from its definition:

$$m_1 = \int_0^{\infty} v n(v) dv. \tag{1.7}$$

Moments of fractional order can also be defined. For example, the fractional moment of order 2/3 of the volume distribution is proportional to the total surface area (the constant of proportionality being a shape factor):

$$m_{2/3} = \int_0^{\infty} v^{2/3} n(v) dv. \tag{1.8}$$

For a distribution in terms of measure of particle length, ℓ, such as particle diameter or radius, the second moment is proportional to total surface area:

$$m_2 = \int_0^\infty \ell^2 n(\ell) d\ell, \tag{1.9}$$

while the third moment is proportional to total volume:

$$m_3 = \int_0^\infty \ell^3 n(\ell) d\ell. \tag{1.10}$$

Shape factors will also have to be introduced in Eqs. 1.9 and 1.10 in order to obtain the surface area and volume.

Joint moments of various orders can be defined for a multivariate discrete or continuous distribution. Again, certain moments have a physical meaning; for example, in the case of a bivariate continuous distribution in terms of particle volume and surface area, α, the first moment with respect to volume will be the total volume of the particles:

$$m_{1,0} = \int_0^\infty \int_0^\infty v n(v, \alpha) d\alpha dv. \tag{1.11}$$

Certain important properties of distributions are defined in terms of moments. It is instructive to define them first in context of probability density functions. The *mean* (or *expectation*) of a PDF, μ, is defined as the first moment – for example, for the PDF of particle volume, $f(v)$, we have:

$$\mu = \int_0^\infty v f(v) dv = m_1. \tag{1.12}$$

The *variance* of a PDF, σ^2, is the second central moment (i.e. the second moment about the mean):

$$\sigma^2 = \int_0^\infty (v - \mu)^2 f(v) dv. \tag{1.13}$$

It can be shown[1] that:

$$\sigma^2 = m_2 - m_1^2. \tag{1.14}$$

The *standard deviation*, σ, is the square root of the variance and a measure of the spread of a distribution. Similar definitions hold for discrete distributions.

For a number density function, the mean and variance have the same values as those of the corresponding PDF and can be obtained by normalising the above equations with the zeroth moment:

[1] $\int_0^\infty (v-\mu)^2 f(v) dv = \int_0^\infty v^2 f(v) dv - 2\mu \int_0^\infty v f(v) dv + \mu^2 \int_0^\infty f(v) dv = m_2 - m_1^2.$

$$\mu = \frac{1}{m_0}\int_0^\infty v n(v) dv = \frac{m_1}{m_0}, \tag{1.15}$$

$$\sigma^2 = \frac{1}{m_0}\int_0^\infty (v-\mu)^2 n(v) dv = \frac{m_2 - m_1^2}{m_0}. \tag{1.16}$$

1.3 Choice of Distributed Variables

The choice of the variables whose distribution is sought is very important for the formulation of the population balance. Some common choices, along with the considerations involved, are listed below.

- *Number of discrete units.* This choice is appropriate for problems described by the discrete PBE, such as coagulation of initially monodispersed particles or polymerisation. The smallest unit present in the system is called a *monomer*.
- *A measure of particle length.* This could be, for example, the diameter or radius of spherical particles. This choice is particularly suitable for problems involving a surface growth rate independent of length (cf. Section 2.3.2), although it makes the description of aggregation and fragmentation more complicated. Furthermore, experimental measurements often report length distributions. Some problems, such as crystal growth, may involve different faces growing with different rates, and in that case multiple length variables can be employed, thus enabling the prediction of particle morphology.
- *Particle volume or mass.* This choice is particularly suitable for aggregation or fragmentation problems, where the particle volume and mass are conserved. It is also easy to accommodate surface growth with this description, so it will be the default formulation for most of the examples in the present book. The final results may be converted to a length distribution for the purpose of comparison, as that is often reported in experiments. If the particle density is constant, the formulations in terms of volume and mass are equivalent.
- *Particle surface area.* This variable can be used to described the morphology of particles with irregular shapes, in addition with a variable describing particle size (such as length or volume). In that case, particles with the same size may have different surface areas.
- *Number of primary particles.* This is an alternative option for describing particle morphology, where one distinguishes between primary particles and aggregates. It is a special case of the first option, that is, a number of discrete units.
- *Particle velocity.* A population of particles with considerable inertia features a distribution of particle velocities, as the particles deviate from the fluid streamlines to a varying extent depending on their inertia.
- *Temperature.* This is needed for problems such spray combustion, which features a population of droplets with different properties. The droplet temperature is

important for several purposes, such as the determination of the evaporation and combustion rates.
- *Concentration of chemical species.* This is important, for example, in the case of atmospheric aerosols that comprise various chemical species and the distribution of their composition is important.
- *Cell age.* This is a variable relevant for biochemical problems, because important processes such as growth or the probability of a division may depend on it.

In many population balance problems, the distributed property is a measure of particle size (such as diameter or volume). In these cases, we speak of a particle size distribution (PSD). In the case of crystals, this is often referred to as a crystal size distribution (CSD). In polymers, the quantity of interest is the molecular weight distribution (MWD).

1.4 Kinetic and Transport Processes

Having reviewed the basic features of distributions, we now turn to the physical, chemical or biological processes that may be acting to bring changes into a population of particles. The very purpose of a population balance model is to predict how the distribution is shaped by the action of these processes. It is instructive to classify them into two groups: *kinetic processes*, which bring changes to the distribution due to interactions of particles with their environment or between themselves, and *transport processes*, which result in changes of the distribution due to transport of particles in physical space.

Kinetic processes include:

- *Nucleation.* This is the formation of new particles, or nuclei, from precursors in the carrier fluid. The term *inception* is also used in the case of soot.
- *Surface processes.* This includes all processes taking place at the surface of the particles and resulting in a continuous change of their size. Examples of such processes are surface growth of crystals and aerosols, oxidation of soot particles, evaporation of droplets and dissolution of bubbles.
- *Aggregation.* This is the formation of a particle from two other particles that collide. Several other terms such as *coagulation*, *coalescence*, *agglomeration* and *flocculation* are used in the literature for such processes, based on the nature of the particular process (such as whether the resulting particle is spherical or not) or on the field of application, although the definitions of these terms are not universal and varies with subject. In the present book, the term 'aggregation' will be used as a generic term, while other terms will also be used in more specific contexts; more details on this are provided in Section 3.2.1.
- *Fragmentation* (or *breakage*). This is the process where particles break into smaller ones. There are various special cases, such as *binary fragmentation* (breaking into two fragments) or *erosion* (removal of a small fragment).

Transport processes may be described in a simplified way in problems where spatial dependence is not considered. If the PBE is to be coupled with fluid dynamics, however, a detailed account of them is needed. They include the following:

- *Convection.* This is the transport of particles by the carrier fluid.
- *Hydrodynamic forces.* This includes forces such as drag and lift.
- *Body forces.* This includes forces due to external fields, such as the weight and the Coulomb force, and forces due to non-inertial frames of reference, such as the centrifugal and Coriolis forces. Other forces that may be included are particle migration due to temperature gradients (*thermophoresis*) or concentration gradients (*diffusiophoresis*).
- *Thermal motion.* This is a mechanism relevant for small particles and may resemble molecular motion or be induced by collisions with fluid molecules (Brownian motion).
- Momentum exchange between particles or between particles and walls, and momentum transfer from particles to flow.

The mechanisms and models for these processes will be considered in more detail in Chapter 3.

1.5 Population Balance and Fluid Dynamics

Many of the processes that can be described by the population balance take place within fluid flows. In these cases, the distribution is spatially non-uniform and the independent variables include both distributed properties and spatial coordinates. For example, a distribution of particle volume that is space and timedependent would be represented by the number density $n(v, \mathbf{x}, t)$, such that the concentration of particles with volume between v and $v + dv$ contained in a fluid element located at $\mathbf{x}$ at time instant t is ndv. Apart from the distribution, other important variables that affect it, such as temperature or concentration of precursors, may also be spatially variable. An example of a spatially non-uniform population balance problem is the crystallisation process in a T-mixer shown in Fig. 1.5, where the spatial variation of supersaturation results in different nucleation and growth rates and different crystal size distributions throughout the reactor.

In a spatially non-uniform formulation, the number density must be considered in the sense of a concentration, that is, number density of particles per unit volume of fluid. Since this is a continuous function of space, the implicit assumption is that the scale at which the flow is varying is considerably larger than the particles. This requirement is easily satisfied for many problems that involve small particles such as aerosols, soot and crystallisation. For larger particles, the population

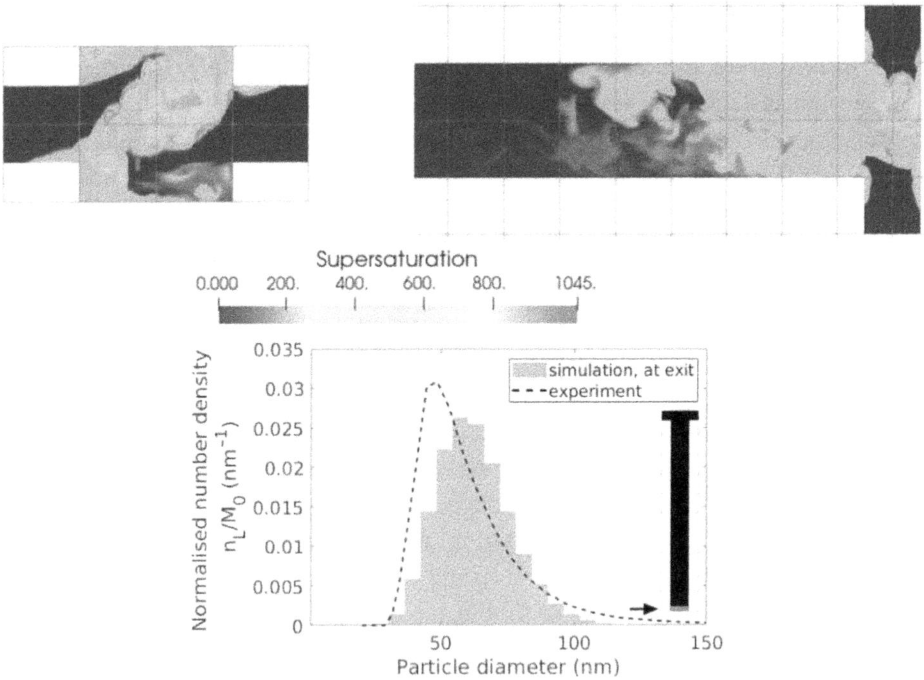

Figure 1.5 Precipitation of $BaSO_4$ in a T-mixer (cf. Section 6.3). Top row: spatial distribution of supersaturation (top left: a cross-section around the inlets, top right: a top view). Bottom: prediction of CSD with coupled CFD-population balance modelling (Tang et al., 2022) and comparison with the experiments of Schwarzer et al. (2006). Reproduced from Tang et al. (2022) under CC-BY 4.0 license.

balance must be applied to a sufficiently bigger volume. This may still be adequate if the flow varies on an even larger scale, such as in the case of atmospheric flows. Otherwise, either the PBE must be applied to a sufficiently large region that can be considered as well mixed (via an ideal reactor or reactor network model as opposed to a coupling with fluid dynamics) or a different approach must be pursued (such as a multiphase flow or discrete element model).

In the case of probability density functions, we will denote spatial dependence by $f(v; \mathbf{x}, t)$ (the symbol ';' is used to indicate that the PDF is normalised with respect to v but not with respect to $\mathbf{x}$ or t). On the other hand, a number density is not normalised with respect to any of its arguments, hence the integration of $n(v; \mathbf{x}, t)$ with respect to v yields the local value of the zeroth moment (which is also spatially dependent):

$$\int_0^\infty n(v, \mathbf{x}, t) dv = m_0(\mathbf{x}, t). \tag{1.17}$$

1.6 Two Examples of Population Balance in Flows

We will now show two examples of population balance models, in order to provide a 'first taste' of the issues to be discussed in the present book. Both cases demonstrate also the role played by fluid dynamics, which will be a major features throughout the book. For now, no equations will be shown, as the mathematical framework and physical models have not been presented yet; this is only a first look at these problems focussing on what the population balance approach can do for each of these cases. Both problems will be discussed in detail in Chapter 6.

1.6.1 Precipitation of Crystals

Crystallisation is a process of producing a crystalline material from a solution. The crystal size distribution (CSD) of the product is of great importance for its properties and its suitability to particular applications. Here, we are considering the case of reaction crystallisation, or precipitation, which is driven by a chemical reaction between species that are fed separately and mix into a reactor. For fast reactions, the mixing is the controlling factor for the outcome of the process, thus presenting a problem featuring strong coupling between population balance and fluid dynamics.

The case shown here involves two reactants are fed from two opposing streams into a T-mixer. As the reactants are brought into contact, a supersaturated solution is formed. Supersaturation is the driving force for crystallisation, and the processes of nucleation and growth depend on it in a nonlinear manner, with nucleation featuring a stronger dependence. Aggregation may also be present under certain conditions. The balance between these two processes determines the CSD; a high nucleation rate results in a large number of crystals and rapid consumption of the reactants, thus yielding small crystals at the end, while a low nucleation rate is followed by longer growth into bigger crystals. The population balance describes the delicate balance between these processes, while the equations of fluid dynamics and mass transfer describe the mixing of reactants. Fig. 1.5 shows the spatial distribution of supersaturation and the CSD at the exit as measured by experiments and predicted by population balance modelling. It is evident that the supersaturation is highly non-uniform and strong gradients are present; thin zones of high supersaturation can also be identified, which give rise to nucleation bursts. This problem will be revisited in Section 6.3, where the methods developed in the book will be employed to analyse the interaction between the processes of nucleation and growth, as well as their interaction with turbulence.

1.6.2 Formation of Soot and Carbonaceous Nanoparticles

The simulation of soot formation is needed to assist in the design of combustion devices. The soot particle size distribution is of increasing importance, as new regulations require control of the number of particles (particularly the smaller ones, which have far more harmful effects) as opposed to the mass of it, as was the case in the past. Similarly, the production of carbonaceous nanoparticles can be aided by modelling and simulation that permits tailoring of the product to applications. The dynamics of carbonaceous nanoparticles is a complex problem that involves fluid mechanics, turbulence, transport phenomena, chemical kinetics and population balance modelling.

Soot precursors are polyaromatic hydrocarbons (PAHs), which are themselves formed from smaller compounds found in flames via complex chemical mechanisms. Soot nucleation, or inception, involves heavy PAH molecules and is poorly understood, partly due to the experimental difficulties in observing these molecules or the soot nuclei. The evolution of soot particles involves surface growth by acetylene and consumption by oxidation, and the detailed description of these processes is also complex as it involves surface chemistry. Aggregation and fusing of soot aggregates also play a role in determining the distribution. Since chemistry is involved in most of these processes, the soot particle size distribution is spatially dependent and heavily influenced by the local concentrations of precursors and by temperature. In flames, concentrations and temperature exhibit large gradients and many species are present in thin zones, due to the interaction of fast chemical reactions with turbulence. The prediction of soot in a turbulent flame is thus one of the most complex challenges for population balance modelling, requiring coupling of a CFD code with chemical kinetics, a turbulence–chemistry interaction model and a population balance model.

Most practical combustion processes involve turbulent flows, which introduces several further implications that will be studied further in Chapter 5. Fig. 1.6 shows results from a CFDPBE simulation of a turbulent sooting flame that will be studied in the case study of Section 6.2. The figure shows contour plots of temperature and of two particle formation processes, nucleation and growth (more processes will be shown in Fig. 6.20). A full discussion of these results will be undertaken in Chapter 6 but, for now, we note this as an example of complex spatial distribution of the physical and chemical processes that are involved in the population balance and shape the particle size distribution.

1.7 A Brief Historical Survey

The concept of describing physical systems with a probability density function and linking the evolution of the latter to physical processes originates in Boltzmann and his famous equation. However, in the present book we employ the term 'population balance' in the more specific context of number density functions, for systems where the concentration of the particles is important as well as their probability

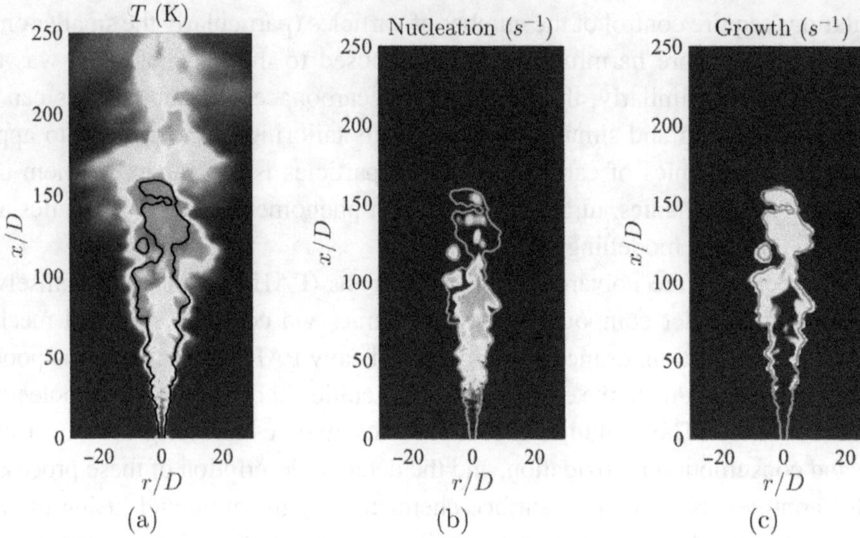

Figure 1.6 Simulation of a turbulent sooting flame with a coupled fluid dynamics – population balance formulation (cf. Section 6.2). Contour plots of the instantaneous (a) temperature, (b) nucleation rate and (c) growth rate. The iso-contour of stoichiometric mixture fraction (cf. Section 5.4) is shown as a solid line in the instantaneous plots. Reproduced from Sun and Rigopoulos (2022) under CC-BY 4.0 license.

distribution. In this sense, the origins of the subject can be traced to Smoluchowski (1917), who formulated a discrete PBE for the coagulation of colloidal particles. Smoluchowski was also able to derive the first solution for his equation, which will be shown in Chapter 4.

Other notable works in the early twentieth century include those of Müller (1928) and Schumann (1940), who derived the continuous PBE for coagulation, although the term 'population balance' was not used at that time. Schumann's work was motivated by atmospheric science and the need to predict the size distribution of cloud and fog droplets. This was the case also for several of the works that followed, including those of Scott (1968), who derived analytical solutions for coagulation equations, and of Bleck (1970), who derived an early numerical method. Melzak (1957a,b) extended the equation to include fragmentation; these two works are not concerned with any particular application but rather with the mathematical properties of the equation. Friedlander and Wang (1966) extended the coagulation equation to include condensation, evaporation and homogeneous nucleation; his work was motivated by the problem of the formation of aerosols. Friedlander and his co-workers are also responsible for the early developments of similarity solutions (Swift and Friedlander, 1964; Friedlander and Wang, 1966). The early works in the colloid, atmospheric and aerosol science communities have been reviewed by Hidy and Brock (1970) and Drake (1972).

Another important stream of research originates in the pioneering works of Randolph and Larson (1962) and Hulburt and Katz (1964), which were motivated by chemical engineering problems. The first of these works was concerned with the problem of crystallisation and formulated an equation for a continuous reactor with inlet and outlet; the term 'population balance' appeared, possibly for the first time, in that paper. The second work emphasised the importance of spatial dependence of the distribution and formulated a continuous PBE (termed a form of Liouville equation) that included spatial transport as well as nucleation, growth and agglomeration; that paper was also responsible for some of the earliest developments in the method of moments for the solution of the PBE. The introduction of the population balance as a general transport equation that includes spatial transport was also brought forward in a short note by Randolph (1964). Several works followed, and this line of research was summarised in the book by Randolph and Larson (1971) (see also Randolph and Larson, 1988, for the latest edition).

The application of the population balance to biochemical engineering commenced at about the same period with works such as Fredrickson and Tsuchiya (1963) and Fredrickson et al. (1967). Early work is summarised in the review of Tsuchiya et al. (1966), while Fredrickson et al. (1967) mention the term 'population balance' and the link of their work in this field with those of Randolph (1964) and Hulburt and Katz (1964). Later reviews of this literature can be found in Ramkrishna (1985, 2000).

With the fundamentals of the subject established, research in the following decades focussed on expanding the range of applications and developing methods of solution. Apart from the work in the communities of aerosol science and crystallisation, which largely fostered the subject's initial growth, a wide range of applications appeared in fields such as multiphase flow, fluidisation, granulation, comminution and biochemical engineering; a comprehensive list of applications up to 1985 can be found in the review by Ramkrishna (1985).

The seminal book of Ramkrishna (2000) was the first one to focus on the population balance as a methodology, while at the same time showcasing many of its applications. Several other books include an account of the population balance (in one of its many guises) in the context of a particular field of application. Williams and Loyalka (1991) is a treatise on aerosol science but includes an extensive treatment of the PBE (or GDE, in the context of that book) and a detailed account of analytical solutions. The books by Friedlander (2000) and Seinfeld and Pandis (2016) on aerosol science also include extensive discussions of the PBE (GDE). Treatises focussing on dispersed multiphase flows from the population balance point of view include Marchisio and Fox (2013) and Yeoh et al. (2014); the former, furthermore, includes valuable material on kinetic theory and moment methods. Hjortsø (2004) deals with biochemical problems. Jakobsen (2014) is a comprehensive treatise on chemical reactors that includes also an extensive section on the population balance. Litster and Ennis (2004) is focussed on granulation and includes a discussion of the PBE, while Litster (2016) deals with design and processing of particulate products and includes a treatment of the PBE with focus is on its application to particulate processes. Several review papers of a more general nature are also available, including Hounslow (1998), Ramkrishna and Mahoney (2002), Sporleder et al. (2012), Ramkrishna and Singh (2014) and Solsvik and Jakobsen (2015).

While the link between population balance and fluid dynamics had been established as early as in the work of Hulburt and Katz (1964), the computational power at the time did not allow the application of numerical methods to the coupled fluid dynamics-PBE problem. Therefore, applications prior to the nineties were based on simplified flow models such as ideal reactors. CFD was developed in the seventies and eighties and enabled the numerical solution of fluid dynamics and scalar transport equations. The development of coupled CFD–PBE methods commenced in the mid-nineties, initially with moment methods in order to minimise the storage and computation time requirements and later with discretisation methods; more information on these methods will be provided in Chapter 4. Methods for dealing with the coupling of population balance and turbulent flow are among the most recent developments at the time of writing. The focus on the coupling of fluid dynamics and population balance is reflected in several recent books and reviews, such as Rigopoulos (2010), Marchisio and Fox (2013), Yeoh et al. (2014), Jakobsen (2014), Raman and Fox (2016) and Rigopoulos (2019).

1.8 Summary

In the present chapter, the population balance has been introduced as an approach for modelling problems involving a population of particles with a distribution of one or more properties. The distribution is determined by a number of processes that can be classified into two groups: kinetic processes, which change the distribution due to interactions of particles with their environment or between themselves, and transport processes, which involve motion of particles in physical space. A number of problems to which the population balance approach can be applied have been identified, as well as possible choices of distributed variables. In many problems, the population balance must be coupled with fluid dynamics in order to determine the spatially varying distribution.

Approaching a problem with the population balance framework involves formulating the PBE for the distribution of the properties of interest, selecting models for the kinetic and transport processes involved, coupling the PBE with a flow model (either a simplified one or a detailed solution of the equations of fluid dynamics), and applying a solution method. The details of how these steps are to be carried out will be the subject of the following chapters. In particular, the formulation of the PBE and coupling with fluid dynamics will be addressed in Chapter 2, with the case of turbulent flow receiving a more extensive treatment in Chapter 5. Models for kinetic and transport processes will be discussed in Chapter 3, while solution methods will be treated in Chapter 4. Finally, Chapter 6 will demonstrate the overall population balance – fluid dynamics approach via three extensive case studies.

2
Formulation of the Population Balance

The objective of Chapter 2 is to present a unifying framework for the various population balance formulations. Only the general mathematical form of the kinetic and transport processes involved will be shown in this chapter; the models for them that serve as constitutive equations will be discussed in Chapter 3. Solution methods will be covered in Chapter 4.

2.1 Overview of Population Balance Formulations

The earliest form of PBE is the Smoluchowski equation (Smoluchowski, 1917), which describes the aggregation of spherical particles. The Smoluchowski equation is a discrete PBE, as the particles are composed of an integer number of monomers (the smallest particles present in the system), and can be extended to a more general formulation that includes processes such as nucleation and condensation. The discrete PBE is often impractical for computations, however, because it involves solving a very large system of equations. Furthermore, there are populations that feature an inherently continuous size spectrum, such as droplets and bubbles. These considerations led to the development of the continuous PBE, which is more often employed in problems and will often be referred to in this book as simply 'PBE' when a distinction from the discrete form is not essential. While the continuous PBE is a single equation, it is a partial integro-differential equation and its solution requires specialised numerical methods.

Certain problems require more than one variable for their description. Examples are multicomponent aerosols, which are particles composed of several chemical species, crystals with a distribution of size in directions and fractal aggregates exhibiting a distribution of size and morphology. Such problems can be addressed via an extension of the PBE to several dimensions or by the coupling of multiple PBEs.

In problems where the entities are transported by a carrier fluid, the PBE must be coupled with the equations of fluid dynamics and transport phenomena. These equations also provide the evolution of the concentration of precursors. Furthermore, a feedback effect in the form of precursor consumption or an effect on the flow may need to be considered. In some cases, it may be possible to simplify the problem by assuming that it takes place within a spatially homogeneous environment, such as a perfectly stirred reactor. In other cases, a solution of the coupled PBE-fluid dynamics problem is essential. The numerical solution of the coupled problem requires combining the PBE discretisation methods with Computational Fluid Dynamics (CFD).

When a population balance problem takes place within a turbulent flow, the population is subjected to fluctuations arising from turbulent transport and fluctuations in kinetic processes such as nucleation and growth (due to fluctuations in the participating chemical species' concentrations and/or temperature). This results in additional modelling issues that must be addressed with suitable closure models.

Fluctuations can also appear in very small populations due to the small size of the sample. The formulation of the population balance so as to account for such fluctuations has been discussed by Ramkrishna (2000). In recent years, with the increased computational power available, small populations can be modelled by tracking every individual member. In the present book, we will consider applications of the PBE to populations sufficiently large for such fluctuations to be negligible.

2.2 Discrete Population Balance Equation (PBE)

The discrete form of the PBE was developed first, having been proposed by Smoluchowski (1917) in its initial form, which is often referred to as the Smoluchowski equation.[1] We first consider this initial form, in which aggregation (coagulation[2]) is the only process being present, and subsequently describe its extension to include more kinetic processes.

2.2.1 Smoluchowski Equation

The Smoluchowski equation describes an aggregation process in a population of particles, each of which comprises an integer number of units. We define a particle

[1] Also known as Smoluchowski coagulation equation, which aids in distinguishing it from Smoluchowski's formulation of the diffusion equation.

[2] While the term 'coagulation' is traditionally employed in the context of the Smoluchowski equation (since it was used by Smoluchowski himself), here we use the term 'aggregation' for consistency with the remainder of the present section. See also the discussion of terminology in Section 3.2.1.

size class i as the set of particles comprising i units and n_i as the number of particles[3] in class i. The Smoluchowski equation is based on the physical argument that the rate of change of n_i is determined by a birth term due to aggregation of all pairs of smaller particles yielding particles in class i and a death term due to aggregation of particles in class i with particles in any other class. The rate of collisions between particles in classes i and j that lead to aggregation is proportional to the product $n_i n_j$ and a coefficient which is called the *kernel* and contains the physics of the collision process. The collision mechanisms and corresponding kernels will be discussed in Chapter 3. The coefficients for all combinations of classes can be arranged into a symmetric matrix β_{ij}.

The rate of aggregation events between particles belonging to different classes, $\beta_{ij}n_i n_j$, must be subtracted from the rates of change of n_i and n_j and added to the that of n_{i+j}. The rate of aggregation events between particles in the same class, $\beta_{ii}n_i^2$, must be subtracted from the rate of change of n_i, while the rate $\frac{1}{2}\beta_{ii}n_i^2$ must be added to the rate of change of n_{2i}. Based on the above, we first write the rate equation for the monomers:

$$\frac{dn_1}{dt} = -n_1 \sum_{j=1}^{\infty} \beta_{1j} n_j. \tag{2.1}$$

The equations for n_2 and n_3 are:

$$\frac{dn_2}{dt} = \frac{1}{2}\beta_{11}n_1^2 - n_2 \sum_{j=1}^{\infty} \beta_{2j} n_j, \tag{2.2}$$

$$\frac{dn_3}{dt} = \beta_{12}n_1 n_2 - n_3 \sum_{j=1}^{\infty} \beta_{3j} n_j. \tag{2.3}$$

From the above, it can be inferred that the following equation can be written for $i > 1$:

$$\frac{dn_i}{dt} = \frac{1}{2} \sum_{j=1}^{i-1} \beta_{j,i-j} n_j n_{i-j} - n_i \sum_{j=1}^{\infty} \beta_{ij} n_j. \tag{2.4}$$

For pairs where $i \neq j$, the factor $1/2$ in Eq. 2.4 is needed to correct for their double counting. For pairs of identical particles, this factor is also dictated by the kinetics as shown above.

The Smoluchowski equation is therefore a system of ODEs that resemble chemical kinetics. Together with initial conditions for the number density, these

[3] In physical problems, this will be a concentration of particles, that is, number of particles per unit volume of fluid.

ODEs constitute an initial value problem whose solution yields the particle size distribution at all time instants.

2.2.2 Formulation of the Complete Discrete PBE

The Smoluchowski equation was extended by Friedlander (1960) in the context of aerosols to include condensation, evaporation and homogeneous nucleation. Condensation is the addition of a single unit, resulting in growth of particles from class i to class $i + 1$, while evaporation is the reverse process. Homogeneous nucleation can be represented as the formation of monomers. As the nuclei may be molecular clusters comprising more than one units, a more general approach is to represent the nuclei as belonging to a class $k \geq 1$ (Seinfeld and Pandis, 2016). These processes, together with aggregation to spherical particles and fragmentation (to be mentioned below), are schematically depicted in Fig. 2.1. Fragmentation is more often encountered in the context of the continuous PBE, although some authors such as Ziff and McGrady (1985) and Vanni (2000) have employed the discrete equivalent.

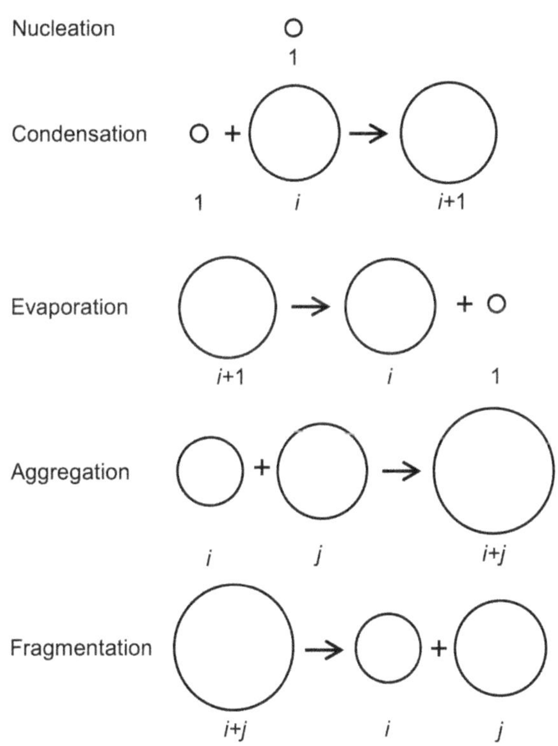

Figure 2.1 Schematic depiction of the processes in the discrete PBE. The numbers indicate the number of monomers in a particle.

In addition to the kinetic processes, we may also include transport processes if the motion of particles in physical space is to be considered. In that case, the number density concentration is space- and time-dependent, i.e. $n_i(\mathbf{x}, t)$. This requires the addition of a convective flux of particles due to the fluid motion and a diffusive flux due to Brownian motion. More information on particle transport will be provided in Chapter 3. Furthermore, inertial particles require further treatment to account for their distribution of velocities. However, the discrete PBE has been used primarily for nanoparticles that are non-inertial, which implies that they follow the flow, hence these considerations will be deferred until the discussion of the more general framework of the continuous PBE. For non-inertial particles, convection and diffusion terms similar to those employed for passive scalars can be added.

We can now write the discrete PBE including the processes that are more commonly encountered in its context, namely aggregation, condensation, evaporation and homogeneous nucleation, in addition to transport in physical space, as follows[4]:

$$\frac{dn_i}{dt} + \frac{\partial(u_j n_i)}{\partial x_j} = \frac{\partial}{\partial x_j}\left(D_{p,i}\frac{\partial n_i}{\partial x_j}\right) + B\delta_{ik} + C_{i-1}n_{i-1} - (C_i + E_i)n_i + E_{i+1}n_{i+1}$$
$$+ \frac{1}{2}\sum_{j=k}^{i-k}\beta_{j,i-j}n_j n_{i-j} - n_i\sum_{j=k}^{\infty}\beta_{ij}n_j + \sum_{j=i+1}^{\infty}\gamma_j b_{ij}n_j - \gamma_i n_i, \qquad (2.5)$$

where $\mathbf{u}$ is the flow velocity[5] and $D_{p,i}$ is the particle diffusion coefficient, which is sizedependent (cf. Section 3.6.4). In addition to aggregation, the following kinetic rates appear at the right-hand side:

- B: Homogeneous nucleation of particles in class k (where $k = 1$ if the smallest cluster comprises one unit).
- C_i: Condensation of monomers on particles in class i, resulting in death of particles in that class and birth of particles in class $i+1$.
- E_i: Evaporation of monomers from particles in class i, resulting in death of particles in that class and birth of particles in class $i-1$.
- γ_i: Fragmentation of particles in class i, resulting in death of particles in that class.
- b_{ij}: Number of particles in class i generated from the fragmentation of a particle in class j, such that $\gamma_j b_{ij}$ is the rate of birth of particles in class i from particles in class j.

[4] In this, as in most spatially dependent equations, the arguments are omitted for brevity.
[5] If thermophoresis is present, the thermophoretic velocity is added to the flow velocity, as discussed in Section 3.6.3.

2.2.3 Applications of Discrete PBE

While most PBE applications employ the continuous form, the discrete form has also found applications in several fields. Examples include:

- Aerosol formation and growth. A discrete PBE with nucleation, condensation, evaporation and aggregation (Eq. 2.5) has been employed by Friedlander (1960). Most applications in aerosol science, however, are based on the continuous PBE.
- Soot formation (Jensen, 1974; Frenklach and Harris, 1987). A discrete PBE with nucleation, surface growth, condensation, aggregation and sometimes aggregation (Kazakov and Frenklach, 1998) has been employed in conjunction with moment (Frenklach and Wang, 1991; Kazakov and Frenklach, 1998) or Monte Carlo (Balthasar and Kraft, 2003) solution methods. Alternatively, the continuous PBE is employed and solved with moment or discretisation methods. These solution methods will be discussed in Chapter 4.
- Nanoparticle synthesis. Landgrebe and Pratsinis (1989) employed the discrete PBE with nucleation and aggregation terms to describe nanoparticle formation. The size range employed was small and the discrete PBE was used for generating reference solutions to validate solution methods for the continuous PBE (Landgrebe and Pratsinis, 1990) and for investigating the range of conditions in which the distribution reaches a self-preserving state (to be discussed in Chapter 4). The continuous PBE is more frequently employed in applications in this field.
- Polymerisation (Ziff, 1980). The growth of macromolecules can be modelled as an aggregation–fragmentation process starting from a population of monomers and is thus an application of the discrete PBE with aggregation and fragmentation.
- Flocculation of suspended matter (Thomas et al., 1999; Partheniades, 2009) and asphaltene (Rahmani et al., 2003; Vilas Bôas Fávero et al., 2017). The description of these processes is based on an aggregation equation that is often the original Smoluchowski equation.
- Astrophysics (Sheth and Pitman, 1997; Lee, 2000). The Smoluchowski equation has been employed to describe clustering of astrophysical entities such as planets, stars and galaxies.

2.2.4 Limitations of Discrete PBE

There are several drawbacks in the discrete PBE. In an aggregation process, the number of variables involved is very large. For example, for an aerosol ranging from 1 nm–1 μm in diameter, the number of discrete size classes that must be accounted for is of the order of 10^9, and this is also the number of ODEs that must be solved. Surface processes may involve the addition or subtraction of a very small unit – in the case of soot, for example, that could be a C_2H_2 unit – and this would have to be taken as the monomer, thus increasing tremendously the number of discrete size classes needed. Furthermore, in some processes such as the evaporation of droplets or dissolution of bubbles, the smallest discrete unit being added or removed cannot even be defined. These considerations led to the development of the continuous PBE.

The discrete PBE can be extended in a straightforward manner to include transport in physical space. In that case, the PBE becomes a set of partial differential equations in space and time for each discrete size class. However, such a formulation is impractical from the point of view of numerical solution as the number of discrete size classes must be multiplied with the number of spatial locations. For this reason, spatially dependent problems are usually approached via the continuous PBE.

2.3 Continuous PBE

The continuous PBE was first derived by Müller (1928) and Schumann (1940) in the context of aggregation.[6] There are two ways of deriving the continuous PBE. The first one is to take the discrete PBE as a starting point and derive the continuous PBE via a limit process. This derivation assumes that the problem can in principle be described by the discrete PBE, regardless of whether it is preferable to work with the continuous form. The second one starts from a continuum principle and is more relevant to problems that cannot be described by a discrete PBE. We will show both approaches here, but it must be emphasised that the final equation is the same in both cases and applicable to any population balance problem.

2.3.1 Derivation Based on Discrete PBE

In the following, we will assume that the particles can be described by a single independent variable indicative of particle size. The generalisation to more independent variables is straightforward and will be discussed in Section 2.4.1. Particle volume or mass is preferable for the description of aggregation or fragmentation because they are invariant with respect to these processes; in the following, we will employ the volume, v. Instead of the number of particles in a given class, i.e. $n_i(t)$, the distribution is now represented by a number density function, $n(v,t)$, and instead of a system of ODEs we will now obtain a single partial integro-differential equation. The derivation in the present section will account for nucleation, growth

[6] One may consult Drake (1972) and Williams and Loyalka (1991) for a discussion of early works on this subject.

and aggregation, while fragmentation will be considered in Section 2.3.2. We first write the general form of the continuous equation as follows:

$$\frac{\partial n(v,t)}{\partial t} = S_{nuc} + S_{growth} + S_{agg}, \tag{2.6}$$

where S_{nuc}, S_{growth} and S_{agg} denote the source terms due to nucleation, growth and aggregation, respectively. The next task is to determine the expressions for these rates in a way consistent with the discrete equation.

Aggregation

If dv_i is a finite interval around the volume of particles with i monomers, then the discrete and continuous number densities are related as follows:

$$n_i(t) = n(v,t)dv_i. \tag{2.7}$$

In the same vein, if we denote the aggregation source term for particles in class i in the discrete equation by $S_{agg,i}$, we have:

$$S_{agg,i} = S_{agg}dv_i, \tag{2.8}$$

where $S_{agg,i}$ is given by (cf. Eq. 2.4):

$$S_{agg,i} = \frac{1}{2}\sum_{j=1}^{i-1} \beta_{j,i-j}n_j n_{i-j} - n_i \sum_{j=1}^{\infty} \beta_{ij}n_j. \tag{2.9}$$

We therefore have:

$$\begin{aligned} S_{agg,i} = S_{agg}dv_i = \frac{1}{2}\sum_{j=1}^{i-1} &\beta(v_j, v_i - v_j)n(v_j,t)dv_j n(v_i - v_j,t)d(v_i - v_j) \\ &- n(v_i,t)dv_i \sum_{j=1}^{\infty} \beta(v_i,v_j)n(v_j,t)dv_j. \end{aligned} \tag{2.10}$$

To obtain the passage to an integral equation, we now substitute $v = v_i$, $w = v_j$ and $\beta(v,w) = \beta_{ij}$ and take the limit as $dw \to 0$, which implies that $d(v-w) = dv$ for fixed v. Hence:

$$\begin{aligned} S_{agg}dv = dv\frac{1}{2}\int_0^v &\beta(w, v-w)n(w,t)n(v-w,t)dw \\ &- n(v,t)dv \int_0^{\infty} \beta(v,w)n(w,t)dw. \end{aligned} \tag{2.11}$$

Note that dv is still a *finite* interval. Dividing by dv, we have:

$$S_{agg} = \frac{1}{2}\int_0^v \beta(w, v-w)n(w,t)n(v-w,t)dw$$
$$- n(v,t)\int_0^\infty \beta(v,w)n(w,t)dw, \tag{2.12}$$

which is the source term containing the birth and death rates due to aggregation in the continuous PBE.[7]

Nucleation

The nucleation source term in the discrete PBE (Eq. 2.5) is $B\delta_{ik}$. The equivalent expression in the continuous formulation is readily obtained as follows:

$$S_{nuc} = B\delta(v - v_0). \tag{2.13}$$

The use of the delta function yields a spike of particle formation with volume v_0, which is the volume of a nucleus particle, such that the rate of formation is B.

Growth

The rate of change in the discrete PBE due to condensation and evaporation, or the net growth rate, is:

$$S_{growth,d} = C_{i-1}n_{i-1} - (C_i + E_i)n_i + E_{i+1}n_{i+1}. \tag{2.14}$$

The equivalent continuous form is[8]:

$$S_{growth} = \frac{\partial(G(v)n(v,t))}{\partial v}, \tag{2.15}$$

[7] The aggregation source term can also be written in the following form (Williams and Loyalka, 1991):

$$S_{agg} = \frac{1}{2}\int_0^\infty\int_0^\infty \delta(u+w-v)\beta(u,w)n(u,t)n(w,t)dudw$$
$$- \int_0^\infty \delta(u-v)n(u,t)\int_0^\infty \beta(u,w)n(w,t)dwdu.$$

In this form, we are considering the birth of particles of size v due to aggregation of all pairs u, w such that $u + w = v$ and death of particles of size v due to aggregation of all pairs u, w such that $u = v$.

[8] It is also possible to include a second-order term as follows:

$$S_{growth} = \frac{\partial(G(v)n(v,t))}{\partial v} + \frac{\partial^2(G'(v)n(v,t))}{\partial v^2},$$

where:

$$G(v) = v_0(C_i - E_i)$$

$$G'(v) = \frac{v_0^2}{2}(C_i + E_i)$$

The second-order term can usually be neglected (Brock, 1972; Ramabhadran et al., 1976).

where the function $G(v)$ is the net growth, that is, the difference of the condensation and evaporation rate:

$$G(v) = v_0(C_i - E_i). \tag{2.16}$$

Therefore, the contribution of this term is:

$$S_{growth} = \frac{\partial(G(v)n(v,t))}{\partial v}. \tag{2.17}$$

Note that the growth rate is, in general, a function of the size coordinate in the PBE (in this case volume). In some problems, it may be independent of a measure of particle length such as diameter, and a description of the size distribution in terms of such a variable may then be advantageous.

Formulation of PBE

Putting all together, the PBE for nucleation, growth and aggregation is:

$$\frac{\partial n(v,t)}{\partial t} + \frac{\partial(G(v)n(v,t))}{\partial v} = B\delta(v - v_0) + \frac{1}{2}\int_0^v \beta(w, v-w)n(w,t)n(v-w,t)dw - n(v,t)\int_0^\infty \beta(v,w)n(w,t)dw. \tag{2.18}$$

If a spatially dependent PBE is required, the transport processes present in Eq. 2.5, namely convection and diffusion of particles, can be added to the continuous PBE in the same way as shown in Section 2.2.2, as long as the particles are non-inertial. A more general treatment of transport in physical space will be discussed in the next section and later in Section 2.7, where the case of inertial particles will also be considered.

2.3.2 Derivation from Continuum Principles

The derivation of the PBE as a continuous equation, rather than as an approximation of the discrete PBE, was first shown by Hulburt and Katz (1964). This derivation is based on the continuity equation in phase space. It is inherently applicable to problems that feature a continuous spectrum of the independent variable and is easily generalised to a multidimensional PBE. A similar equation was postulated by Berry (1969) in the context of cloud modelling.

Let us consider a population of particles with n distributed properties, each of which is described by a continuous variable. These properties may be conveniently represented by the vector $\boldsymbol{\xi} = [\xi_1, \ldots, \xi_n]^T$ defining a phase space. The population of particles can be described by a number density function, $n(\boldsymbol{\xi}, t)$, such that the number of particles at time t with properties between $\boldsymbol{\xi}$ and $\boldsymbol{\xi} + d\boldsymbol{\xi}$

is $n(\boldsymbol{\xi},t)d\xi_1 \ldots d\xi_n$. For the time being, we are considering a spatially uniform system, hence physical space is not included in the above.

Each of the coordinates has an associated rate of change, or generalised velocity, with a physical meaning particular to that coordinate. For example, if we are considering a distribution of particle volume, v, the corresponding rate is the volumetric growth rate:

$$\xi \equiv v, \tag{2.19a}$$

$$\dot{\xi} \equiv \frac{dv}{dt}. \tag{2.19b}$$

Another example is a distribution of concentrations of chemical species in aerosol particles, in which case the rate of change is the chemical reaction rate.

We now write a continuity equation for the number density in phase space, in the same manner as the continuity equation for mass and chemical species (Appendix A). The equation is:

$$\frac{\partial n}{\partial t} + \frac{\partial(\dot{\xi}_k(\boldsymbol{\xi})n)}{\partial \xi_k} = S(\boldsymbol{\xi},n), \tag{2.20}$$

where the source term, S, represents the remaining kinetic processes, nucleation, aggregation and fragmentation.

In order to extend the above equation to a spatially dependent PBE, we now consider the population of particles by a number density function that depends on spatial coordinates[9] as well as time, denoted by $n(\boldsymbol{\xi},\mathbf{x},t)$. It should be noted that, in this case, the number density is a concentration and its dimensions include per unit volume of physical space, as well as per unit of the other independent variables.[10] Therefore, the concentration of particles at physical location $\mathbf{x}$ and time t with properties between $\boldsymbol{\xi}$ and $\boldsymbol{\xi}+d\boldsymbol{\xi}$ is $n(\boldsymbol{\xi},\mathbf{x},t)d\xi_1 \ldots d\xi_n$. Also, let $\mathbf{u_p}$ denote the velocity of the particles in physical space. We now write a balance equation for the accumulation of number density:

$$\frac{\partial n}{\partial t} + \frac{\partial(u_i n)}{\partial x_i} + \frac{\partial(\dot{\xi}_k(\boldsymbol{\xi})n)}{\partial \xi_k} = \frac{\partial}{\partial x_i}\left(D_p(\boldsymbol{\xi})\frac{\partial n}{\partial x_i}\right) + S(\boldsymbol{\xi},n). \tag{2.21}$$

A diffusion term has been included to represent diffusion of particles in physical space due to Brownian motion (if applicable). For the other phase space coordinates, a kinetic process giving rise to diffusional transport is rarely encountered.[11]

[9] In the literature, the particle properties and the dimensions of physical space are often referred to as *internal* and *external coordinates*, respectively.

[10] If particle volume is used as a distributed variable, its units must be distinguished from those of physical volume.

[11] As discussed in Section 2.3.1, a second-order term has been considered in the context of condensation and evaporation of aerosols but is usually neglected. One of the rare cases where it has been encountered is the representation of growth rate dispersion in crystals (Randolph and White, 1977; Tavare, 1985).

Eq. 2.21 refers to non-inertial particles, in which case the particle volume fraction is approximately zero and the convection velocity is the fluid velocity; the latter can be obtained from the solution of the continuity and Navier–Stokes equations (which are shown in Appendix A). By contrast, inertial particles exhibit a distribution of velocities, as will be further discussed in Section 2.7. In that case, the particle velocity should be regarded as an independent variable with an associated rate of change, which is the particle acceleration. We may denote this velocity by $\mathbf{u_p}$, to distinguish it from the carrier flow velocity, $\mathbf{u}(\mathbf{x},t)$, which is a field to be computed, rather than an independent variable. We can now augment the PBE as follows:

$$\frac{\partial n}{\partial t} + \frac{\partial (u_{p,i} n)}{\partial x_i} + \frac{\partial (\dot{u}_{p,i} n)}{\partial u_{p,i}} + \frac{\partial (\dot{\xi}_k(\boldsymbol{\xi}) n)}{\partial \xi_k} = S(\boldsymbol{\xi}, n). \tag{2.22}$$

The diffusion term has been omitted in Eq. 2.22 because it does not usually apply to inertial particles owing to their larger size (the dependency of the diffusion coefficient on particle size will be shown in Section 3.6.4), except for some problems featuring simultaneous diffusion and inertial effects where special treatment is required (see, e.g., Drossinos and Housiadas, 2017). Finally, we note that the extension to a spatially-dependent PBE raises questions with respect to the size of the particles and flow scales, which will be discussed in Section 2.7.

We will now show some concrete examples of PBEs derived from these general balance equations. The examples in the present section will be one-dimensional PBEs that involve kinetic processes, only, that is, they apply to a fully mixed system where the physical space coordinates and transport terms vanish. Multidimensional PBEs and transport processes will be considered in the following sections.

Nucleation and Growth

Let us consider a population of particles with a distribution of volume, v, undergoing nucleation, growth and aggregation. Therefore, v is the only coordinate and $G = dv/dt$ is the corresponding rate of change. In general, G is a function that depends on v – for example, if growth is proportional to surface area, then $G(v) \sim v^{2/3}$. The PBE, therefore, is:

$$\frac{\partial n(v,t)}{\partial t} + \frac{\partial (G(v) n(v,t))}{\partial v} = B\delta(v - v_0), \tag{2.23}$$

where, as in Section 2.3.1, nucleation is represented as a delta function while v_0 is the size of a nucleus particle.

Nucleation and growth problems are often formulated using a measure of particle length as the independent variable, such as the diameter or radius of spherical particles. This is particularly convenient when growth is independent of length, which is often the case in crystal growth and also when the volumetric growth rate is proportional to surface area.[12] If we denote the length measure by ℓ, the PBE is:

$$\frac{\partial n(\ell,t)}{\partial t}+\frac{\partial(G_\ell(\ell)n(\ell,t))}{\partial \ell}=B\delta(\ell-\ell_0), \tag{2.24}$$

where ℓ_0 is the length of a nucleus particle. For size-independent growth, we have:

$$\frac{\partial n(\ell,t)}{\partial t}+G_\ell\frac{\partial n(\ell,t)}{\partial \ell}=B\delta(\ell-\ell_0). \tag{2.25}$$

This is a common form of the PBE in problems where no aggregation or fragmentation is present and has the form of a one-dimensional hyperbolic equation with a source term. The earliest use of this equation was by Randolph and Larson (1962) and Hulburt and Katz (1964), with application mainly in the field of crystallisation. Note that the growth term is typically written on the left-hand side, alongside convection in physical space (if present), to indicate the fact that growth is a convection term in the internal coordinate. In such a formulation, the boundary conditions would be the initial distribution and the condition that $n(0)=0$. An alternative approach is to formulate the PBE without a source term and introduce nucleation as a boundary condition at the lowest end of the volume domain (Randolph and Larson, 1988).

Aggregation and Fragmentation

The aggregation source term can be readily written as in Eq. 2.12. The fragmentation equation appeared first in the works of Melzak (1957a,b) and Valentas et al. (1966). Both works postulated the equation in the continuous form, rather than deriving it from the discrete one. The continuous fragmentation equation is:

$$\frac{\partial n(v,t)}{\partial t}=\int_v^\infty \gamma(w)b(v,w)n(w,t)dw-\gamma(v)n(v,t), \tag{2.26}$$

where $\gamma(v)$ is the fragmentation rate and $b(v,w)$ is the number density of particles of size v being formed from fragmentation of a particle of size w. The first term on the right-hand side represents the birth rate of particles of size v due to fragmentation events of particles of sizes $w>v$, while the second term represents the death rate of particles of size v. The forms of the functions $\gamma(v)$ and $b(v,w)$ must be based on physical considerations and will be considered in Chapter 3.

12 Let the volumetric growth rate be $dv/dt=kv^{2/3}$ and the length be $\ell=\alpha v^{1/3}$, where k is a proportionality constant and α is a shape factor. Then the linear growth rate is $d\ell/dt=(1/3)\alpha v^{-2/3}dv/dt=(1/3)\alpha k$, hence constant.

Complete PBE

We can now write the complete one-dimensional PBE for a spatially homogeneous system, with particle volume as the coordinate:

$$\begin{aligned}\frac{\partial n(v,t)}{\partial t} + \frac{\partial (G(v)n(v,t))}{\partial v} &= B\delta(v - v_0) \\ &+ \frac{1}{2}\int_0^v \beta(w, v-w)n(w,t)n(v-w,t)dw - n(v,t)\int_0^\infty \beta(v,w)n(w,t)dw \\ &+ \int_v^\infty \gamma(w)b(v,w)n(w,t)dw - \gamma(v)n(v,t). \end{aligned} \tag{2.27}$$

While the formulation in terms of volume simplifies the description of aggregation, it may sometimes be desired to use a measure of particle length instead. In that case, the aggregation kernel and integral assume a more complex form. An example of such an equation with nucleation, growth and aggregation can be found in the work of Hounslow et al. (1988), where the crystallisation of calcium oxalate monohydrate crystals was studied in the context of the formation of kidney stones, and aggregation was important due to its potential for rapidly generating large crystals.

2.3.3 Applications of Continuous PBE

The continuous PBE is the formulation employed in most applications. For systems where the size spectrum is inherently continuous, it is the only applicable formulation. Such problems include droplets (when evaporation is considered), bubbles, crystals and cell populations. The distribution of velocities in inertial particles is also treated via the continuous PBE.

For systems described by a discrete size spectrum, the continuous PBE offers the advantage of being a single partial integro-differential equation rather than a very large system of ODEs. The equation offers considerable challenges for its solution, and methods will be reviewed in Chapter 4. For now, we note that one of the main families of methods, the sectional methods, effectively transform the continuous PBE into a set of ODEs (discretised PBE) whose integration in time is similar to the discrete PBE. The difference is that a non-uniform grid can be employed to discretise the PBE in the domain of the internal variable, such that a small number of discretised number densities is required to cover a wide range of particle sizes.

For problems that can be described by both the discrete and continuous PBE, the choice is often dictated by the solution method. For some methods, such as Monte Carlo and certain moment methods, the choice is not important because the solution method applies to both formulations. Monte Carlo methods approximate the solution of the PBE with a notional ensemble of particles and scale with the number of notional particles employed. In moment methods, one is solving only for the moments of the distribution and therefore computes only a few integral properties in both cases. Sectional methods, on the other hand, are based on the continuous formulation. These issues will be discussed in detail in Chapter 4.

2.4 PBE with Multiple Distributed Properties

In some problems, there are more than one distributed properties that are significant. Some examples are listed below:

- In crystallisation, varying crystal morphologies may develop due to different growth rates along different directions. This behaviour can be described by considering the distribution of crystal size along different coordinates.
- Aerosols can have varying morphologies due to different degrees of sintering. This can be taken into account by considering a distribution of surface area or of number of primary particles in an aggregate, along with the aggregate size.
- Atmospheric aerosols may contain several chemical components and thus have a distribution of concentration of each component, as well as size.
- Particles with significant inertia have a distribution of velocity, as well as size. This problem will be discussed in the context of the spatially dependent PBE (Section 2.7).
- In spray combustion, one must consider the distribution of temperature, droplet size and velocity.

The most general approach to modelling such problems is to consider the joint distribution of all variables. The consideration of a multivariate distribution results in a multidimensional PBE. The generalisation of the PBE to multiple dimensions is straightforward, and some examples will be shown in Section 2.4.1. With respect to numerical solution, a multidimensional PBE is much more expensive to solve via discretisation methods. This is not a problem for spatially homogeneous or one-dimensional problems, but can be prohibitive if the PBE is to be coupled with a three-dimensional CFD computation. In that case, one possible remedy may be offered by the fact that the distribution may not be of similar importance with respect to all variables. For example, in a problem involving a distribution of size and surface area, the distribution of the latter may be of interest only as an intermediate process variable. In this case, a moment transformation with respect to some of the variables can be used to reduce the dimensionality of the PBE. As the moment equations are generally unclosed, a modelling assumption will likely be needed to obtain a closed model. This is also the concept underlying multi-fluid models of inertial particles, where a mean velocity is computed for each size class (cf. Section 2.7.4).

A second approach is to employ a system of one-dimensional PBEs for the distribution of each distributed quantity. This approach is more economic in terms of numerical solution via sectional methods, but requires some modelling assumptions. Some examples of this formulation will be shown in Section 2.4.2.

It must be mentioned that the introduction of spatial transport renders the PBE multidimensional in any case, due to the addition of the physical space coordinates, but this will be treated in Section 2.7. The formulations in the present section will refer to a spatially homogeneous system.

2.4.1 Multidimensional PBE

All of the terms in the PBE can be extended to several dimensions in a straightforward manner. It is better to formulate a multidimensional PBE directly for a specific problem, after identifying the need for introducing multiple dimensions, the processes that are present and the availability of sufficient kinetic information in order to make the most of the formulation. Physical insight about these points is very important, because a multivariate distribution contains a lot of information and introduces more parameters to the model. If the kinetic information about these parameters is not sufficiently accurate, the model will not be more accurate than a one-dimensional PBE or coupled PBEs. We will now show some examples of multidimensional PBEs.

Multidimensional Crystal Growth

Crystals can develop varying shapes due to different growth rates at different faces. One example of crystals featuring growth in two dimensions is shown in Fig. 2.2. Predicting the shape, as well as the size, of such crystals requires a multidimensional population balance.

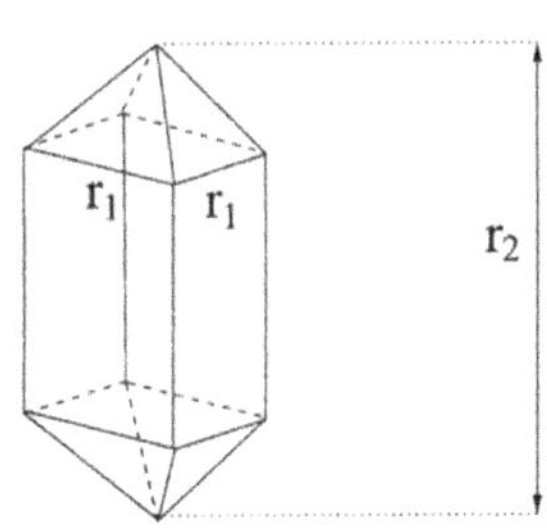

Figure 2.2 Geometry of KDP crystals exhibiting two-dimensional crystal growth; reprinted from Ma et al. (2002) with permission from American Chemical Society. In their notation, r_1 and r_2 are the two coordinates along which the multidimensional growth is formulated.

We denote the directions along which a crystal grows by a vector, ℓ_i, and define a growth rate $G_i(\boldsymbol{\ell})$ along each direction. The PBE for multidimensional crystal growth can then be written as follows:

$$\frac{\partial n(\boldsymbol{\ell},t)}{\partial t}+\frac{\partial (G_i(\boldsymbol{\ell})n(\boldsymbol{\ell},t))}{\partial \ell_i}=0. \tag{2.28}$$

This model describes a problem where growth is the only process present, as e.g. in the case of seeded growth. Information is required about the growth rate along different directions.

Examples of studies of multidimensional crystal growth are Ma et al. (2002) and Qamar et al. (2007). In the former, the following equation is used to simulate nucleation and two-dimensional growth of potassium dihydrogen phosphate (KH_2PO_4 or KDP):

$$\frac{\partial n(\boldsymbol{\ell},t)}{\partial t}+\frac{\partial (G_i(\boldsymbol{\ell},c,T)n(\boldsymbol{\ell},t))}{\partial \ell_i}=B(c,T), \tag{2.29}$$

where the nucleation and growth rates depend on concentration and temperature, and the PBE must be coupled with species and energy balances in an ideal reactor whose solution yields $c(t)$ and $T(t)$ (cf. Section 2.6). Fig. 2.2 shows the geometry of the KDP crystals featured in Ma et al. (2002).

Aggregate Morphology

Aggregate particles are composed of primary particles that may be fused to a certain extent. Examples of such particles are shown in Fig. 2.3. In this case, a distribution of aggregate sizes may not suffice for characterising the particles. For example, the surface area is often important, either as an intermediate variable (for determining

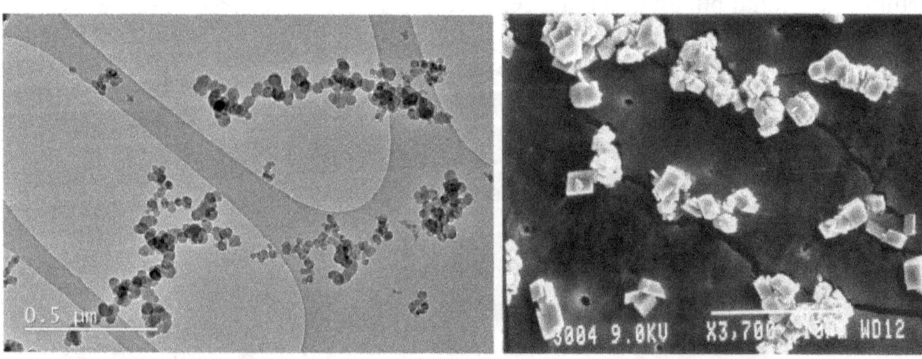

Figure 2.3 Examples of aggregate morphologies. Left: soot aggregates, courtesy of Carlos E. Garcia Gonzalez (Garcia Gonzalez, 2018). Right: agglomerated $CaCO_3$ crystals. Reprinted from Rigopoulos and Jones (2003b) with permission from American Chemical Society.

the rates of surface processes) or as a property of the final product. Aggregates of the same size may exhibit different surface areas, and one way of introducing this feature into the model is via a surface area distribution. Another way of viewing the problem is to consider the properties of the primary particles that compose the aggregate, such as their size and number, in addition to the size distribution of aggregates, an approach discussed in Section 2.4.2.

A two-dimensional discrete PBE that employs particle size and surface area was proposed by Koch and Friedlander (1990). The fusing of primary particles within an aggregate will be further discussed in Chapter 3, but the main underlying idea is that the surface area of a partially fused aggregate ranges between that of a completely unfused aggregate, whose surface area is the sum of those of the constituting particles, and that of a sphere of equal mass that results when fusing is complete. The number density function in the Koch and Friedlander formulation is denoted by $n(v,\alpha,t)$, where α is the aggregate surface area. Two rates are involved: the volumetric growth rate, G_v, and the rate of change of surface area due to particle fusing due to a mechanism such as sintering, $S_{f,\alpha}$ (in Section 3.2.3, we will see that this is a function of surface area). The kinetic expressions needed for these rates will be discussed in Chapter 3; for now, we just note that the former depends (in general) on particle volume while the latter depends on particle surface area. The two-dimensional PBE for growth and aggregation is:

$$\begin{aligned} \frac{\partial n(v,\alpha,t)}{\partial t} + \frac{\partial (G_v(v)n(v,t))}{\partial v} + \frac{\partial (S_{f,\alpha}(\alpha)n(v,t))}{\partial \alpha} &= \frac{1}{2}\int_0^v H(\alpha - \alpha_s(w) - \alpha_s(v-w)) \\ \cdot \int_{\alpha_{min}}^{\alpha_{max}} \beta(w, v-w, \alpha', \alpha-\alpha')n(w,\alpha',t)n(v-w,\alpha-\alpha',t)d\alpha dw \\ - n(v,\alpha,t)\int_0^\infty \int_{\alpha_{min}}^{\alpha_{max}} \beta(v,w,\alpha,\alpha')n(w,\alpha',t)d\alpha dw. \end{aligned} \tag{2.30}$$

In this equation, H is the Heaviside step function that takes the value 1 if the surface area of an aggregate is greater than the sum of the minimum surface areas of the initial particles, while $\alpha_s(w)$ denotes the surface of a sphere with volume w. The purpose of this function is to prevent unphysical aggregation events. The limits of the inner integrals are the minimum and maximum surface areas given by:

$$\alpha_{min} = \left(\frac{w}{v_0}\right)^{\frac{2}{3}} \alpha_0, \tag{2.31}$$

$$\alpha_{max} = \frac{w}{v_0}\alpha_0, \tag{2.32}$$

where v_0 and α_0 are the volume and surface area of the smallest primary particle, respectively.

The solution of the PBE in terms of two or more dimensions via discretisation increases substantially the CPU and memory requirements. One way of circumventing this issue is via the coupled PBE approach that will be presented in the following section.

Multicomponent Aerosols

In a multicomponent aerosol, one is interested in predicting the concentration of various chemical species in the particles. Therefore, the most complete description of this problem is by considering the concentrations as well as the particle size to be the independent variables in the number density function. Gelbard and Seinfeld (1978b) presented such a formulation and derived some analytical solutions for it. However, the development of numerical solution methods for this approach is formidable due to the large number of independent variables. This problem can also be approached via coupled PBEs, as will be discussed in the next section.

2.4.2 Coupled PBEs

A coupled population balance formulation is a system of two or more PBEs, each of which involves the distribution of a different property of the same population. These problems can also be described by a multidimensional PBE. Compared with that, the coupled PBE approach offers a significant computational advantage when discretisation methods are to be employed, because the discretisation of a system of k PBEs with n nodes for each independent variable requires kn nodes as compared with n^k needed by an equivalent k-dimensional formulation. However, the description of a problem in terms of multiple univariate distributions is not equivalent to the description in terms of a multivariate distribution, as it involves a loss of information. In many problems, however, this loss is negligible and the assumptions involved are justified.

In the following, we will focus mainly on the problem of aggregate morphology, for which the coupled PBE approach will be used in Chapter 6 in the case studies on soot and silica synthesis (Sections 6.2 and 6.1, respectively). We will also discuss some alternative approaches for describing aggregate morphology. Finally, we will briefly consider the discrete-continuous PBE and the coupled PBEs for multicomponent particles, both of which have found applications in aerosols.

Two-PBE Formulation for Aggregate Morphology

The problem of aggregate morphology was discussed in Section 2.4.1 from the point of view of the multidimensional PBE. Such formulations are often employed in the context of Monte Carlo methods but are very expensive to use with discretisation methods, especially if the latter are to be coupled with CFD. By contrast, a coupled PBE formulation offers a computationally efficient alternative, especially since, as we will see in Section 4.2.11, its numerical solution can be accomplished with a straightforward extension of discretisation methods for single PBEs.

Two formulations of coupled PBEs for aggregate morphology can be found in the literature. The first one, proposed by Rogak (1997) employs an additional PBE for the number of primary particles per aggregate. The second one, introduced by Tsantilis and Pratsinis (2000), formulates the second PBE for the particle surface area. The latter approach draws an analogy with the two-dimensional PBE where particle volume and surface area are the independent coordinates of the distribution (Eq. 2.30). In the coupled PBE formulation, however, an average surface area corresponds to each particle size, in contrast to the two-dimensional description where a distribution of surface areas corresponds to each value of particle volume. The impact of this assumption is usually not critical; for example, Jeong and Choi (2001) used a surface area-based coupled PBE approach for silica and titania nanoparticles and found the results to be in good agreement with the two-dimensional PBE while requiring considerably lower resources.

We will now focus on the approach that employs the number of primary particles in the second PBE, which has been used extensively in aerosol synthesis and soot formation[13] and will be employed in the case studies in Sections 6.2 and 6.1. The following description is based on the presentation of the method by Sun et al. (2021). We consider a population of particles that includes both primary particles and aggregates, the latter being composed of a number of primary particles. The smallest particles are assumed to instantly coalesce into spherical ones (which are also regarded as primary particles). Above a cut-off size, the fusing process is not fast enough for coalescence to occur before new particles join in, resulting in a fractal shape. This shape and its characterisation will be discussed further in Section 3.2.3. For now, the important point is that the number of primary particles per aggregate is the variable to be tracked by the second PBE. Thus, the problem is described in terms of two number density functions: one for the particles (this includes both spherical ones consisting of a single primary particle and aggregates), $n(v,t)$, and one for the primary particles, $n_p(v,t)$. A consequence of this formulation is that all aggregates of the same size are assumed to have the same number of primary particles per aggregate, which is to be regarded as an average.

[13] See Park and Rogak (2003) for aerosol synthesis and Park et al. (2005), Wen et al. (2005, 2006), Zhang et al. (2008, 2009a,b) for soot formation.

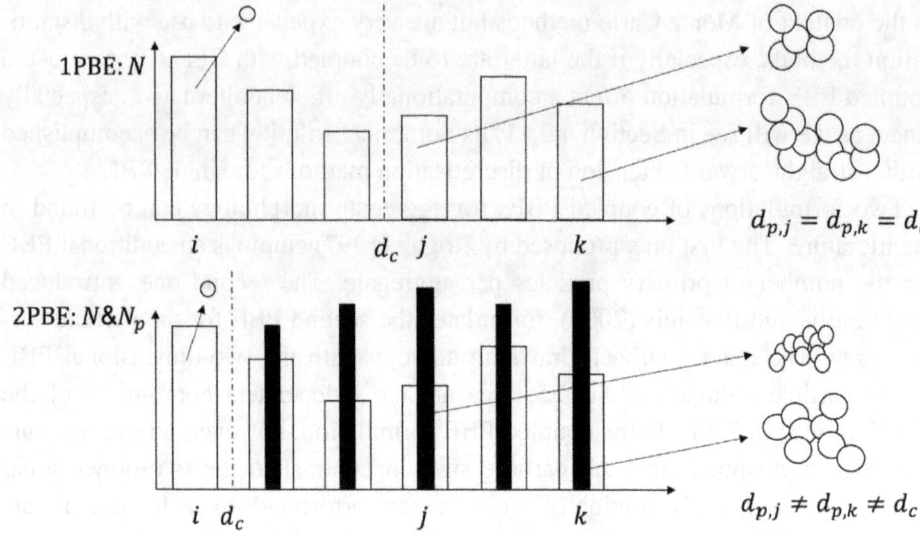

Figure 2.4 A comparison of the concepts involved in the one-PBE and two-PBE models for aggregation. In the lower figure, the white histogram represents number of aggregates, while the black histogram represents number of primary particles. N and N_p are numbers of particles and primary particles, respectively (in the notation used by Sun et al., 2021), while d_c is the diameter at the cut-off point. Reproduced from Sun et al. (2021) under CC-BY 4.0 license.

It should be noted that one-PBE formulations have also been used for problems involving fractal aggregates. This is accomplished by imposing a cut-off size between spherical particles and fractal aggregates and assuming that all primary particles within aggregates have the same size as the largest primary particles (i.e. the cut-off size). In this approach, the PBE for aggregation needs no modifications in terms of overall form and the only change involved is in the aggregation kernel, which should be written in terms of the collision diameter (cf. Section 3.2.3). While simple and computationally efficient, this approach has several shortcomings. The cut-off size needs to be prescribed empirically. Also, it is not possible to account for finite-rate fusing of primary particles. Finally, surface growth, if present, is calculated in an inconsistent manner, because the size of primary particles is assumed to be 'frozen' once they have entered an aggregate, while the size of the whole aggregate is changing due to surface growth. However, this approach can still yield reasonable results when the prediction of morphological parameters is not desired; an example of this will be shown in Section 6.2. Fig. 2.4 illustrates and compares the main concepts of the one- and two-PBE approaches.

To simplify things, we consider only aggregation and fusing of primary particles (by a process such as sintering). The first equation of the model is the PBE for the number density of particles and is identical to a conventional one-dimensional PBE, since the fusing process does not change this number density:

$$\frac{\partial n(v,t)}{\partial t} = \frac{1}{2}\int_0^v \beta(w,v-w)n(w,t)n(v-w,t)dw - \int_0^\infty \beta(v,w)n(v,t)n(w,t)dw. \tag{2.33}$$

The PBE for $n_p(v,t)$ can now be obtained based on the following considerations. We first note that the average number of primary particles contained in aggregates of volume v is $n_p(v,t)/n(v,t)$. Therefore:

- If two particles with volumes w and $v-w$ combine to form an aggregate of volume v, the number of primary particles in the daughter particle is $n_p(w,t)/n(w,t) + n_p(v-w,t)/n(v-w,t)$. Therefore, the birth rate of primary particles is:

$$B_{n_p} = \beta(w,v-w)n(w,t)n(v-w,t)\left(\frac{n_p(w,t)}{n(w,t)} + \frac{n_p(v-w,t)}{n(v-w,t)}\right).$$

- When an aggregate with volume v is combined with any other particle, the resulting death rate of primary particles can be written as:

$$D_{n_p} = \beta(v,w)n(v,t)n(w,t)\frac{n_p(v,t)}{n(v,t)} = \beta(v,w)n(w,t)n_p(v,t).$$

- A source term should be introduced to represent the rate at which the primary particles fuse. The physics and form of this term will be discussed in Section 3.2.3; for now, it suffices to say that it is a function of n and n_p and can thus be denoted as $S_{f,np}(n,n_p)$.

We can thus write the second PBE as follows:

$$\frac{\partial n_p(v,t)}{\partial t} = \frac{1}{2}\int_0^v \beta(w,v-w)n(w,t)n(v-w,t)\left(\frac{n_p(w,t)}{n(w,t)} + \frac{n_p(v-w,t)}{n(v-w,t)}\right)dw - n_p(v,t)\int_0^\infty \beta(v,w)n(w,t)dw + S_{f,np}(n,n_p). \tag{2.34}$$

Various morphological parameters can be extracted from knowledge of n and n_p, as we will see in Sections 6.2 and 6.1.

Extended One-PBE Formulation for Aggregate Morphology

An extended one-PBE approach that incorporates some features of the two-PBE approach was proposed by Tsagkaridis et al. (2023). In this approach, the PBE is complemented by an additional equation for the total number of primary particles. This is obtained with the assumption that the primary particle size is monodisperse, a concept that is the basis of the monodisperse models discussed in Section 2.5. Here, however, it is only the primary particles that are assumed to be monodisperse. If only aggregation and fusing is considered, then only the fusing affects the total number of primary particles and the additional equation is:

$$\frac{\partial N_p}{\partial t} = S_{f,np}(n, n_p), \tag{2.35}$$

where N_p is the total number of primary particles. The PBE, in this case, is Eq. 2.33. The main advantages of the extended one-PBE model over the single PBE are that it allows for variable size of primary particles and for introducing a kinetic model for the fusing process. The model was proposed for application to inorganic nanoparticle synthesis, where nucleation, aggregation and sintering are the main kinetic processes. Its application to problems with surface growth may be more questionable. An application to silica synthesis will be shown in Section 6.1.

Discrete-Continuous PBE

The discrete-continuous PBE is a coupled population balance formulation that has been employed for the description of certain problems in aerosol dynamics, where it is important to describe with accuracy the processes between the molecular clusters at the lower end of the size spectrum and their interaction with the aerosol particles. The molecular clusters comprise an integer number of monomers and are thus better described by the discrete PBE. On the other hand, the particles exhibit a very wide size spectrum and would thus require a huge number of size classes if the discrete PBE were to be applied throughout. The discrete-continuous PBE does not correspond to any two-dimensional formulation, as the molecular clusters and particles are two different populations, which however interact as mass is transported between them.

The discrete-continuous PBE was first proposed by Gelbard and Seinfeld (1979) and solved numerically with a method developed in Gelbard and Seinfeld (1978a), while further numerical methods for its solution were developed by Wu and Flagan (1988) and Landgrebe and Pratsinis (1990). It comprises three equations: a discrete PBE for description of molecular clusters, a continuous PBE for the description of the aerosol particles and a discrete-continuous PBE for the processes between molecular clusters and particles. The discrete and the continuous PBEs are variations of Eqs. 2.4 and 2.12, respectively, while the discrete-continuous PBE is shown in Gelbard and Seinfeld (1979).

Multicomponent Aerosols

Multicomponent aerosols constitute another problem that can be approached by coupled PBEs, as well as by a multidimensional PBE (cf. Section 2.4.1). In such problems, one is interested in predicting the concentration of various chemical species in an aerosol, apart from particle size. Gelbard and Seinfeld (1980) put forward a discretisation method based on the concept of two-dimensional quadrature approaches that accounts for aggregation, nucleation, growth, deposition and intraparticle chemical reaction in a multicomponent aerosol. The equations derived are extensions of the single PBE, with a separate equation written for the mass of each chemical species in particles within a certain size section. They make the assumption that the mean species mass fraction is the same for all particles within the same size section (equivalent to the assumption that the number of primary particles or the surface area is the same in all aggregates of the same size that was made in the two-PBE model of aggregate morphology). This assumption was questioned in a later work (Kim and Seinfeld, 1990).

2.5 Monodisperse PBE

A monodisperse PBE is a formulation that assumes that all particles are identical. The population is described by means of integral properties, such the total number and total volume of particles. Monodispersed PBEs are therefore similar to solving the PBE for the moments of the distribution, an approach that will be considered in Chapter 4. However, monodisperse PBEs are not derived from the PBE via a moment transformation, but rather formulated directly based on modelling assumptions, the most important of which is that the particles are monodisperse. Moment formulations also require closure assumptions, but those are usually weaker than the monodisperse assumption. For example, in a moment method it may be assumed that the distribution has always the same form (typically is a standard one, such as the log-normal distribution).

The advantages of monodispersed PBEs are simplicity and computational efficiency, as the formulation involves only a few variables, and this is particularly beneficial for coupling with fluid dynamics. However, the distribution is not being resolved and, most importantly, the modelling assumptions will likely constrain the range of application of the model. It is therefore important that the model be tested against complete PBE solutions or experiments to ensure that these assumptions are satisfied under the range of conditions in which it will be used. Since monodisperse PBEs have modelling assumptions built into the formulation, it is not possible to write a general form and the equations are postulated for the particular problem.

One example of such a formulation includes the monodisperse soot models that were encountered in early soot research, such as the model of Leung et al. (1991). These models are usually formulated in terms of two variables, soot number density and volume fraction. Their use was motivated by their simplicity and the need for coupling with fluid dynamics, as soot must be resolved in the context of a flame simulation. As more information on the physics and chemistry of soot formation has become available, it is now known that the assumptions of monodisperse soot models are not very realistic. However, these models have found a number of applications in CFD simulations of turbulent sooting flames (e.g. Kronenburg et al., 2000).

Another case of a monodisperse PBE is the model of nanoparticle synthesis via aggregation and sintering (Kruis et al., 1993). This formulation is frequently employed in applications because its assumptions have been demonstrated to be valid under a range of conditions. The model describes nanoparticles using three variables, namely the number concentration, N, surface area per aggregate, α, and volume per aggregate, v. The number concentration is reduced by aggregation, and therefore the equation is:

$$\frac{dN}{dt} = -\frac{1}{2}\beta N^2. \tag{2.36}$$

The equation for surface area of a single aggregate includes a source term due to aggregation and a sink term due to fusing (sintering):

$$\frac{d\alpha}{dt} = -\frac{1}{N}\frac{dN}{dt}\alpha + S_{f,\alpha}(\alpha), \tag{2.37}$$

where the first term on the right-hand side represents the rate of increase by addition of particles without sintering while $S_{f,\alpha}$ is the rate of change of surface area due to sintering, the form of which will be discussed in Section 3.2.3; it has already been encountered in Eq. 2.30 and plays the same role as $S_{f,np}$ in Eq. 2.34. Finally, the rate of change of the volume of an aggregate is given by the following equation:

$$\frac{dv}{dt} = -\frac{1}{N}\frac{dN}{dt}v, \tag{2.38}$$

which indicates that the volume is changing due to aggregation but not sintering.

The monodisperse formulation of the aggregation-sintering problem can be compared with the multidimensional and coupled PBEs employed for the same problem and discussed in Sections 2.4.1 and 2.4.2, respectively. The applicability of the monodisperse model for nanoparticle synthesis was investigated by Mühlenweg et al. (2002) and further developed to account for particle formation from a precursor by Tsantilis and Pratsinis (2004). An equivalent formulation that employs the total number of primary particles, rather than the surface area, will be shown in Section 6.1 and compared with an extended one-PBE and a two-PBE model.

2.6 Coupling of the PBE with Ideal Reactor Models

Most population balance problems take place within fluid flows. The most rigorous way of accounting for spatial transport is via the spatially dependent PBE, which will be discussed in Section 2.7. If, however, the timescale for mixing is much smaller than that of the physical and chemical processes in the PBE, it may be possible to employ a lumped model where the spatially homogeneous PBE is applied within the whole domain or a number of fully-mixed regions. This approach is much easier to deal with computationally and has been used in a big part of the PBE research to date. Ideal reactors are also used for experiments with the purpose of extracting kinetics for the PBE processes.

Ideal Reactor Models

A perfectly stirred reactor (PSR) is a model for a reactor in which, due to rapid mixing, the concentrations and temperature are assumed to be homogeneous. With no flow in or out, the model is called a batch PSR, while with an inlet and an outlet it is called a continuous stirred-tank reactor (CSTR) model. A plug flow reactor (PFR) is a model for a tubular reactor, where it is assumed that the timescale for lateral mixing is much smaller than the timescale of the kinetic processes in the PBE, so that the concentrations and temperature are uniform at every cross-section. These models are shown schematically in Fig. 2.5. Networks of ideal reactors can also be employed to describe more complex flow configurations where it is still possible to identify ideal flow patterns in parts of the system.

Formulation of PBE in Ideal Reactor Models

The PBEs given in Sections 2.2 and 2.3 are directly applicable to a batch reactor. They can also be applied to a PFR with a change of variable from time to distance, since in the PFR model we are effectively tracking the evolution of a fluid particle behaving as an ideal batch reactor. If u is the velocity in the PFR and x is the distance along the tube, then $u = dx/dt$, while number density function, $n(v,x)$, depends on particle volume and distance. In a PFR with incompressible flow, u is

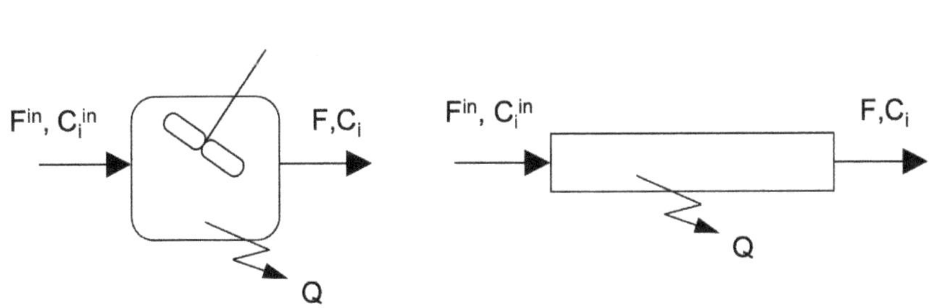

Figure 2.5 Schematic depiction of ideal reactor models. Left: perfectly stirred reactor with flow in/out and heat exchange. The properties (concentrations, temperature, number densities) are assumed to be uniform within the reactor and vary with time. Right: plug flow reactor with flow in/out and heat exchange. The properties are assumed to vary with distance along the reactor.

constant. The one-dimensional PBE with growth and a source term, for example, in the context of a steady-state PFR is the following:

$$u\frac{\partial n(v,x)}{\partial x}+\frac{\partial (G(v)n(v,x))}{\partial v}=S(v,n). \tag{2.39}$$

For a CSTR, that is, a PSR with flow in and out, Eq. 2.39 must be augmented as follows:

$$\frac{\partial n(v,t)}{\partial t}+\frac{n(v,t)-n_{in}(v,t)}{\tau_{res}}+\frac{\partial (G(v)n(v,t))}{\partial v}=S(v,n), \tag{2.40}$$

where $n_{in}(v,t)$ is the inlet number density function and τ_{res} is the residence time in the reactor. Eq. 2.40 describes the dynamical evolution of the distribution in the reactor. If it can be assumed that steady state has been reached, then the following equation is obtained:

$$\frac{n(v)-n_{in}(v)}{\tau_{res}}+\frac{d(G(v)n(v))}{dv}=S(v,n). \tag{2.41}$$

These concepts can be applied to any of the population balance formulations derived so far. For example, the aggregation of particles in steady state in a flow reactor can be described by the CSTR analogue of the Smoluchowski equation (Eq. 2.4):

$$\frac{n_i-n_{i,in}}{\tau_{res}}=\frac{1}{2}\sum_{j=1}^{i-1}\beta_{j,i-j}n_jn_{i-j}-n_i\sum_{j=1}^{\infty}\beta_{ij}n_j. \tag{2.42}$$

When nucleation and growth are present, the PBE must be coupled with the species mass balance equations and possibly the heat transfer equation. The formulation of these equations for ideal reactors can be found in any book on chemical reaction engineering, such as Froment et al. (2011).

Compartmental Models

More complex flow configurations may be described by a network of ideal reactors, if it is possible to identify fully mixed zones. An example of such a model used for population balance modelling of crystallisation is shown in Fig. 2.6. A systematic method for determining such networks has been proposed by Rigopoulos and Jones (2003a). Furthermore, various methods have can be employed to deal with non-ideal mixing, such as the residence time distribution (RTD), for which the reader may refer e.g. to Chapter 12 of Froment et al. (2011). An in-depth discussion of the phenomenological mixing models that can be used in the context of ideal reactors and reactor networks can be found in Chapter 8 of Bałdyga and Bourne (1999). Further discussion of compartmental models particularly for the PBE can be found, for example, in Menz et al. (2014) and Kotalczyk and Kruis (2020).

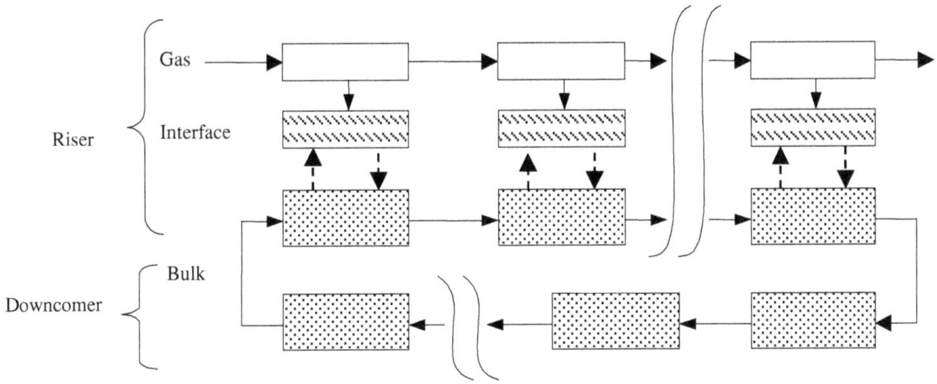

Figure 2.6 Compartmental model for a draft-tube bubble column reactor employed for population balance modelling of crystallisation. Reprinted with permission from Rigopoulos and Jones (2001).

Partially Stirred Reactor Models

The partially stirred reactor (PaSR) is a stochastic mixing model that can be used to account for non-ideal mixing without coupling with a CFD model. It can be viewed as a special case of the transported probability density function (PDF) approach that will be discussed in Chapter 5. As such, the equation describing the PaSR will be shown in Section 5.5.1.

Deposition

When problems involving aerosols or particles in a liquid phase are modelled via ideal reactors, it may be necessary to account for deposition. In the context of the spatially dependent PBE, deposition will appear only at the boundary layer. In the context of ideal reactors, however, deposition must be accounted for via an additional term in the PBE. For example, a continuous PBE for aggregation and deposition within a batch reactor would be:

$$\begin{aligned}\frac{\partial n(v,t)}{\partial t} &= \frac{1}{2}\int_0^v \beta(w,v-w)n(w,t)n(v-w,t)dw \\ &\quad - n(v,t)\int_0^\infty \beta(v,w)n(w,t)dw - D(v)n(v,t), \qquad (2.43)\end{aligned}$$

where the deposition rate, $D(v)$, will depend on the deposition mechanism.

Applications of PBE in Ideal Reactor Models

The PBE in the context of ideal reactor models has been the main formulation encountered in the PBE literature for a long time. This has been partly motivated by the complexity and computational cost of coupling the PBE with fluid dynamics. Tavare (1986) has reviewed ideal reactor models in crystallisation. An example from nanoparticle synthesis can be found in Akhtar et al. (1991). In the field of atmospheric aerosols, networks of ideal reactors are routinely used to model the very large scales of atmospheric phenomena and are sometimes called 'box' models. In such models, the dimensions of the box are allowed to vary with time. More details on the formulation of such models can be found in Chapter 25 of Seinfeld and Pandis (2016).

Apart from the computational convenience in modelling spatially distributed systems, ideal reactor models are employed for extracting the kinetics for the processes in the PBE. In this case, an experiment is designed so as to minimise the effect of mixing. An example is the study of Bramley et al. (1996), where a method for extracting kinetics of nucleation, growth and aggregation in crystallisation is presented. The extraction of population balance kinetics from experimental data (or microscopic-level simulations) is called the *inverse population balance problem* and will be discussed in Section 3.5.

2.7 Spatially Dependent PBE and Coupling with Fluid Dynamics

2.7.1 Preliminary Considerations

We now consider problems where the distribution is spatially dependent and transport terms must be considered in the PBE. In these cases, the latter should also be coupled with the equations of fluid dynamics and possibly heat and mass transfer. The interaction of the PBE with the flow field is important for several reasons:

- The local environment may affect the processes that determine the distribution. For example, crystals are formed from dissolved species, aerosols are formed from precursors, and temperature affects the rate of formation processes. Flow also determines the spatial distribution of the particles and therefore the local surface area, on which processes such as growth and evaporation depend. Furthermore, aggregation and fragmentation processes depend on local variables such as temperature and shear stress.
- Feedback effects on the flow and other variables may have to be considered, such as consumption of reactants by particle formation or changes in temperature due to radiation from particles.
- The particles may move with velocities that depend on the distributed property. For example, particles with significant inertia will experience a drag force that depends on their size.

An example of a case where the coupling of the PBE with fluid dynamics is crucial is the simulation of $BaSO_4$ precipitation in a T-mixer shown in Fig. 1.5.

The driving force for precipitation is supersaturation, which controls nucleation and growth. High supersaturation favours nucleation, and thus control of supersaturation allows control of CSD. Mixing is induced via two impinging jets and generates thin zones of high supersaturation with very high nucleation rates. This effect can only be captured by coupling the PBE with fluid dynamics. This and more cases featuring strong coupling of fluid dynamics with the PBE will be shown in Chapter 6.

Several ways of formulating the spatially dependent PBE are possible, and a number of considerations must be made before choosing the right approach.

First of all, the importance of spatial effects must be considered before deciding to employ a spatially dependent formulation. If the timescales of the kinetic processes are much larger than the flow timescale, spatially homogeneous zones may be identified and the PBE may be applicable in the context of a lumped model such as an ideal reactor or reactor network.

Furthermore, the particles must be considerably smaller than the characteristic scale of the flow in order to be described locally in a statistical manner, that is, by a local number density and size distribution. For small particles, such as aerosols and nanoparticles, this is usually the case, even in small-scale flows. This condition is also easily satisfied in problems with large-scale flows, such as the dispersion of atmospheric particles. If it is essential to resolve the flow around individual particles, then a different approach is required, and such approaches are discussed in multiphase flow treatises (see e.g. Crowe et al., 1998).

The particle concentration must be also considered to determine whether the flow is dilute or dense, and therefore, whether interparticle interactions and feedback effects on the flow are present. Dense flows with significant interparticle interactions and feedback effects on the flow, such as fluidised beds, are also treated with the methods of multiphase flow.

If the particles have considerable inertia, their velocities will be different from the local flow velocity. The most general way to account for inertia in the population balance is to consider a distribution of particle velocities. This leads to a form of Boltzmann equation which calls for the methods of kinetic theory (Reeks et al., 2017; Reeks, 2021). If the velocity distribution is not of importance per se, then resolving it in detail is not necessary and a simpler model is sought, which however may require closure assumptions. Multi-fluid models employ a discretised particle size distribution and solve a momentum equation for the number density of each size class. As such, they are a form of coupled PBE model (Section 2.4.2) based on the assumption that all particles of the same size will move with an average velocity. A further reduction results in two-fluid models, where an average velocity (the first moment of the distribution with respect to velocity) is utilised for the entire particulate phase. Another approach is the stochastic simulation of

Lagrangian trajectories of parcels of particles with different inertia. While these approaches are also treated in multiphase flow treatises, the links between them and the population balance framework will be discussed. The focus of the methods presented in the present book, however, will be on the problems where the particle velocity does not enter as an independent variable in the PBE. This is the case for non-inertial particles and also for particles of low inertia, in which case an approximation can be made for obtaining their velocity as a function of the local fluid velocity (cf. Section 2.7.3).

The effect of turbulence on the population balance is manifested in three ways. Fluctuations in the fluid velocity result in turbulent dispersion of particles. Fluctuations in other variables, such as temperature and precursor species, result in nonlinear effects in the particle formation mechanisms such as nucleation and growth. Turbulence has also effects on the physical mechanisms of aggregation and fragmentation. These issues will be discussed in Chapter 5.

While both the discrete and the continuous PBE can be extended to include spatial transport, the latter is more suited for this purpose. This is mainly because the numerical solution of a spatially dependent PBE requires coupling with CFD, as mentioned in Section 2.2.4, a discrete formulation would result in an extremely large number of variables. Therefore, the continuous PBE will be used as the basis for the developments in this section. Also, the extensions will be based on the PBE with one variable (particle volume) for the sake of brevity, but it should be noted that the multidimensional and coupled PBE formulations can be extended in the same way.

2.7.2 PBE for Non-Inertial Particles

If the particles have negligible inertia, their velocity can be assumed to be equal to the flow velocity.[14] This can be ascertained in a quantitative way via the Stokes number, which indicates the importance of inertial effects on particle motion and is defined as follows:

$$St = \frac{\tau_r}{\tau_f}, \tag{2.44}$$

where τ_r is the *relaxation time*, which is indicative of the time it takes for the particle to respond to changes in fluid velocity, and τ_f is a characteristic timescale of the flow. The physical meaning and calculation of these parameters is further discussed in Section 3.6.5 but, for now, it will suffice to note that the relaxation time typically scales with the square of the particle diameter, which means that the Stokes number

[14] If thermophoresis is accounted for, the thermophoretic velocity is added to the flow velocity; see Section 3.6.3.

is very small for small particles and usually close to zero for sub-micron ones. It must be emphasised, however, that the Stokes number depends on both the particle properties and flow structures. In the limit of $St \to 0$, the particles are non-inertial and follow the streamlines of the fluid flow. Another effect that must be considered for sub-micron particles is Brownian motion, which will be discussed in Section 3.6.4; for now, we note that it gives rise to a diffusion term in physical space. The PBE for non-inertial particles is then Eq. 2.21, shown below for reference:

$$\frac{\partial n}{\partial t} + \frac{\partial (u_i n)}{\partial x_i} + \frac{\partial (\dot{\xi}_k(\boldsymbol{\xi}) n)}{\partial \xi_k} = \frac{\partial}{\partial x_i}\left(D_p(\boldsymbol{\xi})\frac{\partial n}{\partial x_i}\right) + S(\boldsymbol{\xi}, n), \tag{2.45}$$

where $n(\boldsymbol{\xi}, \mathbf{x}, t)$ is the spatially dependent number density.[15] As before, the source term includes the kinetic processes. The fluid velocity must be obtained by a solution of the continuity and Navier–Stokes equations, which are shown in Appendix A (Sections A.1.1 and A.2, respectively). It is likely that the kinetics of nucleation and growth will involve the temperature and concentrations of reactants, in which case the heat and mass transport equations must also be included in the model; these are shown in Sections A.3 and A.1, respectively, of Appendix A.

The PBE for non-inertial particles is appropriate for many problems such as atmospheric aerosols, nanoparticle synthesis, soot and crystallisation. An example of an application to crystallisation was shown in Fig. 1.5, where the PBE with nucleation and growth terms has been coupled to a direct numerical simulation of turbulent flow in a T-mixer to simulate the precipitation of $BaSO_4$ crystals. This and more cases will be discussed in Chapter 6.

2.7.3 PBE for Inertial Particles with Low Stokes Number

When the inertia of the particles is not negligible, their motion starts to deviate from the fluid streamlines and further models are needed for the transport terms in the PBE. In the most general case, the distribution of particle properties must be extended to include the particle velocity components. Before discussing that case, we first consider an intermediate case applicable to low Stokes number inertial particles, with St less than about 0.2 (Balachandar and Eaton, 2010). In that case, it has been shown (De La Mora and Rosner, 1982; Maxey, 1987; Druzhinin, 1995; Ferry and Balachandar, 2001) that the particle velocity can be obtained as a function of the fluid velocity (cf. Section 3.6.5, Eq. 3.145). In that case, there is no need to consider a distribution of velocities and the PBE can be written as in Eq. 2.22, but with the acceleration term omitted:

[15] In this and the following equations, variable dependence will be omitted for brevity.

$$\frac{\partial n}{\partial t} + \frac{\partial (u_{p,i}n)}{\partial x_i} + \frac{\partial (\dot{\xi}_k(\boldsymbol{\xi})n)}{\partial \xi_k} = S(\boldsymbol{\xi}, n), \tag{2.46}$$

where $\mathbf{u_p}(\mathbf{x}, t)$ is to be furnished as a function of $\mathbf{u}(\mathbf{x}, t)$ by Eq. 3.145. While the diffusion term is usually neglected for inertial particles due to their larger size (see Section 3.6.4 regarding the size dependency of particle diffusion coefficient), there are some problems that feature simultaneous diffusion and inertial effects that require special treatment, for which the reader may refer to the review of Drossinos and Housiadas (2017). The PBE with the low Stokes number approximation has been applied to problems involving particles (Pilou et al., 2011), droplets (Aiyer et al., 2019) and bubbles (Yang et al., 2016).

2.7.4 PBE for Inertial Particles with Arbitrary Stokes Number

For inertial particles where the approximation shown in the previous section cannot be applied, the distribution of particle velocity must be taken into consideration. In that case, the particle velocity should be regarded as an independent variable and the number density can be written as $n(\boldsymbol{\xi}, \mathbf{u_p}, \mathbf{x}, t)$, where $\boldsymbol{\xi}$ denotes the distributed variables other than velocity. It can be seen that the dimensionality of the problem increases considerably, since velocity includes three components (for three-dimensional flows). Furthermore, an additional term (the acceleration) is required for transport in the particle velocity coordinates. The PBE for inertial particles has already been shown in the form of Eq. 2.22, and is shown again below with $\mathbf{u_p}$ taken out of the derivative to emphasise the fact that it is now an independent variable (unlike the case of the low Stokes number approximation in the previous section, where it was a field to be computed from the fluid velocity):

$$\frac{\partial n}{\partial t} + u_{p,i}\frac{\partial n}{\partial x_i} + \frac{\partial (\dot{u}_{p,i}n)}{\partial u_{p,i}} + \frac{\partial (\dot{\xi}_k(\boldsymbol{\xi})n)}{\partial \xi_k} = S(\boldsymbol{\xi}, n), \tag{2.47}$$

where $\dot{\mathbf{u}}_\mathbf{p}$ is the particle acceleration, which is equal to the net force per unit mass acting on the particles and includes forces such as drag and gravity. The diffusion term that accounts for Brownian motion is not considered here because inertial particles of higher *St* are large enough for it to be negligible.

In a problem involving particle collisions, the source term requires consideration of collisions of particles with different sizes and velocities, in the manner of kinetic theory. Reviews of the theoretical framework of this approach can be found in Reeks et al. (2017) and Reeks (2021).

Eq. 2.47 is coupled with the continuity and Navier–Stokes equations via an expression for the force on the particles, which includes the fluid velocity, as will be shown in Section 3.6.5. If the system is dense and there is a feedback effect on the flow, the Navier–Stokes equations will have to be modified accordingly with additional source terms.

As an example, we consider the formulation for sprays proposed by Williams (1958, 1961). The distribution of droplet radius, r, needs to be considered, in addition to that of velocity. Eq. 2.47 then takes the following form:

$$\frac{\partial n}{\partial t} + u_{p,i}\frac{\partial n}{\partial x_i} + \frac{\partial (G_r n)}{\partial r} + \frac{\partial (F_i n)}{\partial u_{p,i}} = S(r,n), \tag{2.48}$$

where $\mathbf{F}$ is the net force per unit mass acting on the particles, which is equal to their acceleration. The radial growth rate, G_r, represents droplet evaporation in this case.

The numerical solution of a PBE with velocity distribution via discretisation in all coordinates is very computationally expensive due to its high dimensionality. However, it is not usually required to retrieve such a multidimensional distribution with accuracy. In practice, such formulations are often solved with methods where one computes the Lagrangian trajectories of a number of stochastic entities, each representing a number of particles. More information on such methods can be found in Chapter 7 of Crowe et al. (1998), and examples of such simulations are Dukowicz (1980), Gosman and Ioannides (1983), Jones and Sheen (2000) and Bini and Jones (2008).

2.8 Summary

The objective of this chapter was to present the various formulations of the PBE under a unifying framework and to relate them to the problems that they can be applied to. Several formulations were identified:

- The discrete PBE is appropriate for populations described by particles comprising an integer number of units. Such problems include colloid flocculation, aerosol formation and polymerisation.
- The continuous PBE is appropriate for problems where the distribution is inherently continuous or as a more computationally efficient description of discrete populations. Examples of the first type are droplets (when evaporation is considered), bubbles, cells and crystals, while examples of the second type are aerosol and soot formation.
- Problems involving multivariate distributions, such as multicomponent aerosols and particles with varying morphology, can be described by a multidimensional PBE or by coupled PBEs under certain assumptions. The discrete-continuous PBE is a coupled PBE for aerosol formation where the distribution of a discrete property (molecular clusters) and a continuous one (particle size) must be combined.

- Monodisperse PBEs rely on modelling assumptions. When they can be applied, they are very computationally efficient.
- PBEs can be coupled with ideal reactor models when the timescales of the physical and chemical processes in the PBE are of different order of magnitude than those of the flow processes.
- Spatially dependent problems require coupling of the PBE with fluid dynamics. This can be accomplished at several levels. The most general one involves a distribution over space and velocity, as well as the internal properties. This formulation has been used in spray combustion.
- A moment transformation with respect to velocity leads to the multi-fluid and two-fluid models. This approach is often used in multiphase flow problems, such as dispersed flows with bubbles and droplets.
- When inertial effects are present but not strong, the low Stokes number approximation can be employed to yield the particle velocity as a function of the fluid velocity. In this case, the velocity does not enter the PBE as an independent variable and the latter can be readily coupled with the equations of fluid dynamics. This approach can be used for particles and droplets, depending on their size and the scale of the flow structures.
- When inertial effects can be neglected, the particles follow the streamlines of the flow. As in the last case, the PBE can be readily coupled with the equations of fluid dynamics. This approach is suitable for soot, nanoparticle synthesis and many aerosol and crystallisation problems.

The formulation of a population balance model requires physical insight into the problem, in order to select a model that is both accurate and economic in terms of its solution. Chapters 3 and 6 will present the models and solution methods, respectively, while Chapter 5 will discuss both aspects for problems involving turbulent flow. Finally, in Chapter 6, we will see some case studies of formulation, model and method selection for particular problems.

3
Kinetic and Transport Processes

3.1 Scope and Classification of Models

The PBE includes kinetic and transport processes, and must thus include models that play the role of constitutive equations, The aim of the present chapter is to discuss the incorporation of physical and chemical models with the population balance. While the range of problems that can be described by the population balance approach is very broad, the physicochemical models employed are problem specific. Hence, the coverage here will be focussed on a class of physical systems that share common features so as to allow a reasonably unified treatment. The models presented in this chapter are also featured in the case studies presented in Chapter 6. The general theory and formulation of the population balance as presented in Chapter 2 is, however, applicable to numerous other systems.

The models presented here are mostly relevant to a family of problems that includes aerosols (small solid particles or droplets dispersed in a gas) or colloidal particles dispersed in a liquid (such as the crystals formed in a crystallisation process). These problems are characterised by small particle sizes, sometimes so small that non-continuum effects are important, as in the case of aerosol nanoparticles. With respect to motion within a flow field, such particles can usually be described as non-inertial or by the low Stokes number approximation. Thermal motion (free molecular or Brownian) is often important in this class of problems and provides a diffusive transport process as well as an aggregation mechanism. Aggregation, nucleation and growth are the main kinetic processes present, while fragmentation can also occur in the larger particles. When coupled with transport processes, these problems can be described by a coupled fluid dynamics – population balance approach. This is also the class of problems featured in the case studies in Chapter 6.

Larger droplets and bubbles constitute another family of problems with common characteristics. Inertial forces are important for these problems, the dominant ones being drag and buoyancy, while other hydrodynamic forces such as lift have to be considered in some cases and Brownian motion can be ignored. Deformation can occur and irregular particle shapes play a role in transport and interaction processes. If the volume fraction of the particulate phase is high, the feedback on the flow may also need to be considered. Break-up is important for most problems in this category. While these problems are described by the population balance, the numerical methods employed for their solution are more often the Lagrangian or multi-fluid methods described in the multiphase flow literature. However, many of the basic principles of the models presented in this chapter, particularly for coalescence and break-up, are relevant in that field.

Other major families of problems described by population balance models were mentioned in Section 1.1.2 and include biological, chemical (polymerisation) and astrophysical problems. While many of the formulations and solution methods covered in Chapters 1, 2 and 4 can still be applied in these fields, the discussion of the pertinent models would enlarge too much the scope of this chapter. More information on these subjects can be found in the references provided in Sections 1.1.2 and 1.7.

It must be stressed that, even for the family of problems to be covered here, the range of physical and chemical models is still too wide to permit a detailed exposition of them within the scope of this book. Therefore, references will be provided to sources in subjects such as aerosol and colloid science where such models are treated in detail. The objective of this chapter is to provide the forms for some of the commonly encountered models and to show how they may be incorporated into the PBE, in order to set the stage for the following chapters dealing with the solution and application of population balance models. One of the main aspects of interest is the dependence of the models on particle size, as it is crucial for the development of solution methods.

An issue related to the choice of kinetic models is the determination of their parameters, which are often unknown. In some cases, there may even be uncertainties about the kinetic mechanisms at play. The population balance may then be used to obtain such information from experimental data, typically conducted in ideal reactor systems. This is called the *inverse problem*, as it involves using the population balance to obtain insight from experiments rather than for simulation. While a discussion of the experimental methods themselves is out of the scope of the present book, some approaches for treating the inverse problem and thus linking models and experiments will be discussed.

The chapter is structured in the following way. We first examine the four main classes of kinetic processes in the population balance, namely, aggregation (including coalescence), fragmentation, growth and nucleation. This is followed by a discussion of the inverse problem, while the last part of the chapter is devoted to transport processes and their incorporation into the spatially dependent PBE.

3.2 Aggregation

3.2.1 Preliminary Concepts

General Features of Aggregation Processes

Aggregation is the process in which two members of a population are combined into one. It is induced by a collision between the parent particles. This collision is bound to occur if the trajectories of the particles are geometrically on a collision course, but may also be prevented or enhanced by particle interaction forces. The resulting aggregates are composed of primary particles which may be loosely connected while retaining their original shapes, resulting in an aggregate with a fractal shape, or fuse (instantaneously or gradually) towards an overall spherical shape. Physical models must thus be provided for all these aspects of the aggregation process.

The range of phenomena that involve aggregation processes is very wide. The ones that will be considered in this book have several elements in common that allow a unified treatment. In particular, they all involve particles of some form and aggregation is induced by their relative motion, which may be random or deterministic. Other processes such as polymerisation, biological and astrophysical processes, also share the same general mathematical description provided by the population balance, but the mechanisms involved may be of a different nature. In polymerisation, for example, the corresponding process is a chemical reaction between two macromolecules.

The aggregation mechanisms in aerosols and colloids have many elements in common. Thermal mechanisms in both non-continuum and continuum regimes, fluid-induced relative motion and sedimentation may be active depending on particle size. The resulting aggregates have usually fractal shapes and may gradually fuse, except for the case of small droplets, which merge very rapidly into standard shapes.

Predicting and controlling aggregation is important for many engineering problems. For example, aggregation must be avoided in the synthesis of primary nanoparticles with carefully controlled size distribution, while it is desirable when better separation properties or larger surface area are required.

Terminology

Different names such as coagulation, aggregation, coalescence, agglomeration and flocculation are employed in different contexts. The convention adopted in the present book is to employ the term 'coalescence' when it is needed to emphasise the merging of the primary particles towards a spherical particle, 'aggregation' when the primary particles are fused to an arbitrary extent and 'agglomeration' when the primary particles retain their identities and shapes (i.e. when the forces holding them together are physical). This definition is in accordance with much of the aerosol literature but not unanimous; for example, 'agglomeration' is the term most frequently employed in crystallisation[1] while 'flocculation' is used in water treatment. Furthermore, in the present book, the term 'aggregation' is employed in most instances in a generic manner, since coalescence and agglomeration may be viewed as asymptotic cases of it with complete fusing and no fusing, respectively. The term 'coagulation', which is typically used for a generic merging process when morphology of fractal particles is not considered, will be used occasionally.

The process of fusing of primary particles will also be considered in the context of aggregation. This term also appears with different names in the literature, with 'sintering' being the common term in the case of inorganic particles. For soot, different terms are used depending on the maturity of soot particles[2]; the term 'coalescence' will be used here for young particles which merge into spherical ones, while the gradual fusing process that takes place within aggregates will be referred to as 'fusing' in a generic way.

Aggregation Equation

The distinction between the discrete and continuous forms of the aggregation PBE is not important for the discussion of physical models for the aggregation kernel. Therefore, the continuous form will be employed as a basis for the following discussion. The equation was derived in Section 2.3 and is shown below for reference:

$$\frac{\partial n(v,t)}{\partial t} = \frac{1}{2}\int_0^v \beta(w, v-w)n(w,t)n(v-w,t)dw - n(v,t)\int_0^\infty \beta(v,w)n(w,t)dw. \tag{3.1}$$

The number density is expressed here as a distribution of particle volume, which is advantageous for aggregation problems (cf. Section 2.3.1). Note also that, for the physical kernels that will be considered, the number density will be a concentration, that is, per unit volume of fluid. Some kernels are more easily derived in terms of particle radius, and both forms will be shown in these cases.

The function $\beta(v,w)$, accounts for the geometrical collision rate. In the presence of interparticle forces, additional factors may be needed to account for them. The morphology of the particles can also be accounted by the kernel, and this will be discussed in Section 3.2.3. However, Eq. 3.1 cannot account for a finite timescale

[1] In Randolph and Larson (1988), aggregation and agglomeration are defined in the reverse way in the context of crystallisation, that is, with aggregation denoting weak physical bonds between primary particles.

[2] See Michelsen et al. (2020) for a discussion of the terminology used in soot formation.

of fusing; for that purpose, the formulations in Sections 2.4.1, 2.4.2 and 2.5 can be employed. More information on this subject will be provided in Section 3.2.3.

Non-Continuum and Continuum Regimes

An important preliminary consideration that is relevant for aerosol particles and has consequences for aggregation but also other kinetic and transport processes is whether the analysis should be based on the non-continuum or continuum regime. Particles of diameter less than 100 nm (often termed nanoparticles) suspended in a gas may have sizes smaller or comparable to the mean free path of the gas molecules and behave effectively like free big molecules. For comparison, the mean free path of air in normal conditions is about 65 nm, while the diameter of a hydrogen atom is about 1 Å or 0.1 nm. Non-continuum effects must be taken into account for describing the transport phenomena and aggregation mechanism of particles smaller than or comparable to the gas mean free path.

A dimensionless group, the Knudsen number (Kn), can be used to quantify the extent to which non-continuum effects are important. The definition of the Knudsen number is:

$$Kn = \frac{2\lambda}{d_p}, \tag{3.2}$$

where λ is the mean free path of the gas and d_p is the particle diameter. For a single component gas whose molecules behave like rigid spheres, the mean free path can be estimated from kinetic theory as follows (Friedlander, 2000):

$$\lambda = \frac{1}{\sqrt{2}\pi d_m^2 N}, \tag{3.3}$$

where d_m is the molecular diameter and N is the number of molecules per unit volume. The latter can be obtained from the ideal gas law, yielding the following expression:

$$\lambda = \frac{kT}{\sqrt{2}\pi d_m^2 p}, \tag{3.4}$$

where k is the Boltzmann constant. More elaborate cases such as binary mixtures, as well as expressions that relate the mean free path to measurable quantities such as viscosity and diffusivity, are discussed in Seinfeld and Pandis (2016).

An aerosol particle is considered to be in the non-continuum regime[3] if $Kn \gg 1$. For $Kn \ll 1$, continuum mechanics apply. Depending on the regime, different expressions must be employed for the drag force, as well as for the thermal aggregation kernel. These two regimes are bridged by a transition regime that may be

[3] An early treatise on the dynamics of such particles is that of Fuchs and Sutugin (1971).

approached with correction formulas and interpolations between the laws that apply at either end. Such formulas must be employed in aerosol formation, where the nuclei have sizes of a few nanometers and are in the non-continuum regime, but surface growth and aggregation can generate a continuous spectrum of particle sizes up to a few hundred nanometers, covering the transition and continuum regimes.

3.2.2 Collision Mechanisms and Kernels

Kernels Used in Analytical Studies

Certain kernels have been employed in early studies because they permit analytical solutions of Eq. 3.1, although they do not correspond to physical collision mechanisms. These kernels are still encountered in theoretical studies or used for benchmarking numerical solutions and are shown below:

Constant kernel:

$$\beta(v,w) = \beta_0. \tag{3.5}$$

Sum kernel:

$$\beta(v,w) = \beta_0(v+w). \tag{3.6}$$

Product kernel:

$$\beta(v,w) = \beta_0 vw. \tag{3.7}$$

These kernels will not be further discussed in this chapter but will be encountered in Chapter 4 where solution methods are discussed.

Overview of Collision Mechanisms

The most commonly encountered collision mechanisms in the physical problems under consideration are the following:

- Thermal motion (perikinetic aggregation) in three regimes:
  - Free molecule regime
  - Transition regime
  - Continuum regime (Brownian motion)
- Convection by the carrier fluid (orthokinetic aggregation) in several cases:
  - Laminar shear flow
  - Turbulent flow, non-inertial particles entrained within the smallest eddies
  - Turbulent flow, inertial particles entrained within the smallest eddies
  - Turbulent flow, particles not entrained within the smallest eddies
- Gravitational force (differential sedimentation)

Other mechanisms are also possible but less common, such as acoustic coagulation of aerosols (Williams and Loyalka, 1991), and will not be discussed in the present book.

Basic Concepts of Collision Models

We will now consider in some detail some of the most important physically derived collision kernels. The derivations share some common features, namely the geometric consideration of a collision sphere that will be outlined below. It is assumed that all particles are spherical, while particle interaction forces are neglected.

The derivation of a collision model is based on the following concepts. First of all, we assume that the particles are spherical and that interactions other than hard-sphere collision can be neglected. Then, we consider a spherical coordinate frame fixed on a particle of radius r_1 which we call the *test particle*, as shown in Fig. 3.1. A collision with a particle of radius r_2 occurs when the centre of that particle reaches a distance $r_1 + r_2$ from the centre of the test particle.

Let $\mathbf{J_{12}}$ be a vector denoting the flux of particles with radius r_2 through an element of area on the surface of the sphere with radius $r_1 + r_2$ centred on the test particle (Fig. 3.1). If n_1 is the average concentration of particles with radius r_1 at the location of the test particle, then the average number of collisions between particles of these two radii, C_{12}, is given by:

$$C_{12} = -n_1 \int_S \mathbf{J_{12}} \cdot \mathbf{dA}, \tag{3.8}$$

where S is a suitably defined surface, while the minus sign gives a positive value to an influx of particles. The cross-sectional area of this collision sphere is called the *collision cross-section.*

To obtain an equation specific to a particular collision mechanism, an expression must be available for the flux $\mathbf{J_{12}}$. Some of the most important cases are:

- Constant relative velocity: The integral can be readily evaluated. This arises in the case of differential sedimentation, where the particles have reached a constant settling velocity, but also in turbulent coagulation with the inertia mechanism.
- Varying relative velocity: In this case, the relative velocity of the impinging particle is a function of the distance from the centre of the test particle. This occurs in coagulation due to laminar shear or turbulent flow with negligible inertia.
- Distribution of velocities: An integration must be carried out to account for all possible velocity pairs. This case occurs in free molecular motion.
- Diffusion: The flux is given by a gradient expression. This occurs in Brownian motion.

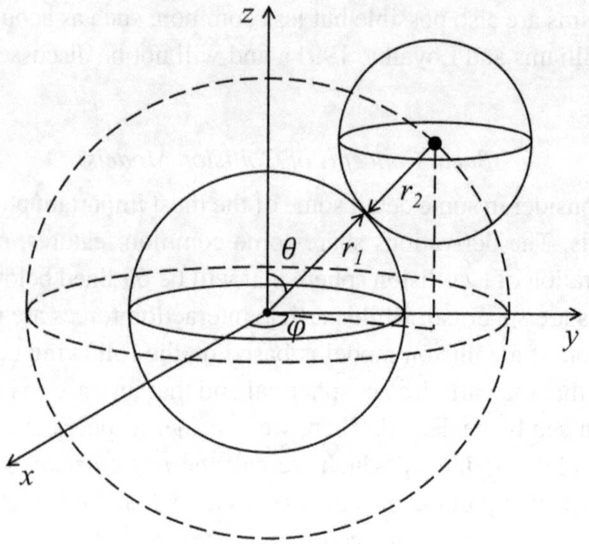

Figure 3.1 Geometry of two-particle collision.

We will now show the kernels for the main physical mechanisms giving rise to coagulation, using the concepts mention above for the particle flux.

Differential Sedimentation

The case of differential sedimentation will be treated first, as it is the simplest one. Let the speeds of the two particles be denoted by $u_{p,1}$ and $u_{p,2}$, with $u_{p,2} > u_{p,1}$. The relative velocity of the incoming particle is then $-|u_{p,1} - u_{p,2}|\mathbf{k}$, and Eq. 3.8 yields:

$$C_{12} = n_1 n_2 \int_S |u_{p,1} - u_{p,2}|\mathbf{k} \cdot \mathbf{dA}, \tag{3.9}$$

where the surface, S, is the upper hemisphere of the collision sphere.

When settling, a particle attains a terminal velocity with the sum of the forces being equal to zero. If the only forces considered are drag and gravity with the former given by Stokes' law, the terminal velocity, $u_{p,t}$, of a particle of radius r_p has the following magnitude:

$$u_{p,t} = \frac{2r_p^2 g(\rho_p - \rho)}{9\mu} \tag{3.10}$$

where ρ_p and ρ are the particle and fluid densities, respectively. The magnitude of the relative velocity is constant over the surface of the collision sphere and can be written as follows:

$$|u_{p,1} - u_{p,2}| = \frac{2g(\rho_p - \rho)}{9\mu}|r_1^2 - r_2^2| \tag{3.11}$$

Substituting into Eq. 3.9, we have:

$$C_{12} = n_1 n_2 \frac{2g(\rho_p - \rho)}{9\mu}|r_1^2 - r_2^2| \int_S \mathbf{k} \cdot \mathbf{dA}. \tag{3.12}$$

Hence[4]:

$$C_{12} = n_1 n_2 \frac{2g(\rho_p - \rho)}{9\mu}|r_1^2 - r_2^2|\pi(r_1 + r_2)^2, \tag{3.13}$$

which is often written as:

$$C_{12} = n_1 n_2 \frac{2\pi g(\rho_p - \rho)}{9\mu}|r_1 - r_2|(r_1 + r_2)^3. \tag{3.14}$$

We can thus write the kernel for aggregation due to differential sedimentation as follows:

$$\beta(r_1, r_2) = \frac{2\pi g(\rho_p - \rho)}{9\mu}|r_1 - r_2|(r_1 + r_2)^3, \tag{3.15}$$

or, in terms of particle volume:

$$\beta(v, w) = \left(\frac{3}{4\pi}\right)^{1/3} \frac{g(\rho_p - \rho)}{6\mu}\left|v^{1/3} - w^{1/3}\right|\left(v^{1/3} + w^{1/3}\right)^3. \tag{3.16}$$

Laminar Shear Flow

The next case to be considered is aggregation in a laminar shear flow, where the particle collisions are brought about by a velocity gradient; this was one of the first cases to be treated (Smoluchowski, 1917).[5] A similar mechanism is aggregation in turbulent flow for particles fully entrained within the smallest scales of fluid motion and with negligible inertia; this will be discussed in the next section, and it will be evident that the kernel has the same dependency on particle size in both cases.

In order to adopt the same coordinate frame as with differential sedimentation and without loss of generality, we assume that the velocity is along the negative z-direction with a uniform and positive shear rate in the y-direction of $G = du/dy$.

[4] The surface integral can be evaluated as follows. The surface element in spherical coordinates is $dS = r^2 \sin\theta d\theta d\phi$. Thus $\mathbf{k} \cdot \mathbf{dA} = \cos\theta dS = \cos\theta r^2 \sin\theta d\theta d\phi$. Integrating over the hemisphere with radius $r_1 + r_2$, we have:

$$\int_S \mathbf{k} \cdot \mathbf{dA} = \int_0^\pi 2\int_0^{\pi/2} \cos\theta (r_1 + r_2)^2 \sin\theta d\theta d\phi = \pi(r_1 + r_2)^2.$$

[5] Smoluchowski's original derivation follows a different approach; the derivation shown here is chosen in order to adopt the same general procedure as for the other collision mechanisms. See Elimelech et al. (1995) for a derivation following Smoluchowski's approach.

The relative velocity of the impinging particle is, therefore, given by $-yG\mathbf{k}$ and, from Eq. 3.8, we have:

$$C_{12} = n_1 n_2 G \int_S y\mathbf{k} \cdot \mathbf{dA}. \tag{3.17}$$

We now need to express y in spherical coordinates. Referring to Fig. 3.1, we have:

$$y = (r_1 + r_2) \sin\theta \sin\phi. \tag{3.18}$$

As in the case of differential sedimentation, the integration is over the upper hemisphere, therefore:

$$C_{12} = n_1 n_2 G \int_0^{\pi} 2 \int_0^{\pi/2} (r_1 + r_2) \sin\theta \sin\phi \cos\theta (r_1 + r_2)^2 \sin\theta d\theta d\phi, \tag{3.19}$$

and finally:

$$C_{12} = n_1 n_2 \cdot \frac{4}{3} G (r_1 + r_2)^3. \tag{3.20}$$

We may thus write the kernel for laminar shear in terms of particle volume as follows:

$$\beta(v, w) = \frac{1}{\pi} G \left(v^{1/3} + w^{1/3} \right)^3. \tag{3.21}$$

Turbulent Flow

In turbulent flow, there is a wide range of spatial and temporal scales of motion. More information on turbulence and its scales will be provided in Chapter 5, which is devoted to the effect of turbulent fluctuations on the population balance. Here, we are interested in how collisions are brought about in turbulent flow. What is important for this purpose is that the collision mechanism depends on the relative size of the particles compared with the size of the smallest scales of motion, called *Kolmogorov scales*. The characteristic size of those scales can be estimated from the kinematic viscosity, ν, and the rate of turbulent kinetic energy dissipation, ϵ (cf. Section 5.2.3 and Eq. 5.15).

The first collision mechanism in turbulent flow is for particles that are smaller than the Kolmogorov scale and thus fully entrained within the smallest eddies, and which are also non-inertial so as to follow the fluid motion. In this case, the collision are due to shear flow within the eddies the following kernel was proposed by Saffman and Turner (1956):

$$\beta(v, w) = 1.3 \left(\frac{\epsilon}{\nu} \right)^{1/2} \left(\frac{3}{4\pi} \right)^{1/3} \left(v^{1/3} + w^{1/3} \right)^3, \tag{3.22}$$

where the dependency of the kernel on particle size is the same as in the case of laminar shear, indicating a similar collision mechanism taking place within the eddies. The deduction of the remaining part and the calculation of the constant is based on assumptions about turbulent flow and can be found in the aforementioned paper.

A second kernel for particles smaller than the Kolmogorov eddies but large enough for their inertia to be considered was also deduced by Saffman and Turner (1956); in that case, collisions are caused by different inertial responses to the change of direction in the flow within an eddy. Larger particles can escape the Kolmogorov eddies, and their collisions are brought about by different mechanisms. The expressions for the kernels for these mechanisms and a discussion of the underlying physical assumptions can be found in Kruis and Kusters (1997).

Free Molecular Motion

Aerosol particles at $Kn \gg 1$ behave like free molecules (cf. Section 3.2.1), and their collision rate can be calculated from kinetic theory. If a stream of particles with radius r_2 moves towards the particle with radius r_1 with relative velocity $|\mathbf{u}_{p,2} - \mathbf{u}_{p,1}|$, the collision rate is given by Eq. 3.9 and the collision cross-section is also the same as in the case of differential sedimentation. However, the relative velocities in this case exhibit the Maxwell–Boltzmann distribution. Therefore, an integration over velocities must be performed to obtain the flux:

$$C_{12} = n_1 n_2 \pi (r_1 + r_2)^2 \int_0^\infty \int_0^\infty f_1(\mathbf{u}_{p,1}) f_2(\mathbf{u}_{p,2}) |\mathbf{u}_{p,2} - \mathbf{u}_{p,1}| d\mathbf{u}_{p,1} d\mathbf{u}_{p,2}, \quad (3.23)$$

where the integration is over all velocity components and $f(\mathbf{u_p})$ is the Maxwell–Boltzmann distribution (Hidy and Brock, 1970). For the details of the derivation, one may consult Brock and Hidy (1965) and Hidy and Brock (1970). The final form of the kernel can be written, in terms of particle volume, as follows (Friedlander, 2000):

$$\beta(v,w) = \left(\frac{3}{4\pi}\right)^{1/6} \left(\frac{6kT}{\rho_p}\right)^{1/2} \left(\frac{1}{v} + \frac{1}{w}\right)^{1/2} \left(v^{1/3} + w^{1/3}\right)^2. \quad (3.24)$$

Brownian Motion

We now consider the case of Brownian motion, which is the mechanism of aggregation due to thermal motion of hydrosols and aerosols with a Knudsen number in the continuum regime. This was the first aggregation mechanism to be investigated, dating back to the original work of Smoluchowski (1917). Brownian motion can be described as a particle diffusion process (cf. Section 3.6.4), hence the flux $\mathbf{J}_{12}$ in Eq. 3.8 is given by:

$$\mathbf{J_{12}} = -D_{p,12} \left.\frac{\partial n}{\partial r}\right|_{r_1+r_2} \hat{\mathbf{e}}_\mathbf{r}, \tag{3.25}$$

where $D_{p,12}$ is the diffusion coefficient for the particles of size r_2 towards those of size r_1 and $\hat{\mathbf{e}}_\mathbf{r}$ is the unit vector in the radial direction. Substituting in Eq. 3.8, we have:

$$C_{12} = n_1 D_{p,12} \left.\frac{\partial n}{\partial r}\right|_{r_1+r_2} \int_S \hat{\mathbf{e}}_\mathbf{r} \cdot \mathbf{dA}, \tag{3.26}$$

or

$$C_{12} = n_1 D_{p,12} \left.\frac{\partial n}{\partial r}\right|_{r_1+r_2} \cdot 4\pi (r_1+r_2)^2. \tag{3.27}$$

To proceed, we need to obtain an expressions for $\partial n/\partial r$ and $D_{p,12}$. To find the former, we need to solve the following diffusion equation in spherical coordinates:

$$\frac{\partial n}{\partial t} = D_{p,12}\left(\frac{\partial^2 n}{\partial r^2} + \frac{2}{r}\frac{\partial n}{\partial r}\right), \tag{3.28}$$

with boundary conditions $n(r,t) = 0$ at $r = r_1 + r_2$ and $n(r,t) = n_2$ at $r \to \infty$, while the initial condition is $n_2(r,0) = n_2$. The result is[6]:

$$n(r,t) = n_2 \left[1 - \frac{r_1+r_2}{r}\,\mathrm{erfc}\left(\frac{r-(r_1+r_2)}{2\sqrt{D_{p,12}t}}\right)\right]. \tag{3.29}$$

Assuming steady state, which is reached for $t >> (r_1+r_2)^2/(\pi D_{p,12})$ and is a reasonable assumption for particles subjected to Brownian motion (see e.g. Seinfeld and Pandis, 2016), we have:

$$\left.\frac{\partial n}{\partial r}\right|_{r_1+r_2} = \frac{n_2}{r_1+r_2}. \tag{3.30}$$

In Section 3.6.4, an expression will be derived for the particle diffusivity due to Brownian motion in the continuum regime (Eq. 3.128). Furthermore, it will be shown that the diffusivity $D_{p,12}$, which relates to the relative displacement of the particles of radii r_1 and r_2, is equal to $D_{p,1} + D_{p,2}$. Using those results, we obtain:

$$D_{p,12} = \frac{kT}{6\pi\mu} \cdot \frac{r_1+r_2}{r_1 r_2}. \tag{3.31}$$

Substituting Eqs. 3.30 and 3.31 into Eq. 3.27, we have:

[6] For the details of the solution, the reader may refer to Crank (1975).

$$C_{12} = \frac{2kT}{3\mu} \cdot \frac{(r_1 + r_2)^2}{r_1 r_2} n_1 n_2. \tag{3.32}$$

Hence the kernel can be written as follows, in terms of particle radius:

$$\beta(r_1, r_2) = \frac{2kT}{3\mu} \left(\frac{1}{r_1} + \frac{1}{r_2} \right) (r_1 + r_2), \tag{3.33}$$

or, in terms of particle volume:

$$\beta(v, w) = \frac{2kT}{3\mu} \left(\frac{1}{v^{1/3}} + \frac{1}{w^{1/3}} \right) \left(v^{1/3} + w^{1/3} \right). \tag{3.34}$$

A notable feature of the Brownian motion kernel is that, for particles of similar size, it can be reasonably approximated by the constant kernel (for given temperature and viscosity), a fact used by Smoluchowski for the derivation of the first analytical solution of his equation (cf. Section 4.2.2) and its application to aggregation of hydrosols. This can be seen by writing from Eq. 3.33 if we let $\alpha = r_1/r_2$ with $r_1 > r_2$, hence:

$$\beta(r_1, r_2) = \beta_0 \left(\alpha + \frac{1}{\alpha} + 2 \right), \tag{3.35}$$

where the dependence on temperature and viscosity has been absorbed into β_0. For equally sized particles, $\beta(r_1, r_2) = 4\beta_0$. For high values of α, the second term in the parenthesis vanishes and the size-dependent part of the expression is approximately equal to $\alpha + 2$. Thus, a constant value can be used for particles of similar sizes. We also observe that the Brownian kernel gives increasingly higher aggregation rates for particles of increasingly unequal sizes.

Eq. 3.33 or 3.33 were derived on the basis of Eq. 3.128 for the particle diffusivity in the continuum regime, which in turn employed Eq. 3.111 for the drag force in creeping flow. The range of applicability of the Brownian aggregation kernel can be extended using the formula for the Cunningham slip factor for the drag (Eq. 3.116) as follows:

$$\beta(v, w) = \frac{2kT}{3\mu} \left(\frac{C_c(v)}{v^{1/3}} + \frac{C_c(w)}{w^{1/3}} \right) \left(v^{1/3} + w^{1/3} \right). \tag{3.36}$$

Further extension into the transition regime can be carried out with an interpolation formula due to Fuchs (1964); for more details and a table of values, the reader may consult Seinfeld and Pandis (2016). Another approach that is often used in practice is to use the following expression involving the kernels for the free molecule regime and Brownian motion with the Cunningham slip factor, denoted here as β_{fc} and β_B, respectively:

$$\beta_t = \frac{\beta_{fm}\beta_c}{\beta_{fm} + \beta_c}, \tag{3.37}$$

where β_t is the kernel for the transition regime.[7]

Comparison of Collision Mechanisms and Rates

Different collision mechanisms may be dominant under different conditions. Some of the factors that must be considered in order to determine the dominant mechanism are listed below.

- If more than one mechanism are physically possible, the particle size plays an important role on the relative magnitude of different mechanisms.
- For aerosols, the Knudsen number determines whether motion is in the free molecule, transition or continuum regime, each of which features distinct mechanisms of thermal aggregation.
- If the aggregation is flow-induced, the type of flow (laminar or turbulent) must be considered in order to determine the appropriate mechanism.
- For aggregation due to turbulent flow, one should consider the size of the particles as compared with that of the Kolmogorov eddies and whether the non-inertial mechanism can be employed.

The particle size is usually the first factor to consider, and, furthermore, the Knudsen number depends on it. Very small aerosol nanoparticles move and collide in the manner of free molecules. In the continuum regime, aerosols up to a few microns collide due to Brownian motion, while the transition regime bridges the free molecule and continuum regimes. Colloidal particles of a similar size range suspended in a liquid may also aggregate due to Brownian motion. As the particles grow larger, the flow-induced and gravitational collision mechanisms begin to dominate; the former depends also on the level of shear. Larger droplets and bubbles coalesce as a result of the fluid motion and gravitational force (e.g. buoyancy for bubbles). More than one mechanism may be present simultaneously, although often one mechanism is dominant.

A comparison of the collision rates due to several different mechanisms, as calculated by Friedlander (2000), is shown in Fig. 3.2. The kernel is computed for particles of 1 μm colliding with particles of different sizes. The value of turbulent energy dissipation used corresponds to conditions in the open atmosphere.

[7] This approach is often referred to as a harmonic mean of the two kernels in the literature, but the harmonic mean is actually twice this expression.

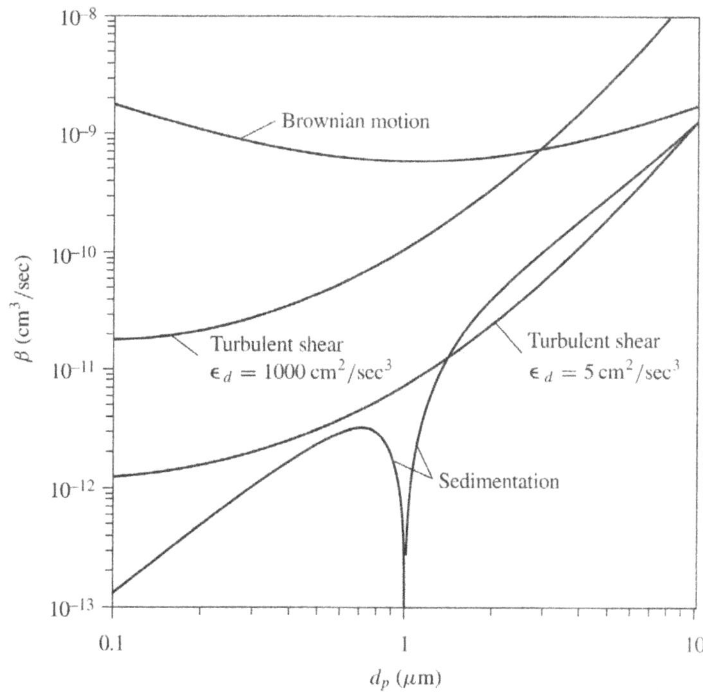

Figure 3.2 Comparison of aggregation kernel values for various mechanisms, for 1 μm particles interacting with 0.1–10 μm particles. Reprinted from Friedlander (2000) with permission from Oxford Publishing Limited.

Particle Interaction Forces

The collision kernel expresses the rate of collisions due to purely geometrical factors, that is, the collisions resulting from the particle trajectories under ideal conditions. This rate does not take into account interparticle forces that alter the geometrical collision rate. Such forces include Van der Waals and Coulomb forces (the latter for electrically charged particles). Another factor to consider is hydrodynamic interaction due to the viscous forces in the carrier fluid as the particles approach each other. These factors can be taken into account via additional factors that modify the ideal collision rates. A discussion of these interactions can be found in Friedlander (2000) or Seinfeld and Pandis (2016) for aerosols and in Elimelech et al. (1995) or Vanni and Baldi (2002) for colloids.

3.2.3 Aggregate Morphology

Description of Morphology

Once a collision occurs, further physical and/or chemical processes take place and determine the morphology of the resulting aggregate. In some cases, the primary

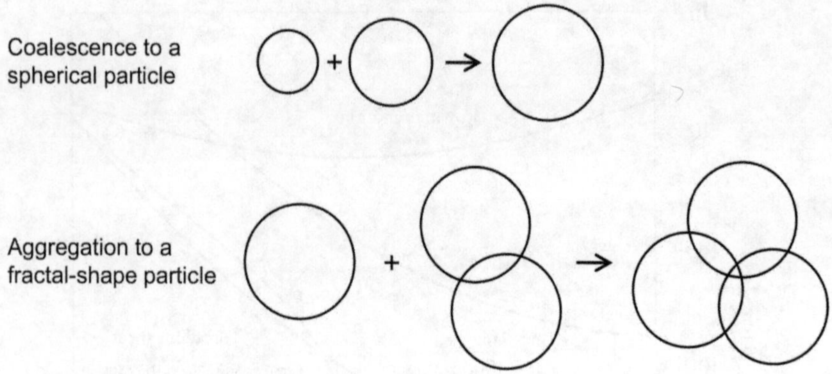

Figure 3.3 Schematic depiction of coalescence into spherical particles (top) and aggregation to form fractal aggregates (bottom).

particles retain their original shape and are held together with loose physical forces, resulting in fractal-shape aggregates. In other cases, the primary particles start to form bridges and the aggregates are gradually being fused. In some cases, the fusing process is so rapid that the new particles coalesce virtually instantaneously. This behaviour is typical of droplets and bubbles, but also of very small aerosols and soot particles. These concepts are illustrated schematically in Fig. 3.3.

The case of instantaneous coalescence is described by the basic aggregation PBE, Eq. 3.1, and the collision models described so far. Fractal aggregates can also be accounted for in the same equation via modifications in the collision kernel, if there is no fusing timescale. The introduction of a finite-rate fusing model requires one of the formulations that were discussed in Sections 2.4.1 and 2.5.

If an aggregate of a characteristic radius r_a is composed of N_p identical primary particles, each with radius r_0, the following equation can be used to describe aggregate structure (Friedlander, 2000):

$$N_p = k_f \left(\frac{r_a}{r_0} \right)^{D_f}, \tag{3.38}$$

where D_f is the *fractal dimension* and k_f is a proportionality constant. For spheres, $D_f = 3$, while for chain-like aggregates, $D_f = 1$. Most fractal aggregates have values of D_f between these two extremes. The characteristic radius of the aggregate can be defined in several ways. One of them is the *radius of gyration*, r_g, which is a geometric concept and is defined as follows:

$$r_g = \sqrt{\frac{\sum_{i=1}^{N_p} (m r_i^2)}{\sum_{i=1}^{N_p} m}}, \tag{3.39}$$

where m is the mass of a primary particle and r_i is its distance from the centre of mass. However, in practice, it is more often possible to measure the *mobility diameter*, which is the diameter of a sphere with the same friction coefficient under similar dynamic conditions. More information about the mobility diameter and its measurement, as well as other measures of aggregate size, can be found in Friedlander (2000).

Collision Kernel

The collision kernels for fractal aggregates can be expressed in terms of the aggregate characteristic size, fractal dimension and primary particle size. For the free molecule regime, we have (Friedlander, 2000):

$$\beta(v,w) = \left(\frac{3}{4\pi}\right)^{\lambda} r_0^{2-6/D_f} \left(\frac{6kT}{\rho_p}\right)^{1/2} \left(\frac{1}{v}+\frac{1}{w}\right)^{1/2} \left(v^{1/D_f}+w^{1/D_f}\right)^2, \tag{3.40}$$

where $\lambda = 2/D_f - 1/2$. For the continuum regime, the kernel is (Friedlander, 2000):

$$\beta(v,w) = \frac{2kT}{3\mu}\left(\frac{1}{v^{1/D_f}}+\frac{1}{w^{1/D_f}}\right)\left(v^{1/D_f}+w^{1/D_f}\right). \tag{3.41}$$

The assumptions involved and limitations of these expressions are discussed by Friedlander (2000). For the shear kernel, the expression is as follows (Elimelech et al., 1995):

$$\beta(r_1,r_2) = \frac{4}{3}Gr_0^3\left(N_{p,1}^{1/D_f}+N_{p,2}^{1/D_f}\right)^3, \tag{3.42}$$

where $N_{p,1}$ and $N_{p,2}$ are the numbers of primary particles in aggregates with radius r_1 and r_2, respectively, and $r = r_0 N_p^{1/D_f}$. The effect of hydrodynamic interaction is considerably reduced in the case of fractal aggregates (Elimelech et al., 1995). These expressions reduce to the forms used for spherical particles in the case of $D_f = 3$.

Aggregate Fusing

When the fusing of the primary particles (sintering in the case of inorganic nanoparticles) within the aggregate takes place at a finite rate, a timescale must be introduced. A general form for the sintering term is the following (Friedlander, 2000):

$$S_{f,\alpha} \equiv \frac{d\alpha}{dt} = -\frac{1}{\tau_f}(\alpha - \alpha_s), \tag{3.43}$$

where τ_f is a characteristic sintering timescale, while the surface areas α and α_s correspond to the aggregate and a sphere of the same volume, respectively. Several models for τ_f have been proposed for different materials with different sintering mechanisms, such as viscous flow and diffusion (Friedlander, 2000), and some of them will be shown in the case studies on silica and soot in Chapter 6.

The term $S_{f,\alpha}$, as given by Eq. 3.43, can be introduced in a PBE formulation with aggregation and sintering that involves the particle surface area, such as Eq. 2.30 or 2.37. In a formulation based on the number of primary particles, such as Eq. 2.34, the corresponding term $S_{f,np}$ can be derived in the following way (Park and Rogak, 2003). The volume of an aggregate comprising N_p primary particles with radius r is $4/3\pi r^3 N_p$, while its surface area is $4\pi r^2 N_p$. Combining these expressions, we obtain the following for the surface area:

$$\alpha = \left(36\pi v^2 N_p\right)^{1/3}. \tag{3.44}$$

Similarly, the surface area of the sphere with equal volume is $\alpha_s = \left(36\pi v^2\right)^{1/3}$. By differentiating Eq. 3.44, we obtain:

$$\frac{d\alpha}{dt} = \left(\frac{4}{3}\pi v^2\right)^{1/3} N_p^{-2/3} \frac{dN_p}{dt}. \tag{3.45}$$

Substituting the above expressions into Eq. 3.43, we finally have:

$$\frac{dN_p}{dt} = -\frac{3}{\tau_f}\left(N_p - N_p^{2/3}\right). \tag{3.46}$$

In a population of particles where n denotes the number density of aggregates and n_p that of primary particles, the average number of primary particles per aggregate of volume v is given by n_p/n. By substituting this ratio for N_p in Eq. 3.46, we can write:

$$\frac{d}{dt}\left(\frac{n_p}{n}\right) = -\frac{3}{\tau_f}\left[\frac{n_p}{n} - \left(\frac{n_p}{n}\right)^{2/3}\right]. \tag{3.47}$$

We thus arrive at the following expression for $S_{f,np}$, that is, the rate of change of n_p due to sintering alone:

$$S_{f,np} \equiv \frac{dn_p}{dt} = -\frac{3}{\tau_f}\left[\frac{n_p}{n} - \left(\frac{n_p}{n}\right)^{2/3}\right] n. \tag{3.48}$$

This is the term that must be substituted in Eq. 2.34.

3.3 Fragmentation

3.3.1 Overview

In fragmentation (also called breakage or break-up), a particle is split into two or more fragments. The kinetics of this process must thus account for two aspects. The first one is the fragmentation rate, while the second one is the number and sizes of the daughter particles.

Fragmentation is present in a wide range of physical processes. Examples include milling and grinding processes, break-up of aerosol aggregates, attrition of crystals, break-up of droplets and bubbles, degradation of polymer chains and splitting of microbial cells. The kinetics involved in these processes are very diverse and problem specific, unlike the case of aggregation where the collision kernels could be traced into a few fundamental mechanisms. As a result, the approaches to fragmentation kinetics are more empirical. Two main routes may be followed to obtain them: either a general-purpose kernel is postulated and its parameters are fitted on the basis of experimental data (as we will see in Section 3.5 on the *inverse problem*), or a phenomenological model is postulated on the basis of the underlying physics.

The literature on fragmentation processes is more often based on the continuous PBE, which takes the form of Eq. 2.26 if no other processes are present. Therefore, this form will be taken as a basis for the following discussion, and is shown below for reference:

$$\frac{\partial n(v,t)}{\partial t} = \int_v^\infty \gamma(w)b(v,w)n(w,t)dw - \gamma(v)n(v,t). \tag{3.49}$$

The discrete form of the fragmentation terms was shown in Eq. 2.2.2 as part of the complete discrete PBE. For fragmentation alone, the equation is:

$$\frac{dn_i}{dt} = \sum_{j=i+1}^{\infty} \gamma_j b_{ij} n_j - \gamma_i n_i. \tag{3.50}$$

The kernel[8] in the integral in Eq. 3.49 includes two factors: the rate (or frequency) of fragmentation events, $\gamma(v)$, and the number density of daughter particles, $b(v,w)$; their equivalents in the discrete equation are γ_i and b_{ij}, respectively. The function $b(v,w)$ is a number density, but only with respect to v, and determines both the number of fragments and their size distribution.[9]

In some cases, it may be desired to specify a minimum size of particles that can break (this has been called *limited breakage*, see Kostoglou, 2007b). This can be accomplished via the following modification to the fragmentation equation:

[8] In the present book, we use the term 'kernel' for the product $\gamma(w)b(v, w)$ in the integral term; either of the two functions is sometimes referred to as a kernel in the literature.

[9] In the literature, this function is sometimes written as the product of two functions, one yielding the number of fragments and the other being the probability density (rather than number density) function of the fragment sizes.

$$\frac{\partial n(v,t)}{\partial t} = \int_v^{\infty} \gamma(w)b(v,w)n(w,t)dw - \gamma(v)n(v,t), \quad \forall v > v_{min}, \tag{3.51a}$$

$$\frac{\partial n(v,t)}{\partial t} = \int_{v_{min}}^{\infty} \gamma(w)b(v,w)n(w,t)dw \quad \forall v < v_{min}, \tag{3.51b}$$

where v_{min} is the critical particle size below which no fragmentation occurs.

The forms of the fragment number density and fragmentation rate functions are the subject of the present section. The focus will be on the overall features of these functions and on general-purpose forms, as the phenomenological models are treated in more specialised literature, a sample of which is mentioned below. A wide range of particle breakage problems such as milling and crushing are discussed in Williams et al. (2007). The break-up of droplets and bubbles is reviewed in Clift et al. (1978), Liao and Lucas (2009), Patruno et al. (2009) and Yeoh et al. (2014), among many other sources. Bubble column reactors in particular are treated in Jakobsen et al. (2005) and in Chapter 8 of Jakobsen (2014). References for problems related to the break-up of cells in biochemical processes include Ramkrishna (2000) and Hjortsø (2004). Fragmentation of polymer chains is discussed by Ziff and McGrady (1985). A general review of models and solutions for the fragmentation equation can be found in Kostoglou (2007b).

Even within the category of general-purpose fragmentation functions, it is possible to find a large number of them in the literature, hence only a selection will be shown here. The focus will be on continuous forms, while more details on the discrete forms can be found in Vanni (2000).

3.3.2 Models for the Fragment Size Number Density

General Features

We first consider the types of function that may be employed to represent the fragment size number density. One property of this function that may be immediately deduced is:

$$b(v,w) = 0, \quad \forall v > w, \tag{3.52}$$

that is, no fragments with volume greater than that of the parent particle should be allowed to form. Furthermore, the moments of this function have the following properties. First, the zeroth moment is equal to the number of fragments resulting from a particle of volume w, $N_f(w)$:

$$\int_0^w b(v,w)dv = N_f(w). \tag{3.53}$$

Second, conservation of mass imposes the following condition for the first moment:

$$\int_0^w vb(v,w)dv = w, \tag{3.54}$$

that is, the total volume of the daughter particles must be equal to that of the parent particle. Note that, given Eq. 3.52, it is sufficient to carry out the integrations up to w. An additional physical constraint that may be imposed on the form of $b(v,w)$ is the following (McGrady and Ziff, 1987):

$$\int_0^u vb(v,w)dv \geq \int_{w-u}^w (w-v)b(v,w)dv, \tag{3.55}$$

for $u < w/2$. This inequality states that, if one of the fragments has size $v \geq w/2$, then the volume of the smaller fragments $(w-v)$ must be present in fragments smaller than $(w-v)$ (i.e. the volume is not unphysically rearranged between the fragments). Kostoglou (2007b) notes that sometimes functions determined empirically by fitting experimental data do not satisfy this constraint. For binary fragmentation, this condition becomes an equality since one particle with volume $(w-v)$ is formed for one particle with volume $v \geq w/2$.

Many fragment size number densities can be written in a self-similar form, in terms of a similarity variable $z = v/w$ (Ramkrishna, 2000). The function is then written in the form:

$$b(v,w) = \frac{b'(z)}{w}, \tag{3.56}$$

where $b'(z)$ is the number density in terms of the scaled variable, which depends on the ratio of sizes rather than on the size of the parent particle. The zeroth and first moment of this number density function are:

$$\int_0^1 b'(z)dz = N_f, \tag{3.57}$$

$$\int_0^1 zb'(z)dz = 1. \tag{3.58}$$

Binary Fragmentation

One of the most important cases is that of binary fragmentation, that is, the splitting of a particle into two fragments. The simplest daughter particle number density is the uniform one, where fragments of every size are equally likely, and is given by the following function:

$$b(v,w) = \frac{2}{w}. \tag{3.59}$$

The fragmentation PBE then takes the following simple form:

$$\frac{\partial n(v,t)}{\partial t} = 2\int_v^\infty \frac{\gamma(w)}{w} n(w,t)dw - \gamma(v)n(v,t). \tag{3.60}$$

Another type of distribution that can accommodate many physical cases is one of a parabolic shape. This can also account for erosion, as we will see below. Hill and Ng (1995) employed the following function:

$$b(v,w) = \frac{24\left(1 - \frac{hw}{2}\right)\left(v^2 - vw + \frac{w^2}{4}\right)}{w^3} + h, \tag{3.61}$$

where h is a parameter such that $0 \leq h \leq 3/w$. For $h = 2/w$, the equation reduces to Eq. 3.59 for uniform binary fragmentation, while for values lower or greater than that the parabola is concave or convex, respectively.

Erosion

Erosion-type fragmentation (or attrition or abrasion) refers to the removal of small fragments during breakage. A review of attrition processes can be found in Bemrose and Bridgwater (1987).

Erosive binary fragmentation can be straightforwardly implemented in the context of the discrete PBE as the removal of a monomer with a conditional function (Vanni, 2000):

$$b_{ij} = \begin{cases} 1 & \text{if } i = 1, j-1, \\ 0 & \text{otherwise.} \end{cases} \tag{3.62}$$

In the continuous PBE, for binary fragmentation, erosion can be implemented via a U-shape function, such as the parabolic function given by Eq. 3.61 for suitable values of the parameters. A simpler way is via the following sum of delta functions (Kostoglou et al., 1997):

$$b'(z) = \delta[z - (1-\epsilon)] + \delta(z - \epsilon), \tag{3.63}$$

where ϵ is the volume of the small fragment.

A more general form of erosion size distribution function, allowing for multiple fragments, has been shown by Kostoglou and Karabelas (2001). Erosion can also be included in the more general approach to fragmentation discussed in the next section.

Multiple Fragments

Fragmentation into N_f equalsized particles can be effected via the following function (Kostoglou et al., 1997), given in the self-similar form (cf. Eq. 3.56):

$$b'(z) = N_f \delta \left(z - \frac{1}{N_f} \right). \tag{3.64}$$

A fragment size number density that yields multiple fragments that was considered by Ziff (1991) and includes as special cases many previously studied functions is the following one:

$$b'(z) = \alpha \gamma' z^{\gamma'-2} + (1-\alpha)\delta z^{\delta-2}, \tag{3.65}$$

where α, γ' and δ are parameters. The number of fragments is determined from their values using Eq. 3.57. Analytical solutions were derived by Ziff (1991) for functions in this class, and a solution for ternary fragmentation is shown in Section 4.2.2.

As an example of an empirically derived function, we may mention the function derived by Austin et al. (1976) on the basis of milling data. This function was shown in a number density form by Hill and Ng (1995), who employed it to compare with numerical results from their discretisation approach, and is of the same form as Ziff's function (Eq. 3.65).

A more general approach to the distribution of daughter particle sizes for the case of multiple fragments was proposed by Hill and Ng (1996b). Their functions were later reformulated and generalised for non-integer parameters by Diemer and Olson (2002), who also showed that many of the previously considered distributions could be included in these forms. They arrived at two functions generalising the Hill–Ng distributions, the first being a power-law product given by:

$$b'(z) = N_f \frac{z^{q-1}(1-z)^{r-1}}{B(q,r)}, \tag{3.66}$$

where B is the beta function, q is a parameter with $q > 0$, the number of fragments must be $N_f \geq 2$ and r is defined as follows:

$$r = q(N_f - 1), \tag{3.67}$$

from which it can be seen that $r > 0$. The second function is a power-law sum:

$$b'(z) = N_f \left[\left(\frac{1}{N_f} \right) \frac{z^{q-1}(1-z)^{N_f-2}}{B(q, N_f - 1)} + \left(1 - \frac{1}{N_f} \right) \frac{(1-z)^{q+N_f-3}}{B(1, q + N_f - 2)} \right]. \tag{3.68}$$

Fig. 3.4 shows a plot of the power-law sum distribution for binary fragmentation ($N_f = 2$). It can be seen that the shapes of the curves for high values of q represent an erosion-type breakage. Plots for other cases can be found in Diemer and Olson (2002).

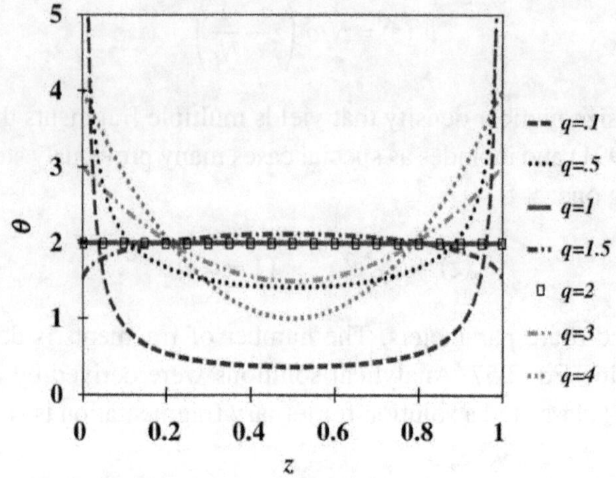

Figure 3.4 The power-law sum distribution (Eq. 3.68) for binary fragmentation (note that, in the notation of the Diemer–Olson paper, $\theta(z)$ corresponds to $b'(z)$ in the present book). Reprinted from Diemer and Olson (2002), with permission from Elsevier.

3.3.3 Models for the Fragmentation Rate

We now turn our attention to the rate of fragmentation events, $\gamma(v)$. Apart from the volume of the parent particle, this function may also involve other variables associated with the local environment of the particles, such as the energy dissipation rate for breakage in a turbulent flow field. Such factors give rise to a wide range of phenomenological models that depend on the physical problem, particularly on whether the particles are solid or fluid (droplets or bubbles). For solid particles, a number of workers have investigated the relationship between particle size and fracture mechanical parameters such as fracture energy or impact velocity, including Yashima et al. (1987), Ghadiri and Zhang (2002), Zhang and Ghadiri (2002) and Vogel and Peukert (2003). For fluid particles, links with flow field and turbulence have been explored by Coulaloglou and Tavlarides (1977), Narsimhan et al. (1979), Kostoglou and Karabelas (2005), Patruno et al. (2009) among many other authors (see Liao and Lucas, 2009 for a review). The details of these phenomenological models and the dependence on fracture or hydrodynamic parameters will not be considered here. Instead, we will concentrate on the general-purpose model forms and the dependence on particle size.

The power law is one of the most commonly employed forms of model for the fragmentation rate:

$$\gamma(v) = \gamma_0 v^{\eta}. \tag{3.69}$$

The exponent may be determined from theoretical arguments or by solving the inverse problem with an approach such as the one described in Section 3.5.3.

While we do not discuss here the dependence of the phenomenological models on fracture mechanical or hydrodynamical parameters, we note that they often involve more complex nonlinear dependence on particle size. As an example, we mention the model of Coulaloglou and Tavlarides (1977) for drop break-up:

$$\gamma(v) = K_1 v^{-2/9} \epsilon^{1/3} \exp\left(-\frac{K_2 \sigma}{\rho_d \epsilon^{2/3} v^{5/9}}\right), \tag{3.70}$$

where ϵ is the rate of energy dissipation, ρ_d is the droplet diameter and K_1, K_2 are dimensionless constants.

3.4 Particle Formation Processes

3.4.1 Overview

We now consider processes that result in formation of particles from their precursors. This includes the generation of new particles and growth of existing ones, but also their shrinking and even removal, as in the case of droplet evaporation or soot oxidation. These processes involve interaction of particles with their local environment and, therefore, involve variables such as the local concentrations of chemical species and temperature. As such, the presence of these processes requires a coupling of the PBE with the equations of transport phenomena.

In the general form of the PBE, these processes are usually depicted via two terms: nucleation, which is the rate of formation of new particles, and growth, which is the continuous rate of change of particle size (positive or negative, thus including evaporation). In practice, however, the modelling of these processes can take various forms; for example, the condensation of aerosols (a form of growth) can be approached as the coagulation of particles with single molecules. Therefore, it is preferable to treat these processes in a unified manner, especially since they all share the common feature of particle–environment interaction.

The models for these processes are quite diverse and heavily depend on the physical system being studied. Furthermore, their description requires an extensive discussion of the underlying thermodynamic and chemical principles. As such, a comprehensive treatment of them is out of the scope of the present book. Rather, the objectives here are to outline some common features and show how the models can be integrated with the population balance. For this purpose, examples drawn from the case studies that will be shown in Chapter 6, while references will be provided to the relevant literature.

While the mechanisms and terminology depend to an extent on the problem under study, certain features are common in different systems. For example, the classical nucleation theory can be used to describe, under idealised conditions, the homogeneous nucleation of crystals and aerosols. Also, in both cases, growth can be limited by diffusion of precursors towards the particle surface or by the processes that occur on the surface. However, the modelling of soot nucleation involves complex chemical steps as one needs to consider the polymerisation of PAHs.

The incorporation of precursor kinetic models in the PBE can be carried out in several ways. If the nucleation rate is given by classical nucleation theory, it can be implemented as a source or nucleus-size particles. In other approaches, it is modelled as a collision process in a way similar to coagulation. Condensational growth can also be modelled as a coagulation of molecules with particles. Other growth processes that involve a surface reaction are incorporated via the growth term, which is flux of particles along the particle size domain.

Of particular importance for the incorporation of rate laws into the PBE is their dependence on particle size, as this has consequences for the formulation and solution method (e.g. size-independent growth without aggregation or fragmentation can be exploited via a formulation in terms of a linear dimension). Nucleation is usually size-independent, while growth limited by diffusion or surface processes gives rise to different size dependent expressions.

The kinetic models for particle formation from precursors are often the biggest sources of uncertainty in the population balance. Nucleation, in particular, is a very complex process that is not fully understood, largely because of the difficulty in obtaining experimental observations of particles at extremely small sizes. These uncertainties must be borne in mind during the comparison of experimental and modelling results. For example, if nucleation cannot be predicted with accuracy but is 'frozen' after an initial stage of the process, it may still be possible to reproduce qualitative trends but not the number or volume fraction of particles. In Section 3.5, we will consider the *inverse problem*, that is, the determination of kinetics from experiments.

In the following, we will consider first crystal nucleation and growth, with an example linked to the case study in Section 6.3. Subsequently, we will discuss the kinetics of aerosol and soot formation, with reference to the case studies in Sections 6.1 and 6.2.

3.4.2 Crystal Nucleation and Growth

The nucleation and growth of crystals is an extensive subject of which we can only provide a brief outline here, with the aid of an example. The focus will be on how the kinetics can be incorporated in the PBE and what are the consequences for the evolution of the particle size distribution. Comprehensive treatments of the underlying thermodynamic and kinetic principles can be found in Mersmann (2001), Mullin (2001), Lewis et al. (2015) and Myerson et al. (2019).

The driving force of crystallisation is *supersaturation*, which is a thermodynamic concept that represents the extent to which a solution deviates from the soluble state. Supersaturation can be expressed in various ways, more information on which can be found in the aforementioned references. We give below an example from the case of precipitation, which is crystallisation induced by a chemical reaction. In Section 6.3, we will show a case study based on the precipitation of barium sulphate ($BaSO_4$), which is a system often employed as a model for precipitation studies. The precipitation of $BaSO_4$ can be described by the following global reaction:

$$BaCl_2 + H_2SO_4 \longrightarrow BaSO_4 \downarrow\ + 2HCl.$$

The supersaturation, S, in this case, is given by the following activity-based expression (Vicum et al., 2003):

$$S = \gamma_\pm \sqrt{\frac{C_{Ba^{2+}_{(\mathrm{free})}} C_{SO^{2-}_{4(\mathrm{free})}}}{K_{SP}}}, \tag{3.71}$$

where K_{SP} is the *solubility product*, which can be found from the literature, and $\gamma_\pm$ are the mean activity coefficients, which are calculated with the modified Debye–Hückel method, as shown in Appendix A of Tang et al. (2022). Eq. 3.71 considers the contributions from the free ions, as $BaSO_4$ tends to form undissociated ion pairs (Vicum et al., 2003).

The description of nucleation and growth problems in crystallisation is typically facilitated by a PBE formulated in terms of length, that is, Eq. 2.24 with the nucleation and growth functions expressed as functions of supersaturation:

$$\frac{\partial n(\ell,t)}{\partial t} + \frac{\partial (G_\ell(\ell,S)n(\ell,t))}{\partial \ell} = B(S)\delta(\ell - \ell_0). \tag{3.72}$$

This equation must be coupled with an expression such as Eq. 3.71 and with mass (and possibly energy) balance equations to compute the supersaturation.

The formation of nuclei from a solution that does not contain existing particles is called *homogeneous nucleation*. On the other hand, *heterogeneous nucleation* is facilitated by existing particles. The above mechanisms constitute *primary nucleation*, to be distinguished from *secondary nucleation* which describes the formation of nuclei via existing fragments or added seed crystals.

The most standard model of homogeneous nucleation is the *classical nucleation theory*, which treats the problem in a thermodynamic manner. A full discussion of this theory can be found in the references provided, as well as in the references in the next section (as it is also relevant to aerosols). The rate predicted by this theory is given by the following expression:

$$B_N = 1.5D_{AB}\left(\sqrt{K_{SP}SN_A}\right)^{7/3}\sqrt{\frac{\gamma_{CL}}{kT}V_m}\exp\left[-\frac{16}{3}\left(\frac{\gamma_{CL}}{kT}\right)^3\frac{V_m^2}{(\nu\ln S)^2}\right], \tag{3.73}$$

where B_N is the rate of generation of particles via nucleation (note that this is related to the B in Eq. 3.72 as $B_N = Bdv_0$ where dv_0 is the interval including the nuclei volume), while N_A, k, D_{AB}, V_m, γ_{CL} and ν are the Avogadro number, Boltzmann constant, apparent diffusion coefficient, volume of a $BaSO_4$ molecule, interfacial energy and dissociation number, respectively. The apparent diffusion coefficient, D_{AB}, is given by the following expression (Mersmann, 2001):

$$D_{AB} = \frac{(z_A + z_B)\,D_A D_B}{z_A D_A + z_B D_B}, \tag{3.74}$$

where D_A and D_B are the diffusivities of the two ions and z_A and z_B are the corresponding charge numbers.

Growth can be controlled by diffusion of species towards the particle (transport controlled) or by surface processes (integration controlled). When growth is transport controlled, growth is inversely proportional to particle length. The following expression can be used for crystallisation:

$$G_L = \frac{k_a}{3k_v}\frac{ShD_{AB}\sqrt{K_{SP}}MW(S-1)}{\rho_C L}, \tag{3.75}$$

where G_L is the linear growth rate and L is the particle diameter, while Sh, MW and ρ_c are the Sherwood number, molecular weight and crystal density, respectively. The Sherwood number stands for the ratio of convection to diffusion mass transfer rate. The parameters k_a and k_v are surface and volume shape factors, respectively; for spherical particles, their values are $k_a = \pi$ and $k_v = \pi/6$. The volumetric growth rate is related to the one given by Eq. 3.75 as follows:

$$G = 3k_v^{1/3}v^{2/3}G_L. \tag{3.76}$$

From Eqs. 3.75 and 3.76, it can be inferred that the volumetric growth rate is proportional to $v^{1/3}$ or particle length. It should also be noted that crystal growth rates may be different at different faces, thus giving rise to a multidimensional population balance formulation as described in Section 2.4.1.

A related process is the precipitation of grains from supersaturated solutions (Lifshitz and Slyozov, 1961). A study of this process focussed on the population balance perspective can be found in Samaras (2006).

3.4.3 Aerosols, Soot and Carbonaceous Nanoparticles

Overview

Aerosols typically involve aggregation and coalescence, therefore a volume-based formulation of the PBE is preferred. Condensation can be treated as a growth process or as coagulation between molecules and particles. In the case of soot, for example, we can distinguish between the condensation of PAHs on the surface and the growth due to surface reaction with acetylene. The PBE can be formulated as follows:

$$\frac{\partial n(v,t)}{\partial t} + \frac{\partial (G(v,\mathbf{Y},T)n(v,t))}{\partial v} = B(v,\mathbf{Y},T)\delta(v - v_0) + C(v,\mathbf{Y},T)$$
$$+ \frac{1}{2}\int_0^v \beta(w, v-w)n(w,t)n(v-w,t)dw - n(v,t)\int_0^\infty \beta(v,w)n(w,t)dw, \quad (3.77)$$

where we have considered nucleation, growth, condensation and aggregation. The dependence of the kinetics on the species mass fractions, $\mathbf{Y}$, and on temperature, as is typical for aerosols and soot, has been indicated.

The classical nucleation theory described in the previous section can also be used to describe the homogeneous nucleation of aerosols, but several modifications and alternative approaches are possible. Growth can be controlled by diffusion or reaction on the particle. More details on the thermodynamic and kinetic expressions employed for aerosols can be found in Friedlander (2000), Seinfeld and Pandis (2016), Drossinos and Housiadas (2017) and Kodas and Hampden-Smith (1999). A review of recent developments in nucleation can be found in Ford (2004).

The formation of soot and carbonaceous nanoparticles is a complex process that has several individual features and involves many chemical steps. Soot kinetics will be employed in the case study in Section 6.2, and therefore a brief introduction to this subject is provided here. The literature on soot kinetics is very large; an overview can be found in Haynes and Wagner (1981) or Glassman (1988), while more detailed reviews can be found in Harris and Weiner (1985), Richter and Howard (2000), Stanmore et al. (2001), Frenklach (2002b), Mansurov (2005), D'Anna (2009) and Wang (2011). The development of models and their incorporation in simulations is covered in Kennedy (1997), Raman and Fox (2016) and Rigopoulos (2019, 2020).

One important feature that differentiates soot and carbonaceous nanoparticles from the other systems considered here is the complex chemistry that leads to their formation. The root to nucleation is via a large number of chemical steps involving the formation of more and more complex poly-aromatic hydrocarbons (PAHs). A multitude of models have been developed over the years, and the complete process is still not fully understood.

Soot formation models may be classified into semi-empirical non-precursor, semi-empirical precursor and detailed models (Rigopoulos, 2019). The first two categories amount to monodisperse PBEs that are formulated to predict one or two soot variables, typically number density and volume fraction. The difference between them is that non-precursor models relate soot to variables such as temperature or mixture fraction, while precursor-based models employ functions of precursor species, typically acetylene (C_2H_2).

Soot Nucleation

Nucleation of soot, also called *inception*, is very difficult to account for in the context of semi-empirical models. While it is usually correlated to C_2H_2, as in the model of Leung et al. (1991), it is now known that the formation of polycyclic aromatic hydrocarbons (PAHs), which are hydrocarbons with two or more aromatic rings, have a key role in bridging gas-phase chemistry and soot nucleation, while also contributing to soot growth via condensation. PAH chemistry is very complex, however, and found only in the detailed soot models. Furthermore, nucleation is fueldependent while, in aliphatic fuels, C_2H_2 appears to be the main species involved, in fuels containing aromatics (such as kerosene), the dominant contribution can come from the aromatic part of the fuel (Vovelle et al., 1994). A notable model is that of Hall et al. (1997), which combines a simplified PAH model with a nucleation model where the nucleation rate is assumed to be eight times the formation rate of $C_{10}{}^{\cdot}H_7$ and $C_{14}H_{10}$.

Detailed models differ in several aspects from semi-empirical ones. First of all, they typically include PAH chemistry, which allows for a more realistic approach to the nucleation step. Reviews of PAH kinetics can be found, for example, in Richter and Howard (2000) or McEnally et al. (2006). The number of possible PAHs is very large, however; furthermore, the transition from PAHs to soot is poorly understood. The main model of PAH growth employed in soot studies is the H-abstraction-C_2H_2-addition (HACA) mechanism (Frenklach and Wang, 1991, 1994; Wang and Frenklach, 1997), which can be represented by the following scheme:

$$
\begin{aligned}
A_i + H &\rightleftharpoons A_i\cdot + H_2,\\
A_i\cdot + C_2H_2 &\rightleftharpoons A_iC_2H_2\cdot,\\
A_iC_2H_2\cdot + C_2H_2 &\longrightarrow A_{i+1} + H,
\end{aligned}
$$

where A_i denotes a PAH with *i* aromatic rings. This sequence can be extended to include PAH oxidation. PAH growth may be terminated at some point such as a four-ring PAH (pyrene) or continued and described as a polymerisation process (Frenklach, 1985; Frenklach and Harris, 1987). Several detailed mechanisms with

PAH chemistry have been developed.[10] Once a PAH formation model has been included in the gas-phase chemistry, nucleation can be considered as a series of PAH reactions. Frenklach and Wang (1991) assumed that soot nuclei are formed by dimerisation of PAHs and employed the method of moments. Another route is to integrate the nucleation steps with the kinetics of a detailed PAH mechanism.[11]

In the case study on soot formation in Section 6.2, the model of Blanquart et al. (2009) is employed, where nucleation is assumed to occur via collision of PAHs to form dimers, which subsequently coalesce to form nuclei or condense on the particle surface. The mechanism considers eight PAHs, from naphthalene ($C_{10}H_8$) up to cyclopenta[cd]pyrene, ($C_{18}H_{10}$), in the dimerisation process. The implementation of this model in the context of a discretisation method for solution of the PBE is described in Section 6.2.3.

Soot Surface Growth and Oxidation

The studies of Harris and Weiner (1983a,b), based on laminar premixed flames, have indicated that acetylene constitutes the primary growth species and that growth is proportional to surface area and acetylene concentration. The semi-empirical precursor-based models are most often acetylene-based, and a widely used one is the two-equation (number density – volume fraction) model of Leung et al. (1991) which employs a square-root dependence of growth on surface area. The model of Brookes and Moss (1999) is also acetylene-based, but assumes growth to be proportional to a power of the surface area, with the power being calibrated from experiments and adopting a value close to unity.

Surface growth is usually incorporated in detailed models via a HACA mechanism analogous to that of PAHs (Frenklach and Wang, 1991), that is,:

$$\begin{aligned}
\mathrm{C(s){-}H} + \mathrm{H} &\rightleftharpoons \mathrm{C(s)}\cdot + \mathrm{H_2},\\
\mathrm{C(s)}\cdot + \mathrm{H} &\longrightarrow \mathrm{C(s){-}H},\\
\mathrm{C(s)}\cdot + \mathrm{C_2H_2} &\longrightarrow \mathrm{C(s){-}H} + \mathrm{H},
\end{aligned}$$

while further surface reaction steps can be added for oxidation. The main difficulty in the detailed modelling of these processes is that they are surface reactions. The initial formulation of Frenklach and Wang (1991) employed the following form of rate equation:

$$R_{g,i} = k_{g,s} C_g \alpha \chi_s S_i N_i, \tag{3.78}$$

[10] See, for example, Marinov et al. (1996), Appel et al. (2000), D'Anna and Kent (2003), Richter et al. (2005), Slavinskaya and Frank (2009), Blanquart et al. (2009) and Saggese et al. (2015).

[11] Such mechanisms are described in Richter et al. (2005), D'Anna and Kent (2008), D'Anna (2008), D'Anna et al. (2010), Sirignano et al. (2010, 2013) and Saggese et al. (2015).

where $R_{g,i}$ is the reaction rate for a soot particle of size class i, $k_{g,s}$ is the per-site rate coefficient, C_g is the concentration of participating gas-phase species, χ_s is number density of surface sites obtained by a steady-state approximation, S_i is the surface area, N_i is the number density and α is a parameter for surface reactivity that accounts for the effect of the morphology of the surface. The prescription of this parameter is the main issue in the application of HACA, and later works have postulated more elaborate models for that parameter where it depends on temperature and particle size (Appel et al., 2000). Soot growth can also occur via condensation of PAHs on the surface of soot particles Frenklach and Wang (1991, 1994), and this is also included in detailed soot models.

Oxidation of soot is assumed to occur via O_2, OH and O. Two of the first O_2-oxidation models were due to Nagle and Strickland-Constable (1962) and Lee et al. (1962), while the model of Neoh et al. (1985) is most commonly employed for oxidation by OH. These models are often employed with modifications in semi-empirical precursor-based models; for example, in the Leung et al. (1991) model, oxidation by O_2 only is included and follows the Lee et al. (1962) model with an adjusted constant, while Liu et al. (2003) consider O_2 and OH with further modifications, particularly for smoking flames.

3.5 Determination of Kinetic Parameters: The Inverse Population Balance

3.5.1 Overview

In Chapter 2, the population balance was formulated as a predictive framework. That is, equations were derived for predicting the evolution of a distribution of a property in a population of particles undergoing certain kinetic and transport processes. In the present chapter, we have so far examined models for the main kinetic processes that appear in the PBE. However, in real problems, there are often uncertainties about the parameters that appear in these models. Even more, the basic mechanisms that come into play in a particular problem may be unknown. For example, it may not be known what type of kernel should be employed in a fragmentation process, or even which combination of basic processes is operating.

The determination of the mechanism and parametrisation of kinetic models is the *inverse population balance problem* and can be seen as complementary to the theory presented so far. Such an inverse problem may be part of an investigation towards generating a predictive model of a process, in order to determine or refine the values of the parameters. But it may also be viewed as a way to employ the population balance to extract mechanistic insight from a set of experimental data.

While we have not examined transport processes yet, it is fit to consider the inverse problem at this stage, as it is associated with the kinetic processes. Furthermore, the kinetic investigations are usually conducted in ideal reactors where the effect of flow is excluded.

The approach to the inverse problem is not as systematic as to the predictive one. First of all, it depends on the amount of information available at the beginning. If the kinetic mechanisms are known and it is simply a matter of determining the parameters, then a suitably formulated PBE can be brought in an appropriate form to perform this estimation, in conjunction with purposely designed experiments. An example of this is the estimation of crystallisation kinetics discussed in Section 3.5.2. If there are uncertainties about the type of mechanism within the context of a certain process, such as, for example, which kernel is at play in an aggregation or fragmentation process, then a model may be postulated and compared with experiments. The uncertainty may extend to the nature of the basic processes themselves, and a more general approach would be to postulate a generic model that, by comparison with experimental data, may yield information about the mechanism itself. An approach along these lines is discussed in Section 3.5.3.

Methods for treating the inverse problem are, to an extent, case dependent. Furthermore, they are often associated with the features of the experimental equipment employed to probe into the particle properties. A treatment of experimental methods is out of the scope of the present text; more information can be found, for example, in Kodas and Hampden-Smith (1999) and Friedlander (2000) for aerosols, or in Lewis et al. (2015) for crystallisation.

One of the earliest examples of an inverse problem is the determination of parameters in crystal nucleation and growth processes from perfectly stirred reactors, as detailed, for example, in Randolph and Larson (1971), one of the first books to feature a systematic treatment of the population balance with focus on crystallisation. Hounslow (1990) and Bramley et al. (1996) developed methods for dealing with crystallisation and precipitation problems that include aggregation. Wright and Ramkrishna (1992) and Sathyagal et al. (1995) presented formulations of the inverse problem for aggregation and break-up, respectively, using the concept of similarity solutions (this will be introduced in Chapter 4, particularly in Section 4.2.3). A review of studies on inverse problems can be found in Ramkrishna (2000).

Since the cases of inverse problems that can be found in the literature are diverse and numerous, the present section will not attempt a comprehensive review. Rather, we will first show the case of crystal nucleation and growth as an example that is simple enough to serve as an introduction. Subsequently, we will discuss a recent approach due to Diemer (2023) that offers a systematic way of treating a range of inverse problems with several kinetic processes.

3.5.2 Crystal Nucleation and Growth

The mixed-suspension mixed-product removal (MSMPR) model, which is used for kinetics extraction in crystallisation, is one of the earliest examples of an inverse population balance problem. A summary of the procedure is given below while more details and worked examples can be found in Randolph and Larson (1988) or Mullin (2001).

The MSMPR describes nucleation and size-independent growth of crystals in a perfectly stirred reactor with inflow and outflow operating at steady state, that is, a CSTR (cf. Section 2.6). With no aggregation or fragmentation present, it is convenient to formulate the PBE in terms of particle length, ℓ. Assuming no particles present at the inlet, the MSMPR equation is:

$$G_\ell \frac{dn(\ell)}{d\ell} + \frac{n(\ell)}{\tau} = 0, \tag{3.79}$$

where G_ℓ is the growth rate and τ is the residence time. Nucleation, rather than being included as a term, can be represented as a boundary condition:

$$n(0) = n_0. \tag{3.80}$$

The solution is:

$$n(\ell) = n_0 \exp\left(-\frac{\ell}{G_\ell \tau}\right). \tag{3.81}$$

To obtain the growth rate using the MSMPR model, an experiment must be carried out to obtain the distribution at steady state. By plotting the logarithm of $n(\ell)$ versus ℓ, the factor $-1/(G_\ell \tau)$ is readily obtained as the slope, from which the growth rate can be found (since the residence time can be fixed).

The kinetic order of nucleation may also be inferred from the one of growth if the latter is known in the following way. If nucleation and growth are both represented by power laws of supersaturation, then:

$$B = k_b S^b, \tag{3.82}$$

$$G_\ell = k_g S^g, \tag{3.83}$$

where S is the supersaturation. By eliminating S, we have:

$$B = k_b k_g^{-b/g} G_\ell^{b/g}. \tag{3.84}$$

Since nucleation was expressed as a boundary condition, we can write:

$$B = \left.\frac{dN}{dt}\right|_{\ell=0} = \left.\frac{dN}{d\ell}\right|_{\ell=0} \frac{d\ell}{dt} = n_0 G_\ell, \tag{3.85}$$

where N is the cumulative distribution, such that $n = dN/d\ell$. We thus have:

$$n_0 = k_b k_g^{-b/g} G_\ell^{b/g-1}. \tag{3.86}$$

Therefore, b/g can be obtained from a plot of the logarithm of n_0 versus that of G_ℓ.

3.5.3 Growth, Aggregation and Fragmentation

We now discuss an approach in which growth, aggregation, and fragmentation are considered simultaneously, due to Diemer (2023). Elements of the mechanism may also be inferred, as compared with presuming the complete form of a model and determining the parameters. This approach is based on the moment transformation of the PBE and utilises steady state and similarity solutions in ideal reactors.[12] The main concepts underlying this method are shown here, while the details can be found in Diemer (2023).[13]

The essence of this approach is to first describe the models for the three processes under consideration in the form of generic power-law kernels. This generic form can accommodate a wide range of possible mechanisms. Afterwards, an equation is derived for the moments of the PBE in ideal reactor models. Under steady state or similarity conditions, features such as the order of the kernels and the values of the parameters can be determined from experimental measurements of selected moments.

General Forms of Kernels and PBE Formulation

The first step is to cast the models into generic power-law forms. Volume is used as the independent variable, and the forms employed are the following:

Growth:

$$G(v) = G_0 v^{\xi}. \tag{3.87}$$

Aggregation:

$$\beta(v, w) = \beta_0 (v^{\alpha} w^{\lambda-\alpha} + v^{\lambda-\alpha} w^{\alpha}). \tag{3.88}$$

Fragmentation:

$$\gamma(v) = \gamma_0 v^{\eta}. \tag{3.89}$$

In these expressions, the coefficients G_0, β_0 and γ_0 may account for the non-volume-dependent part of the kernel. These forms encompass a wide range of models. For growth, as we saw in Section 3.4.2, both the reaction-limited and mass-transfer-limited models in crystallisation are given by power laws, while in Section 3.4.3 we saw that various models for the surface growth of soot are also given by power laws of the surface area (hence of volume). For aggregation, the Brownian motion kernel (Eq. 3.34), for example, can be written in the following way:

[12] While the present section is naturally placed here, it makes use of moment and similarity solution methods for the PBE that will be described in Chapter 4, so it is suggested to study the relevant sections of that chapter first.

[13] The notation employed here differs from the one used by Diemer (2023), for the purpose of consistency with the notation in the present book.

$$\beta(v,w) = \frac{2kT}{3\mu}(2 + v^{-1/3}w^{1/3} + v^{1/3}w^{-1/3}), \tag{3.90}$$

which can be brought to the following form:

$$\beta(v,w) = \beta_0 \left[(v^{\alpha_1}w^{\lambda-\alpha_1} + v^{\lambda-\alpha_1}w^{\alpha_1}) + \kappa(v^{\alpha_2}w^{\lambda-\alpha_2} + v^{\lambda-\alpha_2}w^{\alpha_2}) \right], \tag{3.91}$$

where $\beta_0 = 2kT/(3\mu)$, $\lambda = 0$, $\alpha_1 = 0$, $\alpha_2 = -1/3$ and $\kappa = 1$. This is a linear combination of two terms of the form of Eq. 3.88 and can be treated in the same way. Similar manipulations can be carried out for other kernels, such as the shear kernel in either of its forms (laminar and turbulent). For fragmentation, many general-purpose kernels can be expressed as power laws, as has been discussed in Section 3.3.3. Furthermore, the daughter particle distribution in fragmentation is assumed to be self-similar (cf. Section 3.3.2 and Eq. 3.56).

In order to derive kinetic parameters from experiments, a population balance model must first be brought into a sufficiently simple form, such as the MSMPR model in Section 3.5.2. Two types of models are considered in this approach: steady state and similarity solutions, both in the context of ideal reactors. In Section 2.6, it was shown that the batch and steady-state plug flow reactor are described by effectively the same equation, the former describing an evolution in time and the latter an evolution in space.

The CSTR in steady state, which is described by Eq. 2.41, is described by an algebraic equation, which is particularly advantageous for deriving kinetics. An example of this is the MSMPR model that we saw in the Section 3.5.2.

Similarity solutions will be discussed in Section 4.2.3, and the reader may wish to review that section before proceeding. Briefly, similarity states are reached under certain conditions, such as in a batch PSR with Brownian aggregation after a sufficiently long time. In such states, while the moments of the distribution keep changing (e.g., particles continue to aggregate in the aforementioned case), the shape of the distribution remains unchanged. The similarity solution is expressed in terms of a similarity variable, which is the particle volume scaled by the mean volume. At the similarity state, the mean volume continues to be a function of time, but the distribution in terms of the similarity variable remains constant.

Within the context of these two types of model, there are eight possible combinations of the three kinetic processes that are included in this approach: (1) no processes, (2) growth, (3) aggregation, (4) fragmentation, (5) growth and aggregation (6) growth and fragmentation, (7) aggregation and fragmentation, (8) all three simultaneously. Thus, apart from the trivial case (1), there are seven types of kinetic model that can be investigated. In some of these cases, the problem has a steady-state solution, while in others it has a similarity solution.

Moment Equation for Extracting Kinetics

The approach of Diemer (2023) is based on a moment transformation of the population balance that allows bringing the PBE to a form with few variables. The derivation of moment equations will be shown in Chapter 4 for individual processes (Section 4.2.4 for aggregation and Section 4.3.3 for growth). The moment equation used in this method accounts for a combination of growth, aggregation and fragmentation and is the following:

$$\frac{dm_j}{dt} = jG_0 m_{j-1+\xi} + \gamma_0 (m_{b',j} - 1) m_{j+\eta} + \beta_0 \left[\sum_{k=0}^{j} \binom{j}{k} m_{k+\alpha} m_{j-k+\lambda-\alpha} - m_{\alpha+j} m_{\lambda-\alpha} - m_{\lambda-\alpha+j} m_\alpha \right], \tag{3.92}$$

where $m_{b',j}$ is the jth moment of the self-similar fragment size distribution, $b'(z)$ (cf. Eq. 3.56).

The moments needed in this approach are the zeroth and first moment, which represent the total volume and total number of particles, respectively (cf. Section 1.2.3), and can be used to define a mean particle volume as follows:

$$v_m = \frac{m_1}{m_0}. \tag{3.93}$$

The mean volume is scaled by its initial value:

$$v'_m = \frac{v_m}{v_{m,0}}. \tag{3.94}$$

Furthermore, the moments are also scaled as follows:

$$m'_k = \frac{m_k}{m_0 v_m^k}. \tag{3.95}$$

Note that the scaled zeroth and first moments are both equal to one.

By manipulating the equations for the zeroth and first moments, Diemer was able to bring the equation for the evolution of the scaled mean volume into the following form:

$$\frac{dv'_m}{dt} = \frac{v_m'^{\xi}}{\tau_g} + \frac{v_m'^{\lambda}}{\tau_a} - \frac{v_m'^{1+\eta}}{\tau_f}, \tag{3.96}$$

where τ_g, τ_a and τ_f are characteristic times associated with growth, aggregation and fragmentation, respectively. They are given by the following expressions:

$$\tau_g = \frac{1}{G_0 m'_\xi v_{m,0}^{\xi-1}}, \tag{3.97}$$

$$\tau_a = \frac{1}{\beta_0 m_1 m'_\alpha m'_{\lambda-\alpha} v_{m,0}^{\lambda-1}}, \tag{3.98}$$

$$\tau_f = \frac{1}{(N_f - 1)\gamma_0 m'_\eta v_{m,0}^{\eta}}. \tag{3.99}$$

While these characteristic times depend on the scaled moments which are not, in general, constant, they become constant when a steady or similarity states are reached, and do not vary much for systems evolving towards such states. In the case of the two-parameter aggregation model (Eq. 3.91), the corresponding characteristic time is given by an expression of the same form as Eq. 3.98, shown below[14]:

$$\tau_a = \frac{1}{\beta_0 m_1 (m'_{\alpha_1} m'_{\lambda-\alpha_1} + \kappa m'_{\alpha_2} m'_{\lambda-\alpha_2}) v_{m,0}^{\lambda-1}}. \tag{3.100}$$

As will be discussed in Chapter 4, moment equations are unclosed in the general case, that is, they cannot be brought to a form where the number of equations is the same as the number of moments solved for, except for particular forms of kinetic models. Therefore, closure models may be needed to employ Eqs. 3.97–3.99 and such models are discussed in Chapter 4. In some cases, closed forms can be derived. Diemer (2023) treats cases corresponding to the seven combinations of kinetic processes mentioned above, for either steady-state or similarity states, and the reader is referred to that paper for the resulting expressions.

By comparing these expressions with experimental data, it is possible to extract information such as the combination of kinetic mechanisms, the order of the kernel and the parameter values. In many cases, only one or two kinetic processes are active, hence not all parameters in Eqs. 3.97–3.99 are present. In the simplest case where only one mechanism is active, a logarithmic plot of mean size versus time can be used to infer the order of the kernel. With more than one mechanism being present, the estimation of kernel orders can be more complicated. Therefore, the approach to the inverse problem should be considered together with experiment design in order to isolate mechanisms. Example applications of kinetics extraction shown in Diemer (2023) include the inference of the order of an aggregation and a fragmentation kernel, as well as an inverse problem for aggregation and coalescence of carbon black, based on a monodisperse PBE (cf. Section 2.5).

[14] Diemer, R. B., private communication.

3.6 Transport Processes

3.6.1 Overview

Transport processes are important for problems where the number density is spatially dependent and the PBE must be coupled with the equations of fluid dynamics. Three terms in the PBE are associated with such processes: the particle convection, the acceleration (or transport in velocity space) and the diffusion term. The PBE with transport processes was discussed in Section 2.7. For reference, we show here the two main equations that will be referred to, namely Eq. 2.21 for non-inertial particles:

$$\frac{\partial n}{\partial t} + \frac{\partial (u_j n)}{\partial x_j} + \frac{\partial (\dot{\xi}_k(\boldsymbol{\xi}) n)}{\partial \xi_k} = \frac{\partial}{\partial x_j}\left(D_p \frac{\partial n}{\partial x_j}\right) + S(\boldsymbol{\xi}, n), \tag{3.101}$$

and Eq. 2.47 for inertial particles:

$$\frac{\partial n}{\partial t} + \frac{\partial (u_{p,j} n)}{\partial x_j} + \frac{\partial (\dot{u}_{p,j} n)}{\partial u_{p,j}} + \frac{\partial (\dot{\xi}_k(\boldsymbol{\xi}) n)}{\partial \xi_k} = S(\boldsymbol{\xi}, n), \tag{3.102}$$

where the particle velocity, $\mathbf{u_p}$, is a field to be calculated from the fluid velocity if the approximate model for low Stokes number inertial particles is used (in which case the acceleration term can be omitted, cf. Eq. 2.46) or an independent coordinate if the particle velocity distribution is taken into account.

In the above equations, the quantities that are associated with fluid dynamics and transport phenomena are the fluid velocity, the forces acting on the particles and the particle diffusion coefficient (i.e. u_i, $\dot{u}_{p,i}$ and D_p). Thermophoresis, if present, can be treated as an additional convective velocity. The main transport mechanisms are the following:

- *Convection* of particles by the fluid.
- *Hydrodynamic forces*, such as the drag force.
- *Body forces*, such as gravitational force and thermophoresis.
- *Thermal motion* (free molecular or Brownian motion).

Usually, only a selection of these mechanisms is present in a given problem. For example, the motion of sub-micron aerosols is dominated by convection, thermal motion, drag force and thermophoresis (in the absence of external fields), while the motion of larger droplets and bubbles may involve, apart from convection, various hydrodynamic forces (such as drag force, added mass and lift) and buoyancy. In multiphase flows, momentum exchange between particles or between particles and walls and momentum transfer from particles to flow may be present, but these problems will not be treated in the present book.

In order to determine which forces are important for a particular problem and which models should be employed for these forces, a number of considerations must be made. For aerosols, we must first consider whether the particle motion is in the free molecule, transition or continuum regime. This has implications for the particle motion, as it determines the model to be employed for the drag force, but also for aggregation, as different kernels are used in different regimes. Brownian motion is also an important mechanism for sub-micron particles, such as aerosols (in the continuum regime) and hydrosols. Thermophoresis is important for aerosols. The inertia of the particles must be considered in order to decide whether they follow the fluid streamlines, whether their velocity can be expressed as a function of the local fluid velocity or whether a distribution of velocities must be considered. Finally, the particle concentration determines whether the flow is dilute or dense, and therefore whether particle interactions and feedback effects on the flow must be considered.

Several dimensionless groups can aid in quantifying these considerations, some of which will be introduced in the sections that follow:

- The *Knudsen number* (cf. Section 3.2.1) determines whether continuum or non-continuum models should be employed.
- The *Stokes number* (cf. Section 3.6.5) is representative of the importance of inertial effects and thus indicates how the spatially dependent PBE should be formulated.
- The *relative Reynolds number* (cf. Section 3.6.2) is important for the choice of models for the hydrodynamic forces.
- The *Prandtl* and *Schmidt* numbers are relevant for heat and mass transfer, respectively.

The literature on transport processes in flows with particles, droplets and bubbles is very extensive. Treatises include Hidy and Brock (1970), Williams and Loyalka (1991), Friedlander (2000) and Seinfeld and Pandis (2016) on aerosol science, Elimelech et al. (1995) on hydrosols and Clift et al. (1978), Crowe et al. (1998) and Michaelides and Feng (2017) on multiphase flow, among others. The treatment of this topic here does not aim to be comprehensive; rather, the objectives are to present the fundamental concepts and show how the physical models can be used in the context of population balance formulations. Furthermore, problems involving dense flows and feedback effects by the particles on the flow will not be included, as they are discussed in multiphase flow treatises.

In the following, we will first discuss in some detail the drag force, as it is the main hydrodynamic force that plays a role in the problems considered in the present book, such as the case studies in Chapter 6. Other forces, such as lift and virtual mass, are discussed in Crowe et al. (1998). Body forces such as gravity and thermophoresis are also mentioned, before moving on to Brownian motion. Subsequently, we show how the PBE should be formulated so as to be consistent with the equation of single particle motion.

3.6.2 Drag Force

The drag force is the hydrodynamic force in the direction opposite to the motion of the particle. In order to understand how the expressions for it are derived, we first need to consider the flow relative to the particle.

Relative Reynolds Number

The drag force is best described in a frame of reference fixed on the particle. In this frame and in dimensionless form, the Navier–Stokes equations are:

$$\frac{\partial u_i^*}{\partial t} + u_j^* \frac{\partial u_i^*}{\partial x_j^*} = -\frac{\partial p^*}{\partial x_i^*} + \frac{1}{Re_p} \frac{\partial^2 u_i^*}{\partial x_j^* \partial x_j^*}, \tag{3.103}$$

where the u_i^* is the dimensionless relative velocity of the fluid with respect to the particle and the body force term has been omitted. The scaling has been carried out with the free stream relative velocity, u_∞, and the particle diameter, d_p, as follows: $x_i^* = x_i/d_p$, $t^* = tu_\infty/d_p$, $u_i^* = u_i/u_\infty$ and $p^* = p/(\mu u_\infty/d_p)$. The dimensionless number, Re_p, that is formed in the last term is the *relative Reynolds number* for the particle:

$$Re_p = \frac{u_\infty d_p}{\nu}. \tag{3.104}$$

The value of Re_p must be considered in order to determine the appropriate models for the hydrodynamic forces. For steady flow and in the limit of very small Re_p (*creeping* or *Stokes flow*), the inertia term in Eq. 3.103 can be neglected and the classical Stokes solution for the drag force is obtained; this will be discussed in the next section, along with models for other regimes.

Expressions for the Drag Force

The drag force can be resolved into a component that results from the uneven pressure distribution around the particle, the *form drag*, and a component associated with the shear stress at the fluid–particle interface, the *friction drag*.

To obtain an expression for the drag force, one must first solve the continuity and Navier–Stokes equations for the flow around the particle and derive expressions for the pressure and shear stress at the particle surface. Subsequently, the components of these stresses in the direction of the motion are integrated over the surface to yield the form and friction drag. For example, let us consider a spherical particle and a spherical coordinate system such that the origin is at the centre of the particle and the z axis is aligned with the direction of motion. The component of the pressure force on an element of surface of the particle in the direction of motion is $-p(R,\theta)\cos\theta dA$, where $dA = R^2 \sin\theta d\theta d\phi$. The total pressure force is thus given by:

$$F_p = -\int_0^{2\pi}\int_0^{\pi} p(R,\theta)\cos\theta R^2 \sin\theta d\theta d\phi. \tag{3.105}$$

In a similar vein, the shear force is $\tau_{r\theta}(R,\theta)(-\sin\theta)dA$ and its integral is:

$$F_s = -\int_0^{2\pi}\int_0^{\pi} \tau_{r\theta}(R,\theta)\sin\theta R^2 \sin\theta d\theta d\phi. \tag{3.106}$$

To obtain $p(R,\theta)$ and $\tau_{r\theta}(R,\theta)$, one needs the analytical solution of the equations of flow around a particle, which is possible only in special cases, however. The case of creeping flow is the most well known solution and turns out to be relevant for a wide range of population balance problems where the size of the particles is small. For very small Re_p and steady flow, the inertia and unsteady terms in Eq. 3.103 can be omitted, resulting in the following equations:

$$\frac{\partial u_i^*}{\partial x_i^*} = 0, \tag{3.107}$$

$$\frac{\partial p^*}{\partial x_i^*} = \frac{1}{Re_p}\frac{\partial^2 u_i^*}{\partial x_j^* \partial x_j^*}. \tag{3.108}$$

The first solution of Eqs. 3.107–3.108 equations was obtained by Stokes (1851) for a rigid spherical particle in an infinite medium and the no-slip condition applied at the fluid–particle interface. The solution is obtained by introducing spherical coordinates and an axisymmetric stream function (the *Stokes stream function*), and the details of the derivation can be found in many texts, such as Bird et al. (1960) or White (2006). Expressions are thus obtained for the pressure and shear stress[15]

[15] The equations for the normal and tangential forces in spherical coordinates are:

$$u_r = u_\infty\left[1 - \frac{3}{2}\frac{R}{r} + \frac{1}{2}\left(\frac{R}{r}\right)^3\right]\cos\theta,$$

$$u_\theta = -u_\infty\left[1 - \frac{3}{4}\frac{R}{r} - \frac{1}{4}\left(\frac{R}{r}\right)^3\right]\sin\theta,$$

$$p(r,\theta) = p_\infty - \frac{3\mu R u_\infty \cos\theta}{2r^2},$$

$$\tau_{r\theta}(r,\theta) = -\frac{3\mu R^3 u_\infty \sin\theta}{2r^4},$$

where u_∞ and p_∞ are the velocity and pressure at the free stream, respectively. By substituting into Eqs. 3.105 and 3.106, Eqs. 3.109 and 3.110 are obtained.

which, when substituted into Eqs. 3.105 and 3.106, lead to the following:

$$F_p = \pi \mu d_p u_\infty, \tag{3.109}$$

$$F_s = 2\pi \mu d_p u_\infty, \tag{3.110}$$

where u_∞ is the free stream velocity. The total drag force for creeping flow around a sphere, therefore, is:

$$F_d = 3\pi \mu d_p u_\infty. \tag{3.111}$$

It is customary to represent the drag force as proportional to the product of the kinetic energy per unit volume and a characteristic area, in this case the projected frontal area of the sphere:

$$F_d = c_d \frac{\pi d_p^2}{4} \cdot \frac{1}{2} \rho u_\infty^2, \tag{3.112}$$

where the constant of proportionality, c_d, is called the *drag coefficient*. Combining Eqs. 3.112 and 3.111, we obtain the following expression for the drag coefficient for creeping flow around spheres:

$$c_d = \frac{24}{Re_p}. \tag{3.113}$$

The flow around aerosol particles is usually in the Stokes regime, owing to their small size. Seinfeld and Pandis (2016, page 371) give values of Re_p for particles falling with their terminal velocity in air at 298 K; for $d_p = 10\ \mu$m, $Re_p = 2.5 \cdot 10^{-3}$, while for $d_p = 20\ \mu$m, $Re_p = 0.02$, values for which the Stokes drag formula is accurate. At the upper end, $Re_p = 2$ for $d_p = 100\ \mu$m, at which point the drag force can be represented by small corrections to Stokes' law. Oseen (1910) derived a formula for the drag coefficient by including a linearised convective term to the creeping flow equations, which can be used for Re_p up to 5 (Crowe et al., 1998). The Oseen formula is:

$$c_d = \frac{24}{Re_p} \left(1 + \frac{3}{16} Re_p\right). \tag{3.114}$$

Higher values of Re_p can be treated with the aid of empirical formulas. There are also corrections and formulas for non-spherical and non-rigid particles. Many correlations exist for the drag factor at higher relative Reynolds numbers; more information on this subject can be found in Clift et al. (1978), Williams and Loyalka (1991), Crowe et al. (1998) or Sirignano (2010).

The approach described so far is based on the continuity and Navier–Stokes equations and thus does not take into account non-continuum effects, which can be important for aerosols with sizes below 100 nm. If $Kn >> 1$, kinetic theory must be used (Epstein, 1924). An interpolation formula that can be used for all values of Kn is the following:

$$F_d = \frac{3\pi \mu d_p u_\infty}{C_c}, \tag{3.115}$$

where the slip correction factor, C_c, is given by:

$$C_c = 1 + Kn\left[a + b\exp\left(-\frac{c}{Kn}\right)\right], \tag{3.116}$$

where the coefficients have been determined empirically; $a = 1.257$, $b = 0.4$ and $c = 1.1$ are widely accepted values (Seinfeld and Pandis, 2016).

3.6.3 Body Forces

Gravity must be included in problems involving heavy particles. Buoyancy must also be included when the density of the dispersed phase differs considerably from that of the continuous phase, such as in bubbles. The equation for the net gravitational force is:

$$F_g = m_p g\left(1 - \frac{\rho}{\rho_p}\right), \tag{3.117}$$

where ρ and ρ_p are the densities of the fluid and the particle, respectively, and m_p is the mass of the particle.

Thermophoresis can be modelled as an additional velocity (Friedlander, 2000). Another body force that may be present is the electrostatic force, a treatment of which can be found, for example, in Friedlander (2000) for aerosols and Elimelech et al. (1995) for colloids.

3.6.4 Brownian Motion

Apart from convection by the fluid, the motion of small aerosol and colloidal particles is also subject to fluctuations caused by incessant bombardment by fluid molecules. This is known as Brownian motion and gives rise to a diffusion mechanism of particle transport.

The pioneering works of Einstein, Langevin and Smoluchowski[16] showed that Brownian motion can be treated as both a diffusion process (for a swarm of particles) and a stochastic process (for a single particle). Adopting the diffusion point of view, we can write the following equation for Brownian motion in one dimension:

[16] For English translations of these works see the Dover republication of Einstein (1926), Lemons and Gythiel (1997) for Langevin and Cichocki (2017) for Smoluchowski.

$$\frac{\partial n}{\partial t} = \frac{\partial}{\partial x}\left(D_p \frac{\partial n}{\partial x}\right), \tag{3.118}$$

where $n(x, t)$ is the concentration of monodisperse particles and D_p is the particle diffusivity.

To obtain an expression for D_p, we proceed as follows. We first note that, if the particles have an initial are initially located at $x = 0$ with concentration n_0, the solution of this equation is[17]:

$$n(x, t) = \frac{n_0}{2\sqrt{\pi D_p t}} \exp\left(-\frac{x^2}{4D_p t}\right), \tag{3.119}$$

from which the mean square displacement of a particle as a function of time can be obtained as follows:

$$\overline{x^2}(t) = \frac{1}{n_0}\int_{-\infty}^{\infty} x^2 n(x, t)dx, \tag{3.120}$$

or

$$\overline{x^2}(t) = 2D_p t. \tag{3.121}$$

Another expression for this displacement can now be obtained by considering the motion of a single particle and starting from the following equation:

$$m_p \frac{du_p}{dt} = F_d + F_b(t), \tag{3.122}$$

where F_d is the drag force and $F_b(t)$ is a random force exerted on the particles by the molecules. Eq. 3.122 is therefore not a differential equation in the usual sense but rather a *stochastic differential equation*. Here, however, the objective is not to find the positions of individual particles, but rather their mean displacement. The drag force can be written as:

$$F_d = -fu_p, \tag{3.123}$$

where f is a friction coefficient. By assuming that the mean value of $F_b(t)$ vanishes, the following solution can be obtained (Friedlander, 2000):

$$\overline{x^2}(t) = \frac{2\overline{u_p^2} m_p}{f} t. \tag{3.124}$$

One of Einstein's most crucial assumptions was that, since the particles are subjected to the thermal motion of the fluid molecules, the principle of equipartition of energy can be applied to yield:

[17] For the details of the solution see, for example, Crank (1975).

$$\frac{1}{2}m_p\overline{u_p^2} = \frac{1}{2}kT, \tag{3.125}$$

where k is Boltzmann's constant. We thus have the following expression for the mean square displacement:

$$\overline{x^2}(t) = \frac{2kT}{f}t, \tag{3.126}$$

which can be equated with the one obtained from Eq. 3.121 to yield the following expression for D_p:

$$D_p = \frac{kT}{f}. \tag{3.127}$$

For Stokes drag, $f = 3\pi\mu d_p$ (cf. Eq. 3.111) and we have:

$$D_p = \frac{kT}{3\pi\mu d_p}. \tag{3.128}$$

If non-continuum effects are important, Eq. 3.115 can be used instead for the drag force in order to introduce the slip correction factor.

Finally, we consider the diffusion coefficient for the relative motion of particles with two different sizes. This was encountered when considering aggregation due to Brownian motion (Section 3.2.2). In this case, the coefficient $D_{p,12}$ can be written in terms of the mean square relative displacement of the two particles (Friedlander, 2000) and we can write, in the vein of Eq. 3.121:

$$\overline{(x_1 - x_2)^2} = 2D_{p,12}t. \tag{3.129}$$

The mean square relative displacement can be expanded as:

$$\overline{(x_1 - x_2)^2} = \overline{x_1^2} - 2\overline{x_1 x_2} + \overline{x_2^2}. \tag{3.130}$$

Since the motion of the two particles is independent, $\overline{x_1 x_2} = 0$ and we obtain:

$$D_{p,12} = \frac{\overline{x_1^2}}{2t} + \frac{\overline{x_2^2}}{2t}, \tag{3.131}$$

or

$$D_{p,12} = D_{p,1} + D_{p,2}. \tag{3.132}$$

where $D_{p,1}$ and $D_{p,2}$ are the diffusivities of each of the two types of particles. This result was used to derive Eq. 3.31 in Section 3.2.2.

3.6.5 Incorporation of Transport Models in the PBE

Having considered the models for spatial transport, we now show how the PBE must be formulated in order to be consistent with the equations of particle motion. As mentioned in Section 3.6, we are considering dilute flows where particle–particle interactions are negligible.

The first question is whether particle trajectories deviate from the streamlines of fluid motion and if so, how this should be taken into account. To answer this question, we start from the equation of single particle motion and introduce the concepts of relaxation time and Stokes number. Subsequently, we examine the conditions under which Eq. 2.22 may be simplified. Once the appropriate form of the equation has been determined, the next step is to introduce the transport models into the PBE, and some examples will be discussed in the present section.

Equation of Particle Motion

In order to obtain the equation of motion for a particle, an equation for the particle velocity based on Newton's law is required. The equations, therefore, are:

$$\frac{dx_{p,i}}{dt} = u_{p,i}, \tag{3.133}$$

$$m_p \frac{du_{p,i}}{dt} = F_{p,i}, \tag{3.134}$$

where $x_{p,i}$ is the vector indicating the particle position and $F_{p,i}$ is the total force applied on the particle and m_p is the particle mass.

We first consider the case of a spherical particle where drag is the only force present and motion is in the creeping flow regime. Based on Eq. 3.111, the equation for particle motion is:

$$m_p \frac{du_{p,i}}{dt} = 3\pi\mu d_p (u_i - u_{p,i}), \tag{3.135}$$

or

$$\frac{du_{p,i}}{dt} = \frac{18\mu}{\rho_p d_p^2}(u_i - u_{p,i}). \tag{3.136}$$

This equation may be rewritten as follows:

$$\frac{du_{p,i}}{dt} = \frac{1}{\tau_r}(u_i - u_{p,i}), \tag{3.137}$$

where τ_r is a timescale representative of the time required for a particle to adapt to a change in the flow velocity and is called the *relaxation time*. In this case, it is equal to:

$$\tau_r = \frac{\rho_p d_p^2}{18\mu}. \tag{3.138}$$

In order to get some physical intuition about this concept, we consider a particle starting from rest, for which the solution of Eq. 3.137 is:

$$u_{p,i} = u_i \left(1 - e^{-t/\tau_r}\right). \tag{3.139}$$

In this case, then, the relaxation time can be interpreted as the time for a particle to achieve a value equal to 63% of the free-stream velocity.

For many problems involving inertial particles, the main forces applied are drag and gravity. It is useful to introduce the *drag factor*, f, defined as the ratio of the drag coefficient to the Stokes drag:

$$f = \frac{c_d Re_p}{24}. \tag{3.140}$$

The following equation is thus obtained for the particle velocity:

$$\frac{du_{p,i}}{dt} = \frac{f}{\tau_r}(u_i - u_{p,i}) + g_i \left(1 - \frac{\rho}{\rho_p}\right). \tag{3.141}$$

By using Eq. 3.111, the equation of particle velocity can also be written as follows:

$$\frac{du_{p,i}}{dt} = \frac{1}{2} c_d \frac{\pi d_p^2}{4} \rho (u_i - u_{p,i})|u_i - u_{p,i}| + g_i \left(1 - \frac{\rho}{\rho_p}\right). \tag{3.142}$$

A comprehensive equation for the motion of a rigid spherical particle in creeping flow that accounts for effects such as those associated with transient and nonuniform flow has been developed by Maxey and Riley (1983). For higher Reynolds number, different drag expressions can be introduced via the drag coefficient or drag factor. Further hydrodynamic forces, such as lift, and body forces, such as the electrostatic force, may have to be considered where appropriate. A critical review of analytical expressions for particle motion in flow, as well as for heat and mass transfer, can be found in Michaelides (2003). Complex equations of particle motion that take into account secondary effects are more suited to computation with Lagrangian approaches, where the equation of particle motion is solved directly.

Stokes Number

The Stokes number determines the importance of inertial effects on the particle motion. It was defined in Section 2.7.2 as $St = \tau_r/\tau_f$, where τ_r is the relaxation time, which is representative of the time it takes for the particle to respond to changes in fluid velocity, and τ_f is a characteristic timescale of the flow. The latter should be chosen so as to be relevant to the problem considered; for example, in a flow with mean velocity u_0 over an obstacle with characteristic dimension l, it can be taken as l/u_0. The Stokes number may equivalently be defined as a ratio of two lengthscales, the *stop distance* – the initial velocity multiplied by the relaxation time – and a lengthscale characteristic of the flow.

In the limit of $St \to 0$, the particles adapt instantaneously to changes in the flow and thus follow the streamlines. As St gets larger, the particle trajectories begin deviate from the streamlines. For non-zero but small values of St, an approach will be shown below where the particle velocity can be calculated from the local fluid velocity and its gradient. For larger values of St, the particles exhibit a distribution of velocities that must be accounted for in a population balance that includes the particle velocity as an independent coordinate; alternatively, an equivalent Lagrangian computational method may be employed where the particle trajectories are calculated by solving directly Eqs. 3.133 and 3.134.

Non-Inertial Particles

For St $\to$ 0, the particles follow the streamlines of the flow. Such particles are typically subjected to Brownian motion, so the diffusion term must be retained. The PBE, therefore, written in terms of particle volume and including a growth and a source term (which may incorporate nucleation, aggregation and fragmentation) is based on Eq. 2.21:

$$\frac{\partial n}{\partial t} + \frac{\partial (u_i n)}{\partial x_i} + \frac{\partial (Gn)}{\partial v} = \frac{\partial}{\partial x_i}\left(D_p(v)\frac{\partial n}{\partial x_i}\right) + S. \tag{3.143}$$

This equation can be used for problems involving small colloidal particles in liquid flows, as commonly encountered in crystallisation and precipitation processes, and for many problems in aerosol science such as atmospheric aerosols, nanoparticle synthesis and soot formation. It should be mentioned that, while atmospheric aerosol particles can reach larger sizes, the flow scales in the atmosphere are also larger and therefore the Stokes number is often very small. In some of these problems, such as soot formation and nanoparticle synthesis, thermophoresis should also be included. For example, if we consider particles in the continuum regime and introduce Eq. 3.128, we have:

$$\frac{\partial n}{\partial t} + \frac{\partial}{\partial x_i}\left[\left(u_i - u_i^t\right) n\right] + \frac{\partial (Gn)}{\partial v} = \frac{\partial}{\partial x_i}\left(\frac{kT}{3\pi \mu d_p}\frac{\partial n}{\partial x_i}\right) + S, \tag{3.144}$$

where u_i^t is the thermophoretic velocity and $d_p(v)$ is the particle diameter as a function of its volume. This equation can be used for soot and nanoparticle synthesis in laminar flows, as in the case studies presented in Sections 6.2 (the Santoro flame case) and 6.1.

Low Stokes Number Approximation

For particles of low Stokes number, it has then been shown (De La Mora and Rosner, 1982; Maxey, 1987; Ferry and Balachandar, 2001) that the particle velocity can be obtained as a function of the fluid velocity as follows:

$$u_{p,i} = u_i - \tau_r \left(u_j \frac{\partial u_i}{\partial x_j} - g_i \right) + O(\tau_r^2). \tag{3.145}$$

This equation can be used to provide the particle velocity as a field variable to be introduced into the PBE, which takes the form of Eq. 2.46. The application of this formulation approximation requires a careful consideration of the Stokes number of the fluid–particle system, which depends on both the particles and the flow structures. A typical range of applicability is for Stokes number less than about 0.2 (Balachandar and Eaton, 2010). Higher-order terms can also be derived (Ferry and Balachandar, 2001).

PBE with Velocity Distribution

If the particles are inertial and the low Stokes number approximation cannot be invoked, a distribution of particle velocities must be considered and the acceleration term in Eq. 2.22 must be retained. This formulation is usually appropriate for particles large enough for Brownian motion to be negligible. The acceleration, $\dot{\mathbf{u}}_\mathbf{p}$, must include the appropriate models for the forces on the particles such that Eq. 2.22 is consistent with the equation of particle motion. For example, if only the drag force is considered and we include volumetric growth, we obtain:

$$\frac{\partial n}{\partial t} + \frac{\partial (u_{p,i} n)}{\partial x_i} + \frac{\partial}{\partial u_{p,i}} \left[\frac{1}{2} c_d \frac{\pi d_p^2}{4} \rho (u_i - u_{p,i}) |u_i - u_{p,i}| n \right] + \frac{\partial (Gn)}{\partial v} = S. \tag{3.146}$$

Equations of this form are often used for sprays. An early appearance of such an equation can be found in Williams (1958, 1961), as discussed in Section 2.7.4.

In the case of turbulent flows, there are fluctuations in particle position and velocity. For inertial particles, this problem is usually addressed via a PDF transport equation rather than a PBE. As mentioned in Section 1.2.2, a PDF transport equation is formulated in terms of a probability density function rather than a number density function. This approach originates in concepts from kinetic theory, where the aim is to obtain the probability distribution of molecular velocities arising from the random motion of molecules. A review of this approach can be found in Reeks et al. (2017) and Reeks (2021). The issues associated with turbulent flow will be revisited in Chapter 5.

3.7 Concluding Remarks

In the present chapter, models for the kinetic and transport processes in the PBE were considered. Since the range of population balance problems is very wide, the focus was on the general forms of models and their incorporation into the PBE. The dependency of the models on particle size is particularly relevant, as it has repercussions for the solution methods. More specific examples were drawn from the fields of aerosols and crystallisation that feature in the case studies of Chapter 6. A few further points may be made about the selection of kinetic models.

While some aspects of kinetics, such as collision mechanisms and transport processes, are sufficiently established, other aspects such as particle formation kinetics (particularly nucleation) and fragmentation rates have considerable uncertainty and are largely problem specific. For problems that have not been investigated kinetically, the inverse population balance methodology can be employed to determine kinetics and extract more mechanistic insight from experimental data.

Apart from experiments, simulation methods such as molecular dynamics and Monte Carlo simulations at the microscopic level are expected to play an increasingly important role in the future. The population balance methodology can thus act as a bridge between the molecular and continuum level approaches.

Model selection should be considered together with issues relating to the formulation and solution method. More complex kinetic models typically involve more parameters and greater uncertainty. The availability and reliability of a kinetic model enable the use of a population balance formulation and solution method that are commensurate with the kinetics in terms of complexity and level of detail.

4

Solution of the PBE

4.1 Introduction and Overview

4.1.1 General Features of the PBE

In the present chapter, we will discuss solution methods for the various forms of the PBE. At first, the methods will be developed in the context of the spatially homogeneous PBE, while the later sections will address the coupling of the PBE with CFD for the solution of spatially inhomogeneous problems.

The first analytical solution of the PBE was due to Smoluchowski (1917). It is a solution of the discrete aggregation PBE (known as the Smoluchowski equation) with the constant kernel and monodisperse initial conditions. There are very few other known analytical solutions for discrete PBEs. Numerical integration is possible, but the size of the ODE system (order of 10^9 equations for a particle size spectrum ranging from 1 nm to 1 μm in diameter) renders this approach prohibitively expensive. It is thus usually preferable to solve the equivalent continuous PBE.

The continuous PBE is a nonlinear partial integro-differential equation for which only a few analytical solutions are known. Numerical methods are thus required for its application over a wide range of problems and conditions. The form of the continuous PBE, however, poses certain unique challenges for numerical methods. Part of the reason is the need to accommodate processes of a different nature. Aggregation and fragmentation are described by integral equations. Surface processes such as growth, on the other hand, are described by a hyperbolic partial differential equation whose solution requires methods similar to those employed in gas dynamics. Furthermore, aggregation results in rapid size increase so that the solution covers a wide spectrum of scales. The need to address these issues simultaneously has led to the development of a large number of methods.

Questions of existence, uniqueness, boundedness and positiveness of solutions have been posed and addressed in the literature, mainly in the context of aggregation and fragmentation equations. These questions will not be discussed in the present text, but the interested reader may refer to Drake (1972), Ramkrishna (2000) or Banasiak et al. (2020a,b).

The aim of this chapter is to explain the main concepts underlying the methods for solving the PBE and to discuss in detail a small selection of them which will be used in the case studies presented in Chapter 6. No comprehensive description of all methods will be attempted, as the literature in this field is very extensive, but references will be provided. The emphasis of the treatment will be on the fundamental concepts underpinning the methods, and it is hoped that this will allow the readers to acquire the big picture and negotiate the details of individual methods from the relevant papers. The discussion will often focus on individual kinetic processes, the methods for which constitute the building blocks for solving more complex PBEs combining several processes. We will commence with aggregation and fragmentation, which lend the PBE the character of an integral equation. We then turn our attention to growth, whose solution introduces new concepts, and also to problems where it is coupled with other processes. The solution of multi-dimensional and coupled PBE formulations is based on extensions of methods for the one-dimensional PBE and will be treated next. The last part of the chapter is devoted to the coupling of the various methods with fluid dynamics and CFD. The modelling issues associated with turbulent flow will be the subject of Chapter 5.

4.1.2 Overview of Solution Methods

A very large body of literature has been devoted to solution methods for the PBE, either for individual processes or for combinations of them. In terms of the basic concepts, the methods can be classified into analytical, similarity, Monte Carlo, moment and discretisation methods. Before discussing the methods for each process in detail, it is worth mentioning the main features of each family of methods.

Analytical Solutions

Analytical solutions of the PBE can be obtained only for a few special cases. While restricted in terms of applicability to realistic problems, these solutions provided pivotal insights in early research and are also important for validating numerical methods. In the present book, they will be discussed (if only briefly) in Section 4.2.2 for aggregation and fragmentation and in Section 4.3.2 for growth and problems where it is combined with other processes.

Similarity Solutions

Under certain conditions, the PBE can be reduced into an equation in terms of a single similarity variable, and solution of that equation yields a distribution independent of time and initial conditions (although the similarity variable is itself timedependent and initial conditions affect the time to reach this distribution). If a similarity solution for a certain PBE problem exists, it can be found analytically (with some approximations) or numerically. The problems for which such solutions exist are limited, but they are of considerable physical significance and relevant to a number of practical problems. Similarity solutions have been obtained mainly for aggregation and fragmentation problems, although some solutions for problems including growth can also be found, and will be discussed in Section 4.2.3.

Moment Methods

Moment methods involve solving equations for certain moments of the distribution, rather than resolving the distribution itself. The main motivation for moment methods stems from the fact that the number of variables employed is minimal since only a few moments are being computed, while discretisation methods require more variables for an adequate representation of the distribution. There are two main drawbacks, however, in the moment approach. The first one is that the distribution is not being predicted, and assumptions must be made for reconstructing it from the moments. The second one is that the moment equations are unclosed except for special cases such as aggregation with the constant kernel or growth with size-independent kinetics, as will be shown in Section 4.2.4. Various approximate methods have been proposed to provide closure. These methods can be classified into the following general categories.

- *Presumed distribution.* This approach is very effective in applications where there is evidence that the distribution assumes a wellknown form – some examples are the log-normal distribution, often encountered in aerosols, and the Rosin–Rammler distribution, which is associated with sprays. In many problems, however, the combination of surface processes, aggregation and fragmentation, as well as the effect of mixing, give rise to distributions deviating considerably from standard forms.
- *Series approximation of the distribution.* In this approach, the distribution is approximated as a sum of trial functions, whose coefficients are allowed to change with time. The choice of functions imposes an assumption on the a functional shape of the distribution, so this approach is related to the previous one.
- *Model approximation.* As the reason for the lack of closure lies in the form of the physical models, such as the aggregation kernel, it may be possible to replace them with approximate simplified forms in some cases. Such approximations must be carried out specifically for each problem and tested extensively to determine their range of validity.

- *Moment interpolation.* In this approach, the unknown fractional moments are obtained by interpolation from the integral ones. This approach has found widespread application in soot models.
- *Numerical quadrature methods.* This family of methods originates in the Quadrature Method of Moments (QMOM), where Gaussian quadrature is employed to approximate the unclosed terms inside integrals. A later method based on the same concept is the Direct Quadrature Method of Moments (DQMOM), where ODEs for the quadrature parameters are derived.

Section 4.2.4 will provide more details on the fundamentals of moment methods for aggregation and fragmentation, while Section 4.2.5 will introduce closure models for these processes. Moment methods for problems including growth will be discussed in Section 4.3.3.

Discretisation Methods

Discretisation methods are based on a discrete representation of the continuous distribution. The distribution is divided into intervals, also called sections, hence discretisation methods are often called *sectional methods*.[1] A discretisation method with a uniform grid and an increment of one monomer is computationally equivalent to a solution of the discrete PBE. A discretised continuous PBE, however, allows dividing the domain of the independent variable into intervals of unequal size. This is essential for aggregation problems, where the solution can cover a wide range of scales. Several issues are encountered in devising a successful discretisation method for the PBE, due to the integro-differential nature of the equation and the need to accommodate processes of a different nature:

- Aggregation-dominated problems require expanding grids, as the particle volume can increase by several orders of magnitude. In a uniform grid, which is equivalent to the discrete PBE, an aerosol ranging from 1 nm to 1 μm in diameter would require an order of 10^9 grid nodes in volume space (aggregation is best described with volume or an equivalent quantity, such as mass, as the independent variable – cf. Section 1.3).
- Aggregation involves integral terms whose discrete form is non-local, that is, includes contributions from the entire domain. A large number of operations is thus needed, in contrast with the discretisation of derivatives, which involve only neighbouring intervals.

[1] This term is usually employed for methods where the distribution is approximated with a piecewise constant function, that is, as a histogram. In the present book, we will employ the term 'discretisation method' to encompass a wider family of methods.

- It is not straightforward to obtain a discretisation of the aggregation terms that is conservative with respect to the moments of the distribution in an arbitrary non-uniform grid. The origin of this issue is in the aggregation birth term, which resembles a convolution (and is, in fact, a standard convolution in the case of the constant kernel). Conservation of the first moment with respect to volume is particularly important, as it is an invariant in aggregation processes and represents conservation of mass. It must be noted, however, that when growth is present in addition to aggregation and/or fragmentation, the first moment is not constant and its prediction relies also on the treatment of the growth term, which depends on other properties of the distribution (such as the surface area for reaction-limited growth).
- Growth-dominated problems often result in a distribution with a sharp front whose resolution with discretisation methods poses considerable problems.

Several methods have been proposed for the discretisation of the aggregation terms in the PBE. Some of them are derived in an ad hoc manner based on the physics of the aggregation process, while others proceed by applying formally a numerical approach such as the finite element or finite volume method. It will be shown that the ad hoc methods are special cases of the formal methods. Three families of methods can be identified:

- *Pointwise approximation methods* represent the distribution as concentrated at a finite number of points in the size domain, that is, as a sum of delta functions. In effect, these methods replace the continuous PBE with an equation resembling the discrete PBE. On a non-uniform grid, however, many of the resulting particles cannot be accommodated by the grid nodes, and the challenge for these methods is to redistribute these particles to neighbouring nodes in a manner guided by the conservation of the moments. These methods, therefore, place emphasis on the prediction of the moments rather than on the distribution, and will be discussed in Section 4.2.6.
- *One-dimensional quadrature methods* are derived by applying a finite element or finite volume method to the PBE. Since the latter is an integral equation, the application of these methods results in double integrals. In this class of methods, the inner integral is approximated with one-dimensional quadrature, thus resulting in a polynomial expression that can be readily integrated. One problem faced by these methods, which will be discussed in Section 4.2.7, is the conservation of moments.
- *Two-dimensional quadrature methods* is a family that includes methods some of which were initially derived in an ad hoc manner, but which can all be formally derived via the finite volume method and the treatment of the double integrals that result with two-dimensional quadrature. These methods allow for exact

conservation of one moment (typically the first moment) and accurate prediction of the distribution; they will be discussed in Section 4.2.8.

Fragmentation is also an integral equation, albeit a less challenging one as it does not feature the convolution-like aggregation source term. Discretisation methods for fragmentation are usually extensions of the methods for aggregation and will be discussed together with those.

The treatment of the growth term shares common features with methods found in gas dynamics, since it is an advection equation. In Section 4.3 we will discuss fixed, moving and adaptive grid discretisation methods.

Monte Carlo Methods

Monte Carlo methods employ an ensemble of notional particles that are subjected to random events representing the processes in the PBE, the frequency of these events being consistent with the rates of the processes. Monte Carlo methods are computationally expensive because a large number of particles are needed to obtain convergence. For this reason, Monte Carlo methods have been employed mainly in spatially homogeneous or one-dimensional problems, as well as to ideal reactor networks and partially stirred reactors. However, the CPU effort of Monte Carlo methods does not scale exponentially with dimensionality, as is the case with discretisation methods, which makes them a good option for high-dimensional problems. Other applications include small populations where fluctuations are important and cases where the formulation of a deterministic PBE may not be straightforward. Monte Carlo methods for aggregation will be discussed in Section 4.2.9. Another, recently proposed, use of Monte Carlo methods is for overcoming the numerical diffusion involved in the discretisation of the growth term (cf. Section 4.3.7).

Multidimensional, Coupled and Spatially Dependent Formulations

The multidimensional and coupled PBE formulations that were introduced in Sections 2.4.1 and 2.4.2 can be solved with extensions of the methods described so far, and these will be addressed in Section 4.2.10.

All of the solution methods will be initially developed in the context of the spatially homogeneous PBE, which corresponds to a perfectly stirred reactor. Application to problems that can be approximated by other ideal reactors is straightforward; in Section 2.6, we discussed how the PBE is formulated for each of them.

The numerical treatment of the spatially dependent PBE requires coupling of a numerical method for solving the PBE with a CFD code. The issues related to the coupling of the PBE with CFD will be discussed in Section 4.4. An important distinction must be made between non-inertial and inertial particles. In the case of the former, the particles are transported with the local flow velocity. In the case of the latter, the particle velocity deviates from the local flow velocity. An intermediate case is particles of low Stokes number, whose velocity is not equal to the local flow velocity but can be computed from that.

The treatment of particles with high inertia requires a consideration of their velocity distribution and is typically carried out with Lagrangian methods or Eulerian methods where each size class is treated as a separate phase with its own momentum equations. For dense flows, it may be necessary to account for two-way or four-way coupling between phases. These problems and methods are treated in multiphase flow treatises and will not be covered in detail in the present book, although an introduction will be provided in Section 4.5, focussing on their relationship with the methods and models described here and providing pointers to relevant literature.

Finally, Section 4.4.7 will introduce the application of data-driven methods, including machine learning, to the population balance. These methods are gaining widespread adoption to a wide range of problems, and a rapid growth of their application to the solution of the PBE is expected in the near future.

4.2 Aggregation and Fragmentation

Solution methods for aggregation and fragmentation processes will be discussed first. Problems where these processes are combined with growth will be discussed in Section 4.3. We will also initially assume that the particles are spherical, while formulations for dealing with more complex aggregate morphology will be discussed in Section 4.2.10.

Aggregation which will feature the most extensive presentation, as it is the most complex process in the PBE from the point of view of numerical treatment; several concepts and methods will be introduced there and discussed only briefly when encountered in the subsequent sections. Fragmentation is, in a way, the reverse of aggregation, and solution methods for it have a lot in common with those for aggregation, so their discussion will be brief and draw mostly on the same principles.

4.2.1 General Features

The Aggregation PBE

The discrete and continuous forms of the aggregation PBE were derived in Chapter 2 and are written again here for reference.

Discrete form:

$$\frac{dn_1}{dt} = -n_1 \sum_{j=1}^{\infty} \beta_{ij} n_j, \tag{4.1a}$$

$$\frac{dn_i}{dt} = \frac{1}{2} \sum_{j=1}^{i-1} \beta_{j,i-j} n_j n_{i-j} - n_i \sum_{j=1}^{\infty} \beta_{ij} n_j, \qquad \forall i > 1. \tag{4.1b}$$

Continuous form:

$$\frac{\partial n(v,t)}{\partial t} = \frac{1}{2} \int_0^v \beta(w, v-w) n(w,t) n(v-w,t) dw - n(v,t) \int_0^\infty \beta(v,w) n(w,t) dw. \tag{4.2}$$

Since aggregation is by nature a discrete process, both forms are equivalent descriptions of physical problems, the continuous form being an approximation of the discrete one (as is evident from its derivation in Section 2.3.1). Solution methods for the discrete equation will be discussed in the beginning of Section 4.2.2, while the rest of this section will be devoted to the continuous form.

Two alternative, equivalent forms of the PBE birth term are often preferred for the development of numerical methods, for reasons of computational efficiency. In the original forms, the aggregating pairs are counted twice in the aggregation birth term (except for the case of equal-sized particles in the discrete PBE), and that is compensated by the factor $1/2$. Therefore, in the discrete PBE, the aggregation birth term can be written in the following way. For even i, we have:

$$\frac{1}{2} \sum_{j=1}^{i-1} \beta_{j,i-j} n_j n_{i-j} = \sum_{j=1}^{i/2-1} \beta_{j,i-j} n_j n_{i-j} + \frac{1}{2} \beta_{i/2, i/2} n_{i/2}^2. \tag{4.3}$$

For odd i, let k be the greatest integer less than $i/2$. We then have:

$$\frac{1}{2} \sum_{j=1}^{i-1} \beta_{j,i-j} n_j n_{i-j} = \sum_{j=1}^{k} \beta_{j,i-j} n_j n_{i-j}. \tag{4.4}$$

A similar operation can be applied to the birth term in the continuous PBE[2]:

[2] This intuitive result can be proven in the following way (Williams and Loyalka, 1991):

$$\frac{1}{2} \int_0^v \beta(w, v-w) n(w,t) n(v-w,t) dw = \frac{1}{2} \int_0^{v/2} \beta(w, v-w) n(w,t) n(v-w,t) dw + \frac{1}{2} \int_{v/2}^v \beta(w, v-w) n(w,t) n(v-w,t) dw.$$

$$\frac{1}{2}\int_0^v \beta(w,v-w)n(w,t)n(v-w,t)dw = \int_0^{v/2} \beta(w,v-w)n(w,t)n(v-w,t)dw. \tag{4.5}$$

Finally, a truncation of the infinite domain in the sum or integral that appears in the aggregation death term is essential for some numerical methods:

$$n_i \sum_{j=1}^{\infty} \beta_{ij} n_j \approx n_i \sum_{j=1}^{N} \beta_{ij} n_j, \tag{4.6}$$

$$n(v)\int_0^{\infty} \beta(v,w)n(w,t)dw \approx n(v)\int_0^{v_{max}} \beta(v,w)n(w,t)dw. \tag{4.7}$$

where N and v_{max} denote the maximum number of monomers and maximum particle volume within the simulation domain, respectively. One consequence of this approximation is a loss of mass due to birth of particles outside the maximum allowed size. The domain should be sufficiently large to ensure that this loss is negligible.

Two important features of the aggregation equation with reference to kernels *self-similarity* and *gelation*. The former is the basis of similarity methods and will be discussed in Section 4.2.3. The latter refers to the accumulation of mass into a cluster of infinite size. The behaviour of kernels with respect to gelation has been investigated in a number of works, and the reader may refer to Ziff (1980), Davies et al. (1999), or Ernst and Pratsinis (2006), among others, for further details.

The Fragmentation PBE

Like aggregation, fragmentation is described by an integral equation which, however, does not pose the same severe challenges. This has to do with the facts that the fragmentation equation is linear and does not have the convolution-like term that represents aggregation birth. As will be seen in the rest of the section, much of the challenge in the numerical solution of the aggregation PBE, particularly in the case of discretisation methods, is associated with the treatment of the aggregation

Via a change of variable $u = \frac{v}{2} - w$ in the first integral and $u = w - \frac{v}{2}$ in the second one, this expression becomes:

$$\frac{1}{2}\int_{v/2}^{0} \beta\left(\frac{v}{2}-u, \frac{v}{2}+u\right) n\left(\frac{v}{2}-u, t\right) n\left(\frac{v}{2}+u, t\right) du + \frac{1}{2}\int_{0}^{v/2} \beta\left(\frac{v}{2}+u, \frac{v}{2}-u\right) n\left(\frac{v}{2}+u, t\right) n\left(\frac{v}{2}-u, t\right) du$$

$$= \int_{v/2}^{0} \beta\left(\frac{v}{2}-u, \frac{v}{2}+u\right) n\left(\frac{v}{2}-u, t\right) n\left(\frac{v}{2}+u, t\right) du.$$

The last result follows from the symmetry of the kernel and the fact that the integrand is an even function. Changing back into $w = \frac{v}{2} - u$, we obtain the desired result:

$$\int_{v/2}^{0} \beta\left(\frac{v}{2}-u, \frac{v}{2}+u\right) n\left(\frac{v}{2}-u, t\right) n\left(\frac{v}{2}+u, t\right) du = \int_0^{v/2} \beta(w, v-w)n(w, t)n(v-w, t)dw.$$

birth term. The general fragmentation equation, Eq. 2.26, is shown again here for reference:

$$\frac{\partial n(v,t)}{\partial t} = \int_v^\infty \gamma(w)b(v,w)n(w,t)dw - \gamma(v)n(v,t). \tag{4.8}$$

This is a linear integro-differential equation where only the birth term involves an integral. While a discrete fragmentation equation may also be written, the majority of the solution methods and applications have been based on the continuous form, hence Eq. 4.8 will be the basis for the following discussion. The fragmentation equation includes two functions whose product is the kernel of the integral birth term (while only one of them appears in the death term). Various forms of these functions and their physical meaning were discussed in Chapter 3.

4.2.2 Analytical Solutions

Smoluchowski Equation

Smoluchowski obtained the first solution for the discrete aggregation equation (Smoluchowski, 1917) with the constant kernel and a monodisperse initial condition:

$$\frac{dn_1}{dt} = -\beta_0 n_1 \sum_{j=1}^{\infty} n_j, \tag{4.9a}$$

$$\frac{dn_i}{dt} = \frac{1}{2}\beta_0 \sum_{j=1}^{i-1} n_j n_{i-j} - \beta_0 n_i \sum_{j=1}^{\infty} n_j, \qquad \forall i > 1, \tag{4.9b}$$

with $n_1 = n_0$ and $n_i = 0$ at $t = 0$, where n_0 is the initial number of particles. Smoluchowski's solution is obtained as follows. We first introduce the zeroth moment into Eqs. 4.9:

$$\frac{dn_1}{dt} = -\beta_0 n_1 m_0, \tag{4.10a}$$

$$\frac{dn_i}{dt} = \frac{1}{2}\beta_0 \sum_{j=1}^{i-1} n_j n_{i-j} - \beta_0 n_i m_0, \qquad \forall i > 1. \tag{4.10b}$$

This system of equations can be solved recursively because the equation for n_i involves only $n_1, \ldots, n_{i-1}$ and m_0 at the right-hand side, if an expression for m_0 is available. A differential equation for m_0 can be obtained by summing Eq. 4.9:

$$\frac{dm_0}{dt} = \frac{1}{2}\beta_0 \sum_{i=2}^{\infty}\sum_{j=1}^{i-1} n_j n_{i-j} - \beta_0 m_0^2. \tag{4.11}$$

The first term on the right-hand side can be written in the following way:

$$\frac{1}{2}\beta_0 \sum_{i=2}^{\infty}\sum_{j=1}^{i-1} n_j n_{i-j} = \frac{1}{2}\beta_0(n_1^2 + 2n_1n_2 + n_2^2 + 2n_1n_3 + \ldots)$$
$$= \frac{1}{2}\beta_0(n_1 + n_2 + \ldots + n_n)^2$$
$$= \frac{1}{2}\beta_0 m_0^2.$$

The equation for the zeroth moment, therefore, becomes:

$$\frac{dm_0}{dt} = -\frac{1}{2}\beta_0 m_0^2. \tag{4.12}$$

where $m_0 = n_0$ for $t = 0$. This equation has the following solution:

$$m_0 = \frac{n_0}{1 + \frac{1}{2}\beta_0 n_0 t}. \tag{4.13}$$

Substituting into Eq. 4.10 and solving consecutively the equations for n_1, n_2 and so on, the following general formula can be obtained for n_i:

$$n_i = n_0 \frac{\left(\frac{\beta_0 n_0 t}{2}\right)^{i-1}}{\left(1 + \frac{\beta_0 n_0 t}{2}\right)^{i+1}}. \tag{4.14}$$

By setting $n_i^* = n_i/n_0$ and $t^* = \beta_0 n_0 t$, this equation can also be written in the following dimensionless form:

$$n_i^* = \frac{\left(\frac{t^*}{2}\right)^{i-1}}{\left(1 + \frac{t^*}{2}\right)^{i+1}}. \tag{4.15}$$

Smoluchowski's solution is the most important analytical solution of the discrete aggregation equation. While limited to the constant kernel, it can be approximately applied to the physical problem of aggregation of initially monodispersed particles due to Brownian motion because, as we saw in Section 4.2.1, the constant kernel approximates the Brownian motion kernel when the colliding particles are of similar sizes. A few more solutions have been obtained, such as a solution for the product kernel (McLeod, 1962).

The Smoluchowski equation can be solved numerically for any kernel and initial condition, albeit at considerable computational cost. Such solutions are very accurate and can be regarded as substitutes for analytical solutions. The temporal integration can be carried with standard numerical procedures. The simplest approach is to employ the explicit Euler method:

$$n_i^{t+dt} = n_i^t + dt\left(\frac{1}{2}\sum_{j=1}^{i-1} \beta_{ij} n_j^t n_{i-j}^t - \beta_{ij} n_i^t \sum_{j=1}^{\infty} n_j^t\right). \tag{4.16}$$

Time savings can be obtained by employing the alternative aggregation birth term (Eqs. 4.3 and 4.4). Even so, the number of calculations scales with the square of the number of discrete size classes. For an aerosol in the range of 1 nm–1 μm in diameter, an order of 10^9 classes must be computed. Hidy (1965) showed results with up to 600 classes calculated with an early supercomputer, while Landgrebe and Pratsinis (1989) used 5,000 size classes on a desktop computer of that time. With modern computers, much larger systems can be integrated to obtain numerical solutions with a variety of kernels. Such numerical solutions can be employed for model validation in lieu of an analytical solution. For problems where the Smoluchowski equation must be solved for many different conditions, however, and particularly for spatially dependent problems, this direct numerical approach is not an option.

While the discrete fragmentation equation will not be discussed in detail, it is worth noting that a similar numerical method can be set up for it in a straightforward manner, similar to the method for solving the Smoluchowski equation.

Continuous Aggregation PBE

The analytical solutions that have been obtained for the aggregation equation are very few and restricted to special cases of kernels (mostly the theoretical ones) and initial conditions. Here, we will show two solutions that have been used extensively for validating numerical methods. The analytical solution for the constant kernel was first derived by Schumann (1940). A more comprehensive analytical study of the theoretical kernels was conducted by Scott (1968). The solutions are shown here in the form presented by Gelbard and Seinfeld (1978a) and are obtained with the method of Laplace transforms, but the derivations will be omitted; for those and for more analytical solutions, the reader may consult, apart from the cited papers, Drake (1972) or Williams and Loyalka (1991).

The first solution is for an aggregation process with the constant kernel and an exponential initial distribution:

$$\frac{\partial n(v,t)}{\partial t} = \frac{1}{2}\beta_0 \int_0^v n(v-w,t)n(w,t)dw - \beta_0 n(v,t)\int_0^{\infty} n(w,t)dw, \tag{4.17}$$

$$n(v,0) = \frac{N_0}{v_0}\exp\left(-\frac{v}{v_0}\right), \tag{4.18}$$

where N_0 is the initial total number of particles and v_0 is the initial average particle volume. The solution is:

$$n(v,t) = \beta_0 \frac{4N_0}{(T+2)^2} \exp\left(-\frac{v}{v_0}\frac{2}{T+2}\right), \tag{4.19}$$

where T is a dimensionless time defined as follows:

$$T = \beta_0 N_0 t. \tag{4.20}$$

The second solution is for the sum kernel, with the initial condition being again Eq. 4.18. The PBE is shown below:

$$\frac{\partial n(v,t)}{\partial t} = \frac{1}{2}\beta_0 \int_0^v (v+w)n(v-w,t)n(w,t)dw - n(v,t)\beta_0 \int_0^\infty (v+w)n(w,t)dw, \tag{4.21}$$

and the solution is:

$$n(v,t) = \frac{N_0(1-T)}{vT^{1/2}} \exp\left[-(1+T)w\right] I_1(2wT^{1/2}), \tag{4.22}$$

where I_1 is the modified Bessel function of the first kind of order one. In this case, the dimensionless time is defined as follows:

$$T = \beta_0 N_0 v_0 t. \tag{4.23}$$

Continuous Fragmentation PBE

Several analytical solutions for the fragmentation PBE have been presented in the literature. A couple of solutions that have been used to validate discretisation methods will be shown here, again without derivation. More solutions can be found in McGrady and Ziff (1988), Williams (1990) and Nicmanis and Hounslow (1998) and Kostoglou (2007b), among others.

Ziff and McGrady (1985) derived several analytical solutions for binary fragmentation. We show here the solution for binary fragmentation and $\gamma(v) = v$:

$$\frac{\partial n(v,t)}{\partial t} = 2\int_v^\infty n(w,t)dw - vn(v,t), \tag{4.24}$$

with the exponential initial condition (Eq. 4.18) with $v_0 = 1$ and $N_0 = 1$. The solution is:

$$n(v,t) = (1+t)^2 \exp[-v(1+t)]. \tag{4.25}$$

Ziff (1991) considered a general class of scaling solutions to the fragmentation equation. The special case of ternary fragmentation with a monodisperse initial condition, $n(v,0) = \delta(v - v_{max})$, has the following solution:

$$n(v,t) = \frac{1}{v_{max}} e^{-tv_{max}^2} \delta(v - v_{max}) + 6tv \int_v^{v_{max}} y^{-2} e^{-ty^3} dy. \tag{4.26}$$

4.2.3 Similarity Solutions

Similarity solutions originate in the pioneering work of Swift and Friedlander (1964) and Friedlander and Wang (1966). They are based on a non-dimensional scaling of the PBE that results in the derivation of an equation in terms of a single variable. The solution of that equation is the *self-preserving distribution (SPD)*, a distribution that is independent of time when expressed in terms of the similarity variable, which is timedependent. The SPD is independent of the initial conditions, although the latter determine the time required to reach it. The similarity concept offers an economic way of solving the PBE because, if a similarity solution exists and the SPD is known, the actual distribution can be easily obtained from the latter and the timedependent similarity variable. Before the method can be applied, however, it must be asserted that the problem admits a similarity solution and the time elapsed is sufficient for the SPD to be reached.

Similarity methods for the aggregation PBE (Eq. 4.2) are based on the use of the following similarity variable:

$$\eta(v,t) = \frac{v}{v_m(t)}, \tag{4.27}$$

which is a scaling of the particle volume with the mean particle volume, $v_m(t)$, related to the zeroth and first moments as follows:

$$v_m(t) = \frac{m_1}{m_0(t)}. \tag{4.28}$$

Note that m_0 is timedependent, while m_1 is timeindependent when only aggregation is present.

The basic assumption of similarity methods is that the solution can be described by the SPD, which is a function of η alone. The SPD is a function, $\psi(\eta)$, such that $\psi(\eta)d\eta$ is the fraction of particles whose scaled volume lies between η and $d\eta$. Therefore:

$$\psi(\eta)d\eta = \frac{n(v,t)dv}{m_0(t)}, \tag{4.29}$$

or

$$\psi(\eta)d\left(v\frac{m_0(t)}{m_1}\right) = \frac{n(v,t)dv}{m_0(t)}, \tag{4.30}$$

from which the following relationship between the number density and the SPD, for a given t, can be established:

$$n(v,t) = \psi(\eta)\frac{m_0(t)^2}{m_1}. \tag{4.31}$$

Thus, if $\psi(\eta)$ is a known function and a solution for $m_0(t)$ is available (together with the value of m_1 which is invariant), the number density at any time can be obtained. The similarity approach therefore requires the existence of a solution for $\psi(\eta)$ and of a closed form for the equation for the zeroth moment (Eq. 4.36). If these conditions are satisfied and $\psi(\eta)$ can be obtained, analytically or numerically, then the distribution at any time can be reconstructed. However, these conditions cannot be guaranteed to hold true for every kernel. It has been shown (see e.g. Pruppacher and Klett, 1996) that this is possible when the kernel is a homogeneous function, that is, fulfils the following condition:

$$\beta(\alpha v, \alpha w) = \alpha^k \beta(v,w), \tag{4.32}$$

for some real number k. The question of the closure of the equation for the zeroth moment will be discussed in Section 4.2.4, in the context of moment methods.

If a similarity condition does exist, the next questions to be answered are to derive it and to find the time required to reach it. Two approaches can be followed. The first one is to substitute the similarity variable into the PBE and seek an analytical solution to the resulting equation. This is a complex task and complete analytical solutions are usually not possible, but approximate solutions have been derived (Friedlander and Wang, 1966; Lai et al., 1972; see also Friedlander, 2000 for a summary of these works). The second approach is to solve the discrete PBE numerically for sufficiently long time to reach the SPD. An early computation with 600 size classes for the Brownian kernel was performed by Hidy (1965), while a complete computation was performed by Landgrebe and Pratsinis (1989). While computationally expensive in the past, this task is easily done with today's resources.

Similarity methods have also been explored for fragmentation equations. They will not be discussed in detail here, but comprehensive accounts can be found in Ramkrishna (2000) and Kostoglou (2007b). A similarity solution for a problem featuring joint aggregation (constant kernel) and fragmentation (binary) was put forward by Patil and Andrews (1998) and corrected by Lage (2002); this solution is referred to in Section 4.2.8 for the validation of the conservative finite volume discretisation method. More similarity solutions for several combinations of kinetic processes can be found in Diemer (2023).

4.2.4 Moment Methods: Fundamentals and Closed Forms

The method of moments for the PBE originates in early works such as Golovin (1963), Hulburt and Katz (1964) and Enukashvili (1964).[3] The basic concept of moment methods, introduced in Section 4.1, is to solve equations for certain moments of the distribution, rather than resolving the distribution itself. In the present section, we will discuss the application of this approach to the PBE for aggregation processes. We will introduce the equations for the rate of change of the moments, identify the closure problem and discuss several methods that have been proposed for solving it.

Moment Equations for the Aggregation PBE

The equations for the time evolution of the moments of a discrete distribution are obtained by multiplying the discrete aggregation PBE by v^k and summing over all particle sizes:

$$\frac{dm_k}{dt} = \frac{1}{2}\sum_{i=1}^{\infty}\sum_{j=1}^{i-1} v_i^k \beta_{j,i-j} n_j n_{i-j} - \sum_{i=1}^{\infty}\sum_{j=1}^{\infty} v_i^k \beta_{ij} n_i n_j. \tag{4.33}$$

A similar operation, with integration replacing summation, yields the equations for the moments of the continuous distribution:

$$\begin{aligned}\frac{dm_k}{dt} = \frac{1}{2}\int_0^{\infty}\int_0^{v} & v^k \beta(w, v-w) n(w,t) n(v-w,t) dw dv \\ - \int_0^{\infty}\int_0^{\infty} & v^k \beta(v,w) n(v,t) n(w,t) dw dv.\end{aligned} \tag{4.34}$$

The moment equations in either the discrete or continuous case comprise an infinite set that must be truncated to obtain solution. More importantly, the equations are unclosed in their general form. This is because right-hand sides still contain particle numbers or number density functions. To be able to solve these equations, they must be brought to a form where only moments are present. The closure problem and the methods for dealing with it will be discussed in the rest of the section.

The equations for the zeroth and first moment are of particular importance and can be obtained based on physical considerations. The zeroth moment corresponds to the total number of particles, and thus its rate of change can be obtained by summation or integration over all aggregation events. For the discrete PBE, the resulting equation is:

[3] See Drake, 1972, for an account in English of the Russian works.

$$\frac{dm_0}{dt} = -\frac{1}{2}\sum_{i=1}^{\infty}\sum_{j=1}^{\infty}\beta_{ij}n_in_j, \tag{4.35}$$

where the factor 1/2 is needed for aggregation between identical particles, as well as to prevent double counting of pairs of particles with different sizes (cf. Section 2.2.1). For the continuous PBE, the equation is:

$$\frac{dm_0}{dt} = -\frac{1}{2}\int_0^{\infty}\int_0^{\infty}\beta(v,w)n(v,t)n(w,t)dwdv. \tag{4.36}$$

Aggregation does not change the total volume of the particles, hence the first moment is an *invariant* when aggregation is the only process present:

$$\frac{dm_1}{dt} = 0. \tag{4.37}$$

Moment Equations for the Fragmentation PBE

The continuous fragmentation PBE, Eq. 4.8, can be manipulated in the same way as with the aggregation PBE to yield a set of equations for the moments:

$$\frac{dm_k}{dt} = \int_0^{\infty}\int_v^{\infty} v^k\gamma(w)b(v,w)n(w,t)dwdv - \int_0^{\infty} v^k\gamma(v)n(v,t)dv. \tag{4.38}$$

The condition given by Eq. 4.37 also holds in the case of fragmentation. Eq. 4.38 poses the same truncation and closure problem as the aggregation equation.

Closure Problem and Closed Forms

Before a solution of the moment equations can be sought, it must first be truncated to a finite and closed system of equations. Only for certain special forms of the kernel is it possible to obtain a closed set of equations. For most of the physically realistic kernels, a closure assumption must be applied.

Another important problem is related to the recovery of the distribution from the moments and the conditions under which the moments retain physically realistic values. The problem of moments was first considered by Stieltjes (1894, 1895), and a comprehensive discussion can be found in Shohat and Tamarkin (1970). This issue will not be discussed further in this book, but a discussion in the context of the PBE can be found in Drake (1972). In the following, we will discuss the method of moments from the point of view of calculating the evolution of the moments alone.

We will first consider certain cases of the aggregation equation for which a closed form can be derived. The treatment will be based on the continuous form, Eq. 4.34, which is employed in most cases. We begin with the derivation of a general result due to Thompson (1968), which can be used to obtain closed forms for a class of kernels including the three main theoretical ones (constant, sum and product kernel). First, we note the following identity:

$$\int_0^\infty \int_0^v f(v,w)dwdv = \int_0^\infty \int_w^\infty f(v,w)dvdw. \quad (4.39)$$

This identity is based on the fact that the domain of integration in both sides is the area between the v-axis and the line $w = v$. The first term on the right-hand side of Eq. 4.34 can thus be written as follows:

$$\frac{1}{2}\int_0^\infty \int_0^v v^k \beta(w, v-w)n(w,t)n(v-w,t)dwdv$$
$$= \frac{1}{2}\int_0^\infty \int_w^\infty v^k \beta(w, v-w)n(w,t)n(v-w,t)dvdw. \quad (4.40)$$

A change of variable, $u = v - w$, is now introduced. Noting that the Jacobian of this transformation is 1, we change appropriately the limits of the inner integral and interchange the order of integration to obtain:

$$\frac{1}{2}\int_0^\infty \int_w^\infty v^k \beta(w, v-w)n(w,t)n(v-w,t)dvdw$$
$$= \frac{1}{2}\int_0^\infty \int_0^\infty (u+w)^k \beta(w,u)n(w,t)n(u,t)dudv. \quad (4.41)$$

Owing to the symmetry of the kernel, the second term on the right-hand side of Eq. 4.34 can be written as follows:

$$\int_0^\infty \int_0^\infty v^k \beta(v,w)n(v,t)n(w,t)dwdv$$
$$= \frac{1}{2}\left(\int_0^\infty \int_0^\infty u^k \beta(v,u)n(v,t)n(u,t)dudv\right.$$
$$\left. + \int_0^\infty \int_0^\infty w^k \beta(v,w)n(v,t)n(w,t)dwdv\right). \quad (4.42)$$

By renaming appropriately the variables and assembling the terms, we arrive at the final result:

$$\frac{dm_k}{dt} = \frac{1}{2}\int_0^\infty \int_0^\infty \left[(v+w)^k - v^k - w^k\right] \beta(v,w)n(v,t)n(w,t)dwdv. \quad (4.43)$$

We first note that, for any kernel, Eqs. 4.36 and 4.37 are readily derived from Eq. 4.43. The following equation can now be derived for the moments of order $k > 1$:

$$\frac{dm_k}{dt} = \frac{1}{2}\sum_{i=1}^{k-1} \binom{k}{i} \int_0^\infty \int_0^\infty v^i w^{k-i} \beta(v,w)n(v,t)n(w,t)dwdv. \quad (4.44)$$

Further simplification is possible only for specific forms of the kernel. Martynov and Bakanov (1961) have obtained results for the following general form of the kernel (see also Drake, 1972):

$$\beta(v,w) = A + B(v+w) + Cvw. \tag{4.45}$$

We can see that this corresponds to the constant kernel for $B = 0$ and $C = 0$, to the sum kernel for $A = 0$ and $C = 0$, and to the product kernel for $A = 0$ and $B = 0$. For this kernel, it is possible to express the right-hand side of Eq. 4.44 in terms of moments of order lower than k:

$$\frac{dm_0}{dt} = -\frac{1}{2}(Am_0^2 + 2Bm_0 + C), \tag{4.46}$$

$$\frac{dm_k}{dt} = \frac{1}{2}\sum_{i=1}^{k-1}\binom{k}{i}(Am_i m_{k-i} + Bm_{i+1}m_{k-i} + Bm_i m_{k+1-i} + Cm_{i+1}m_{k+1-i}), \qquad \forall k > 1. \tag{4.47}$$

The most important case is that of the constant kernel, for which $A = \beta_0$ and we have the following analytical solution:

$$\frac{dm_0}{dt} = -\frac{1}{2}\beta_0 m_0^2, \tag{4.48}$$

$$\frac{dm_k}{dt} = \frac{1}{2}\beta_0\sum_{i=1}^{k-1}\binom{k}{i}m_i m_{k-i}, \qquad \forall k > 1. \tag{4.49}$$

More analytical solutions can be found in Peterson et al. (1978). This class of theoretical kernels is the main one that has been studied analytically with the aid of the method of moments.

For the physical kernels, however, it is not possible to obtain a closed form of the moment equations. This will be demonstrated here for the Brownian motion kernel, Eq. 3.34, in which case the equation describing the evolution of the zeroth moment is:

$$\frac{dm_0}{dt} = -\frac{1}{2}\int_0^\infty\int_0^\infty \frac{2kT}{3\mu}\left(\frac{1}{v^{1/3}} + \frac{1}{w^{1/3}}\right)\left(v^{1/3} + w^{1/3}\right) n(v,t)n(w,t)dwdv, \tag{4.50}$$

or:

$$\frac{dm_0}{dt} = -\frac{2kT}{3\mu}\left(m_0^2 + m_{1/3}m_{-1/3}\right). \tag{4.51}$$

Thus the equation involves moments of fractional order.

For the free molecule regime kernel, Eq. 3.24, even the simplification into an equation with fractional-order moments is not straightforward. Lee et al. (1984) derived the following simplified form:

$$\frac{dm_0}{dt} = \left(\frac{3}{4\pi}\right)^{1/6} \left(\frac{6kT}{\rho_p}\right)^{1/2} b\left(m_{2/3}m_{-1/2} + 2m_{1/3}m_{-1/6} + m_{1/6}m_0\right), \tag{4.52}$$

where b is a factor that depends on the distribution. Their calculations showed that b depends on the initial geometric standard deviation and the moment to be solved for, and reaches a constant value at $t \to \infty$. Nevertheless, the equation is still unclosed due to the presence of fractional moments. Lee et al. (1984) used a presumed distribution to obtain closure. As the values of b span a relatively narrow range, an average value was used, on the grounds that the effect of this assumption is smaller than that of the presumed distribution that was employed in that study to obtain closure of the moment equations.

To conclude, the moment transformation of the aggregation equation results in a closed system of equations only for the theoretical kernels, while for the physical kernels it yields an unclosed system.

The moment equations for fragmentation pose a similar closure problem that depends on the kernel. Analytical solutions for some special cases can be found e.g. in Ziff and McGrady (1986). However, the majority of the kernels used in fragmentation are empirical and can take complicated forms.

4.2.5 Moment Methods: Closure Models

The approaches that have been proposed for closure of the moment equations will now be discussed. It must be mentioned that the literature on methods for moment closure is very extensive, and only the basic concepts underlying each class of methods will be introduced here. A comprehensive account of moment methods can be found in Marchisio and Fox (2013), while Kostoglou and Karabelas (2004) have presented a review of moment methods for fragmentation.

Presumed Distribution

If the shape of the distribution can be presumed, then equations for the evolution of the moments in closed form can be derived from the PBE. Analytical solutions are possible for some combinations of presumed distributions and kernels, otherwise numerical quadrature can be employed for treating the resulting integrals. To obtain the closed form, either analytical expressions for the moments are derived from the presumed distribution and substituted into the unclosed equations that involve fractional moments (such as Eqs. 4.51 and 4.52), or the analytical expression for the distribution is written in terms of the moments and substituted into the initial moment equations (Eq. 4.33 or 4.34).

The log-normal distribution has been employed extensively for moment closure, particularly in aerosol dynamics, due to the fact that it is often used for fitting aerosol distributions (Friedlander, 2000). Lee (1983) derived the following equation for the zeroth moment of the log-normal distribution in the case of the Brownian motion kernel:

$$m_0 = \frac{n_0}{1 + \left[1 + \exp(\ln^2 \sigma_0)\right] K n_0 t}, \tag{4.53}$$

where σ_0 is the initial value of the standard deviation and $K = 2kT/(3\mu)$. The derivation involves an approximation for a term involving the standard deviation at time t, σ. The similarity with Eq. 4.13 is notable. Subsequent works (Lee et al., 1984; Lee and Chen, 1984) extended this approach to aggregation in the transition and free molecule regimes. A comprehensive study of the presumed distribution method was presented by Williams (1986). Equations were derived for the log-normal and gamma distributions and for the Brownian, gravitational, laminar shear, turbulent shear and turbulent inertial kernels, while collision efficiency models were also considered.

One situation often encountered in aerosol processes is that the distribution features more than one mode because the particles originate from more than one distinct sources. A presumed distribution moment method proposed to accommodate this problem is the modal aerosol dynamics (MAD) approach by Whitby and McMurry (1997). In this method, separate moment equations are derived for each mode of the distribution and separate terms are included for intramodal and intermodal aggregation. Other processes, such as growth, can also be included. The detailed equations can be found in the aforementioned paper, where application to a fully mixed reactor is also shown.

Presumed distribution methods have found numerous applications, including atmospheric aerosols (Whitby and McMurry, 1997), nanoparticle synthesis (Pratsinis, 1988), nuclear aerosols (Williams, 1986) and soot formation (Dobbins and Mulholland, 1984). When they can be applied, these methods have many advantages such as simplicity of formulation, computational efficiency and robustness. In many problems, however, the distributions cannot be assumed to follow standard forms. This happens particularly when the problem features a combination of kinetic processes and also when there are spatial inhomogeneities and strong mixing effects.

Series Approximation of the Distribution

The approach of series expansion originates in the work of Hulburt and Katz (1964) and Hulburt and Akiyama (1969), who proposed expansion in orthogonal functions. This approach has also been shown to be related to the method of weighted residuals Ramkrishna (1971). The approximation of the distribution in the form of a series is also an assumption regarding the functional form of the distribution, hence this approach is related to presumed distribution methods.

Model Approximation

In Section 4.2.1, it was mentioned that the constant kernel can provide an approximation for the Brownian motion kernel when the particle sizes are similar. A few other kernel approximations have been tried. Frenklach and Harris (1987) used a Taylor series expansion to approximate the kernel in the free-molecule regime (referred to as 'Method I' in that work) and found that the approximation was sufficiently accurate for narrow particle size distributions and no nucleation. More recently, Taylor series expansions have been employed by Yu et al. (2008) for the Brownian motion kernel.

Moment Interpolation

The approach of using interpolation to obtain the fractional moments was first proposed by Frenklach and Harris (1987) in the context of aerosol and soot modelling. It was described as 'Method II' in that work but later named Method of Moments with Interpolative Closure (MOMIC). The main concept is to approximate the fractional moments that result from the application of the moment transformation in the case of the physical kernels, such as in Eqs. 4.51 and 4.52, via a logarithmic interpolation between the integral moments. The derivation must be carried out individually for each kernel, and various numbers of integer moments and interpolation methods can be used. A review of MOMIC can be found in Frenklach (2002a), where example parametrisations for the free molecule and Brownian kernels are also provided. MOMIC has been extended to provide closure for surface processes and aggregation into fractal particles, and has been used extensively for modelling soot formation (e.g. Frenklach and Wang, 1991, 1994; Kazakov and Frenklach, 1998).

A similar concept was proposed by Barrett and Jheeta (1996), motivated by the use of the log-normal distribution for aerosols. The interpolation formula used in that work yields the moments of the log-normal distribution as a special case and provides increasingly accurate approximations when higher-order polynomials are used (requiring more moments). That approach illustrates the link between moment interpolation and presumed distribution methods.

Numerical Quadrature Methods

The methods described so far employ an approximate form of the integrand in the unclosed term in order to bring closure to the moment equations. Numerical quadrature methods, on the other hand, approximate the entire integral by numerical quadrature.

A numerical quadrature rule is a formula for approximating an integral as a weighted sum of values of the integrand at a number of discrete points, as per the following general form:

$$\int_a^b f(x)dx = \sum_{i=1}^{N} W_i f(x_i), \tag{4.54}$$

where W_i are the weights, x_i are the abscissas where the function is being evaluated and N is the number of discrete points employed. Further background on the derivation of quadrature rules can be found, for example, in Heath (2002).

In the application of quadrature methods to moment equations, Gaussian quadrature is employed, which attains the maximum possible accuracy for a given number of points by using non-equidistant points. The points where the function is being evaluated are the roots of orthogonal polynomials, which have been tabulated in the literature. Several choices of orthogonal polynomials exist, but the rule that is most relevant to the aggregation moment equation is Gauss–Laguerre quadrature, which is applicable to integration over a semi-infinite domain, as is the case with the integrals in that equation. The Gauss–Laguerre rule can be written as:

$$\int_0^{\infty} g(x)f(x)dx \approx \sum_{i=1}^{N} W_i f(x_i), \tag{4.55}$$

where the integrand is written as a product of two functions. The weights have been derived and tabulated for certain standard forms of the function $g(x)$, hence many functions can be integrated if they can be written as a product in this way. The method can be straightforwardly extended to multiple integrals.

In the case of the moment equation for the aggregation PBE, the integral to be approximated is the double integral in Eq. 4.43. The quadrature rule cannot be applied in a straightforward manner, however, because the integral contains the unknown number density distribution. Therefore, the number density is treated as the weight function, with the remaining part of the integrand being the function to be approximated. This is shown below:

$$\frac{1}{2}\int_0^\infty \int_0^\infty \left[(v+w)^k - v^k - w^k\right] \beta(v,w)n(v)n(w)dwdv$$
$$\approx \frac{1}{2}\sum_{i=1}^{N_1}\sum_{j=1}^{N_2} W_i W_j \left[(v_i+v_j)^k - v_i^k - v_j^k\right] \beta(v_i,v_j), \quad (4.56)$$

where two sets of weights and two sets of abscissas, comprising N_1 and N_2 values, are employed to approximate the double integral. The problem, now, is that of finding the values of the weights, W_i and W_j, and the locations of the abscissas, v_i and v_j.

Hulburt and Katz (1964) (see also Hulburt and Akiyama, 1969), in their pioneering work on the formulation of the PBE, proposed also a method of solution along these lines, employing the standard Gauss–Laguerre weights and abscissas. The implication of this approach is that the distribution assumes a presumed form yielded by Laguerre polynomials, and therefore the method is equivalent to a presumed distribution approach (with three points, the distribution being presumed is the gamma distribution). A similar approach was proposed by Enukashvili (1964).

In the Quadrature Method of Moments (QMOM), proposed by McGraw (1997), the abscissas on which the integrand is evaluated are not fixed, nor are the weights corresponding to standard functions. Rather, an algorithm for obtaining the weights and abscissas for arbitrary functions (Press et al., 1997) is employed for calculating from the initial distribution. Subsequently, the change in the moments over a time step is calculated using the current values of these parameters. In the next time step, the parameters are recalculated using the same algorithm. This operation involves calculating the eigenvalues and eigenvectors of a Jacobi matrix, and the details can be found in McGraw (1997); note that this paper is about the closure of the growth term, but the method was later extended to aggregation (Barrett and Webb, 1998) and aggregation–fragmentation Marchisio et al. (2003).

The requirement for an inversion of the matrix raises several issues, as it may be ill-conditioned; furthermore, this operation renders the original QMOM unwieldy for extending to the multidimensional PBE or coupling with fluid dynamics. Marchisio and Fox (2005) proposed the Direct Quadrature Method of Moments (DQMOM), where differential equations are derived for the quadrature parameters (rather than the moments), thus eliminating the need for the matrix inversion step. In DQMOM, the distribution is represented in the following way:

$$n(v) = \sum_{i=1}^{N} W_i \delta(v - v_i). \quad (4.57)$$

The link with Gaussian quadrature is evident when we consider the integration of a term involving the distribution and an arbitrary function:

$$\int_0^\infty f(\upsilon) \sum_{i=1}^{N} W_i \delta(\upsilon - \upsilon_i) d\upsilon = \sum_{i=1}^{N} f(\upsilon_i) W_i, \tag{4.58}$$

which is of the same form as Eq. 4.55. Note also the similarity of Eq. 4.57 with the representation in pointwise discretisation methods (Eq. 4.59) to be discussed in the next section. More details about DQMOM and other variants of QMOM can be found in Marchisio and Fox (2013).

4.2.6 Discretisation Methods: Pointwise Approximation

The following three sections are devoted to discretisation methods, which can be classified into three families, as discussed in Section 4.1.2. The first of these comprises methods that approximate the distribution with a finite set of values at discrete points in the size domain.

Basic Concepts

With methods in this category, the distribution is assumed to be concentrated at a finite set of nodes, which is equivalent to a sum of delta functions:

$$n(v) = \sum_{i=1}^{N} n_i \delta(v - v_i), \tag{4.59}$$

where v_i and n_i are the values of the particle volume and number density, respectively, at the discrete nodes. The resulting discretised PBE is of a similar form to the discrete PBE, and identical in the case of a uniform grid with an increment of one monomer. However, a non-uniform grid (typically expanding for aggregation) needs to be employed in order to reduce the number of nodes. In a sense, pointwise approximation methods seek to construct a discrete PBE on a non-uniform grid.

In a non-uniform grid, a particle generated by an aggregation or fragmentation event does not always lead to a particle whose volume lies on a grid node. This is illustrated with an example in Fig. 4.1, which depicts an expanding geometric grid and an aggregation event between particles with volume 2 and 8; clearly the daughter particle cannot be accommodated. The main concept of pointwise approximation methods is to *redistribute the volume of such daughter particles between neighbouring nodes*; in the case of Fig. 4.1, that would be nodes with volumes 8 and 16. However, such a redistribution results in an arbitrary modification of the distribution; furthermore, a straightforward proportional approach does not conserve particle volume. This can be seen again in Fig. 4.1, where the aggregation of 10 particles of volume $2v$ with 10 particles of volume $8v$ would result in 10 particles of volume $10v$. If one-fourth of the resulting particles are allocated to volume $8v$ and three-fourths to volume $16v$, the total volume is $140v$, which differs markedly

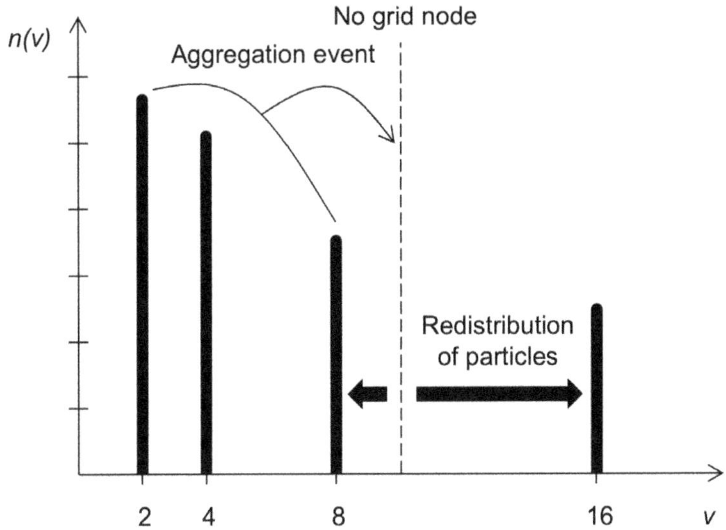

Figure 4.1 In pointwise approximation methods, the non-uniform grid results in aggregation events with no grid nodes to accommodate the particles that are formed. Therefore, the particles must be redistributed among neighbouring nodes.

from the correct value of $100v$. The success of these methods relies on an effective scheme for the redistribution of particle volume.

The following equation describes the methods in this family in a general sense[4]:

$$\frac{dn_i}{dt} = \frac{1}{2} \sum_{j=1}^{M} P_{jk(i,j)} \beta_{jk(i,j)} n_j n_{k(i,j)} - \sum_{j=1}^{N} Q_{ij} \beta_{ij} n_i n_j, \tag{4.60}$$

where $k(i,j)$ is the index of a node such that the sum of volumes $v_j + v_{k(i,j)}$ is close to v_i to have its volume partly redistributed there, β_{ij} is the kernel in its discrete form and N is the total number of nodes, while $P_{jk(i,j)}, Q_{jk}$ and M depend on the method and how it redistributes particle volume. Note that it is possible to retrieve the Smoluchowski equation (Eq. 2.4) by choosing a uniform grid containing all possible sizes and letting $P_{jk(i,j)} = 1$, $Q_{ij} = 1$, $k(i,j) = i - j$, $M = i - 1$ and $N \to \infty$. This fact demonstrates the link of methods based on Eq. 4.60 and the discrete PBE.

A similar equation can be employed to describe fragmentation:

$$\frac{dn_i}{dt} = \sum_{j=i+1}^{N} R_{i,j} \gamma_i p_{i,j} n_j - S_i \gamma_i n_i, \tag{4.61}$$

[4] Note that there are no implied summations of repeated indices in Eqs. 4.60 and 4.61; the summations are explicitly indicated.

where, in a similar vein, γ_i and $p_{i,j}$ are the discrete counterparts of the fragmentation rate and daughter particle distribution, while $R_{i,j}$ and $Q_{i,j}$ are correction factors that depend on the method. By setting these factors to 1 and choosing a uniform grid containing all possible sizes, the discrete equation for fragmentation is recovered.

Numerical Schemes

One type of expanding grid that offers a considerable simplification to the derivation of a discretisation method is the geometric grid with common ratio 2, that is, $v_i = 2v_{i-1}$. Berry (1967) was possibly the first to propose this grid, in the context of a study of cloud droplet aggregation. Several of the pointwise approximation methods are derived so as to exploit the unique properties of this grid, which greatly reduce the number of combinations of particle volumes to be considered. In particular, when considering the aggregation birth within the interval (v_{i-1}, v_i), at least one of the parent particles must lie within (v_{i-2}, v_{i-1}), while the aggregation of two particles within the same interval results in a particle in the next interval (Hounslow et al., 1988). Some typical methods in this family will be briefly described below. It should be mentioned that several of them have been extended to fragmentation, based on the same concepts.

Batterham et al. (1981) proposed the first method of this type, in the context of a study on pelletisation of powders. Their model was also based on the geometric grid with common ratio 2. Particles lying between grid nodes were allocated to the nearest neighbour. Batterham et al. realised that conservation of the first moment should be prioritised, as it is implies conservation of mass, and thus chose the coefficients in Eq. 4.60 so as to ensure this property. The distribution is done such that particles of a median volume between two nodes are distributed evenly between their neighbours. For example, if v_0 is the size of the smallest particle, then the grid comprises volumes $2v_0, 4v_0, \ldots$, and a particle of volume $3v_0$ is divided into $3/4$ of a particle with volume $2v_0$ and $3/8$ of one with $4v_0$. Fragmentation was also considered.

Hounslow et al. (1988) developed a method that ensures conservation of both particle number and volume, as well as accommodating nucleation and growth. As with the Batterham et al. (1981) approach, their formulation is based on the geometric grid with common ratio 2. In their paper, the equations are derived on the basis of a distribution of particle length, hence the objective is to conserve the zeroth and third moments of the distribution. In order to achieve this property, they derived the discretised PBE for the aforementioned grid and introduced a correction factor. They then derived the equations for the rate of change of the zeroth and third moments and showed that the former did not depend on the correction factor, while the latter could be predicted correctly if the factor was equal to $2/3$.

The above methods are bound to the geometric grid with common ratio 2 by construction. A further group of methods aimed to relax this constraint. Marchal et al. (1988) derived a method not bound to a particular grid by considering aggregation as a chemical reaction with stoichiometric coefficients determined in such a way that mass is conserved. Litster et al. (1995) extended the method of Hounslow et al. (1988) so as to allow the use of a grid defined by $v_{i+1}/v_i = 2^{1/q}$ where $q \in \mathbb{N}$ and $q \geq 1$. Hill and Ng (1996a) developed a method applicable to uniform and geometric grids, the latter with any ratio. This was attained by deriving different forms of discretised equation for different ranges of geometric ratio. The same authors also developed a pointwise approach for fragmentation (Hill and Ng, 1995).

A more general approach to the derivation of pointwise approximation methods was proposed by Kumar and Ramkrishna (1996a) (see also Ramkrishna, 2000). This method addresses both aggregation and fragmentation, ensures the conservation of any two moments of the distribution and is applicable in an arbitrary grid. This is accomplished by solving two algebraic equations to obtain the correction factors, rather than prescribing them, such that the two chosen moments are conserved while the particles are distributed to the two adjacent nodes. Their initial method was called a *fixed pivot technique* because the particles in each interval are represented by one pivotal size that remains fixed. In a subsequent paper (Kumar and Ramkrishna, 1996b), they proposed a *moving pivot technique* which improves on the accuracy of the original method by allowing the pivotal sizes to be adjusted. The Kumar and Ramkrishna approach is one of the most widely applied. Kumar et al. (2006) developed a method that improves on the fixed pivot technique, the *cell-average technique*, and Kostoglou (2007a) presented an extension of that method that conserves three moments, although it does not always improve the prediction of the distribution.

Discussion of Pointwise Approximation Methods

In summary, discretisation methods in the category of pointwise approximation are relatively simple to implement and computationally efficient, while they place the emphasis on conservation of moments rather than accuracy in the prediction of the distribution. As such, these methods bridge the gap between moment methods and the finite element/volume discretisation methods that will be described in the following sections and focus more on the prediction of the distribution. The conservation of one or more moments is ensured by the construction of the method. In many applications, it is mostly the moments that can be compared with experimental measurements, and indeed pointwise discretisation methods have found wide application. Compared with moment methods, they offer the advantage that they do not require closure assumptions. In terms of implementation ease and computational efficiency, they stand between moment and finite element/volume discretisation methods. It should be mentioned that the first moment is strictly conserved only in aggregation and/or fragmentation problems; when growth is present, a different property of the distribution (such as the surface area for reaction-limited growth) is crucial for predicting mass change. Kostoglou and Karabelas (1994) and Vanni (2000) have tested and compared discretisation methods, focussing mainly on the methods in this category.

4.2.7 Discretisation Methods: One-Dimensional Quadrature

While pointwise approximation methods construct a discrete PBE introducing correction factors in the process, the methods discussed in the present and next section discretise the continuous PBE with the finite element or finite volume method. These methods do not involve correction factors to impose moment conservation and emphasise accuracy in the prediction of the distribution. However, it will be shown that the straightforward application of the finite element method (the one-dimensional quadrature approach) cannot guarantee the moment conservation due to the convolution-like form of the aggregation birth term. This issue can be addressed with the two-dimensional quadrature methods discussed in the next session.

The General Procedure

In finite element methods, the domain of the independent coordinate is divided into a finite number of intervals (elements). The unknown function is approximated via piecewise functions over each element, which are most often constant functions, but can also be higher-order polynomials such as splines. The equation is then multiplied by a weight function and integrated over each element, and several choices of weight functions result in different variations of the method. The equation is then integrated over each element, and a set of algebraic equations (for steady state problems) or ODEs (for transient problems) results with the unknowns being the coefficients of the trial functions. The general procedure will be shown below, while one may consult e.g. Reddy (1993), Chapter 14, for more details.

First, consider the one-dimensional PBE written in the following form:

$$\frac{\partial n(v,t)}{\partial t} = S(v,n), \tag{4.62}$$

where $S(v,n)$ is a source term including aggregation and fragmentation. The domain of v is now discretised into n intervals. For an interval ranging from v_{i-1} to v_i, the following equation can be written:

$$\int_{v_{i-1}}^{v_i} w(v)\frac{\partial n(v,t)}{\partial t}dv = \int_{v_{i-1}}^{v_i} w(v)S(v,n)dv, \tag{4.63}$$

where $w(v)$ is a weight function for which possible choices will be shown below.

The distribution within each interval is now approximated via a sum of m trial functions, each of which involves a parameter, so that there are m parameters over each of the n elements. These parameters are the unknowns to solve for. As we will see, the parameters can be the nodal values of the discretised number density function. Eventually, a set of ODEs is derived for the temporal evolution of these parameters, and its solution yields the timedependent distribution. With different choices of trial functions, several variations of the finite element method are obtained; the following are some common choices that have been applied to the PBE in the literature:

- *Collocation*: The weight functions are delta functions, i.e. $w(x) = \delta(x - x_i)$. Therefore, collocation methods force the residual to be zero at the nodal points. In this sense, collocation methods resemble pointwise approximation methods.
- *Galerkin*: The weight functions are the same as the trial functions. Galerkin methods are particularly suitable for problems with variational principles (Finlayson, 2014).
- *Subdomain*: The interval is divided into subdomains while the weight function is unity for each subdomain and zero outside it. When the whole interval is taken as a single subdomain and the trial function is constant, the result is the *finite volume method.*
- *Moment*: The weight function is a power of the independent variable, hence the resulting integral is a moment of the distribution. For the zeroth moment (i.e. weight function 1), the resulting method is the same as the subdomain or finite volume method. This method is particularly relevant to the PBE as it can enforce conservation of a single moment.

In the following, only the finite volume method will be considered in detail, as it forms also the basis for the two-dimensional quadrature methods in the next section. For the finite volume method, $w(v) = 1$, and it is also customary to approximate the unknown function as piecewise constant, such that n_i is the value of the number density over interval (v_{i-1}, v_i). In this case, writing out the source term in detail, Eq. 4.63 becomes:

$$\begin{aligned} \delta v_i \frac{\partial n_i}{\partial t} = & \frac{1}{2} \int_{v_{i-1}}^{v_i} \int_0^v \beta(w, v-w) n(w,t) n(v-w,t) dw dv \\ & - \int_{v_{i-1}}^{v_i} n(v,t) \int_0^\infty \beta(v,w) n(w,t) dw dv \\ & + \int_{v_{i-1}}^{v_i} \int_v^\infty \gamma(w) b(v,w) n(w,t) dw dv - \int_{v_{i-1}}^{v_i} \gamma(v) n(v,t) dv. \end{aligned} \tag{4.64}$$

We can see from Eq. 4.64 that the application of a finite element or volume term to an integral equation such as the PBE results in double integrals. The straightforward way to deal with them is to approximate the inner integral with quadrature, which is why methods in the present category are named one-dimensional quadrature methods. The treatment of fragmentation in this way is straightforward. The main problem arises in aggregation, and in particular in the aggregation birth term. Therefore, the following will concentrate in the aggregation equation.

Application of the Finite Volume Method to the Aggregation PBE

We now take as starting point Eq. 4.2 with the aggregation birth term written as in Eq. 4.5, and also truncate the domain to a finite range, (v_{min}, v_{max}). This range should start from the nucleus volume and reach up to a sufficiently high value to contain the distribution as it is expected to evolve in the time period considered. The equation to discretised, therefore, is:

$$\frac{dn(v,t)}{dt} = \int_{v_{min}}^{\frac{v}{2}} \beta(w, v-w)n(w,t)n(v-w,t)dw - n(v,t)\int_{v_{min}}^{v_{max}} \beta(v,w)n(w,t)dv. \tag{4.65}$$

This domain is now discretised into n intervals, such that interval i denotes (v_{i-1}, v_i), and the distribution is approximated with piecewise constant functions, such that n_i is the number density within interval i. After integration over interval i, the resulting discretised equation is:

$$\begin{aligned}\delta v_i \frac{dn_i(t)}{dt} &= \int_{v_{i-1}}^{v_i}\int_{v_{min}}^{\frac{v}{2}} \beta(w, v-w)n(w,t)n(v-w,t)dwdv \\ &- \int_{v_{i-1}}^{v_i} n_i(t)\int_{v_{min}}^{v_{max}} \beta(v,w)n(w,t)dwdv,\end{aligned} \tag{4.66}$$

where $\delta v_i = v_i - v_{i-1}$.

The treatment of the death term is straightforward, as it can be split into a sum over intervals and the number density is constant in each interval:

$$\int_{v_{i-1}}^{v_i} n_i(t)\int_{v_{min}}^{v_{max}} \beta(v,w)n(w,t)dwdv = n_i(t)\delta v_i \sum_{j=1}^{N}\left(n_j(t)\int_{v_{j-1}}^{v_j} \beta(v,w)dw\right), \tag{4.67}$$

where N is the number of intervals. Thus, it only remains to numerically approximate the integral of the kernel over interval i with a numerical quadrature rule, such as Gaussian quadrature. This term can also be tabulated to save computational time.

The aggregation birth term, however, presents a different problem. Splitting as before into a sum over intervals, we obtain:

$$\int_{v_{i-1}}^{v_i}\int_{v_{min}}^{v_{max}} \beta(w, v-w)n(w,t)n(v-w,t)dwdv$$
$$= \delta v_i \sum_{j=1}^{N}\left(n_j(t)\int_{v_{j-1}}^{v_j} \beta(w, v-w)n(v-w,t)dw\right). \qquad (4.68)$$

The inner integral cannot be manipulated further because the number density of the complementary volume $v - w$ lies within a different interval and possibly spans several intervals.

The methods in the present section approximate the inner integral in Eq. 4.68 with numerical quadrature, hence the term *one-dimensional quadrature methods*, unlike the methods in the next section that follow a different approach to the treatment of the entire term. The approximation of the inner integral has, however, two adverse consequences:

- The birth terms are not strictly balanced with the death terms. This is because they are approximated separately, and, furthermore, the integrand of the birth term (Eq. 4.68) includes both number density and kernel while that of the death term (Eq. 4.67) involves the kernel alone. *This is the reason why moment conservation cannot be attained in a strict sense via the methods in this family.*
- The birth term integral cannot be pretabulated, as it involves the number density.

Several methods in this category have been proposed, differing in the choice of finite element approach and quadrature schemes. The earliest one was due to Gelbard and Seinfeld (1978a) who employed spline trial functions. Later works include Nicmanis and Hounslow (1998) who used collocation and Galerkin methods, Rigopoulos and Jones (2003c) based on a collocation method and Roussos et al. (2005) employing a Galerkin method. In the method of Rigopoulos and Jones (2003c), the moment conservation error was analysed and an approach was proposed for overcoming it by generating a grid consisting of the complementary aggregation volumes (i.e. the volumes $v - w$ that complement volume w to form v) and splitting further the integral in Eq. 4.68. This approach reduces the conservation error but does not eliminate it, because the birth and death fluxes are still not balanced in an absolute sense. It also attains computational efficiency by allowing for a pretabulation of the integral terms. We should note also the recent work of Vervisch et al. (2023), where the birth and death terms are split with the method of fractional steps (cf. Section 4.4.4). The death term is then treated separately via an analytical solution (owing to its linearity), while the birth term is evaluated via a high-order approximation of the distribution. More details on this will be shown in Section 4.3.5.

Discussion of One-Dimensional Quadrature Methods

Overall, the finite element and finite volume methods based on one-dimensional quadrature cannot ensure the conservation of moments in aggregation and fragmentation, but rather aim to control it via the accuracy attained in the prediction of the distribution. This is in contrast to the pointwise approximation methods, where moment conservation is attained by construction, although this is accomplished by introducing modifications to the distribution. The discretisation of fragmentation does not pose problems and accurate results for both distribution and moments can be obtained. The discretisation of aggregation, however, poses a particular challenge due to the convolution-like aggregation birth term. This term is responsible for the error in the conservation of moments, and it is not possible to strictly balance birth and death terms if one-dimensional quadrature is applied to the inner integrals. The ease of implementation and computational efficiency in these methods varies depending on the particular formulation; at best it can be comparable to pointwise methods, if the implementation allows a pre-tabulation of the aggregation and breakage integrals. The third category, two-dimensional quadrature methods, is closely related to the methods in the present section since it is based on the finite volume method, which is a special case of the framework presented in this section (using the subdomain trial functions). However, they differ in the treatment of the integral terms, where two-dimensional quadrature is applied to the entire term, and are able to attain the conservation of one moment by construction.

4.2.8 Discretisation Methods: Two-Dimensional Quadrature

Overview

We now discuss a family of methods that are also based on the framework of finite volume methods, but are distinguished from those in the previous section in the treatment of the double integrals that result from the discretisation of the aggregation terms. It will be shown that, with the application of two-dimensional quadrature, it is possible to balance source and sink terms and endow the method with the conservation property for one moment.

These methods were initially developed from an intuitive point of view, based on the physics of aggregation. The first method of this type was proposed by Bleck (1970), where a graphical procedure was used to illustrate the concept on a geometric grid with common ratio 2. The approach of Gelbard et al. (1980) provided a more general and systematic account, and was extended to multicomponent aerosols by Gelbard and Seinfeld (1980) and to the discrete-continuous PBE by

Wu and Flagan (1988). The use of a geometric grid with common ratio 2 simplifies greatly the approach, as an arbitrary grid requires a complex partitioning of the integration domain, but this is needed particularly for problems where aggregation is combined with growth and/or fragmentation. Landgrebe and Pratsinis (1990) developed a discrete-continuous PBE based on a geometric grid with common ratio 2 while anticipating in an appendix the generalisation of the method to arbitrary expanding grids. It can be shown that these methods, all of which employ two-dimensional quadrature, can be formulated using the more general framework of finite volume methods, although this concept was put forward much later by Filbet and Laurençot (2004), who also carried out a comprehensive analysis of the properties of the numerical scheme. Liu and Rigopoulos (2019) developed an approach for partitioning the domain of integration in an arbitrary grid. A more comprehensive exposition was put forward by O'Sullivan and Rigopoulos (2022), where the approach was also extended to fragmentation and validated with a number of test cases. This approach will be used as the basis for the exposition below. The aggregation PBE will be treated in detail, since the aggregation birth term is the main source of the conservation problem. Fragmentation will be discussed afterwards, as it is a straightforward extension of the methodology.

The conservative finite volume approach of Liu and Rigopoulos (2019) and O'Sullivan and Rigopoulos (2022) will be employed in the applications in Chapter 6, so it will be examined in some detail. Before seeing how a strictly conservative scheme can be devised, we first take a look at how conservation is built into finite volume methods.

Conservation of Mass in Finite Volume Methods

The finite volume method was initially developed in the field of computational fluid dynamics (CFD) and quickly became dominant, largely owing to its conservation property. As we saw in Section 4.2.7, however, its straightforward application to the population balance does not conserve mass. It is worth understanding the reasons behind this issue before looking at how the two-dimensional quadrature methods attain conservation.

Finite volume methods work with *fluxes*. Fig. 4.2 shows schematically how a finite volume operates in the context of CFD: a single flux is computed between cells and acts as outflow for one of them and inflow for the other. This way, all inflows and outflows are balanced and conservation of mass is achieved by construction of the method. The same figure shows the fluxes in the discretised PBE. Unlike CFD, however, here the fluxes are not between neighbours but may connect intervals anywhere in the domain. Furthermore, there is not a single flux acting as outflow and inflow; rather, two outflow fluxes from parent intervals combine to form an inflow flux to a daughter interval.

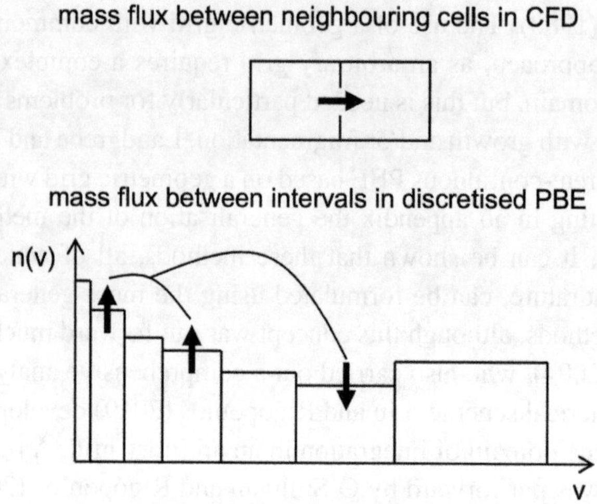

Figure 4.2 Mass fluxes in finite volume schemes.

This consideration provides a direction for the development of a conservative finite volume method for the PBE: the outflow and inflow fluxes must be calculated such that every inflow is the sum of its constituent outflows. Attaining this, however, is not straightforward because of the convolution-like form of the aggregation birth term (cf. Eq. 4.65). While the death term clearly denotes an outflow of particles from an interval, the birth term contains inflows from two intervals combined in a convolution-like integral. Thus, the first step must be to delineate the birth fluxes and identify the parent intervals.

Conservative Finite Volume Discretisation of the Aggregation PBE

The starting point for the following derivations is again the truncated aggregation equation (Eq. 4.65). Furthermore, in order to ensure conservation of mass (or, equivalently, the first moment for constant density particles), the equation is multiplied by v. It should be mentioned that, by multiplying with v^k, a method that conserves any single moment can be derived.[5] The mass-conservative aggregation equation is:

$$\frac{\partial}{\partial t}(vn(v,t)) = \int_{v_{min}}^{\frac{v}{2}} v\beta(w, v-w)n(w,t)n(v-w,t)dw - vn(v,t)\int_{v_{min}}^{v_{max}} \beta(v,w)n(w,t)dw. \qquad (4.69)$$

[5] The method can be called a *moment weighted residual method* in the terminology of Finlayson (2014).

As in the previous section, the number density will be approximated by a piecewise constant function (which is customary with finite volume methods). This equation is now integrated over interval i to yield:

$$\hat{v}_i \frac{dn_i(t)}{dt} = \int_{v_{i-1}}^{v_i} \int_{v_{min}}^{\frac{v}{2}} v\beta(w, v-w)n(w,t)n(v-w,t)dwdv$$
$$- n_i(t) \int_{v_{i-1}}^{v_i} \int_{v_{min}}^{v_{max}} v\beta(v,w)n(w,t)dwdv, \tag{4.70}$$

where:

$$\hat{v}_i = \frac{v_i^2 - v_{i-1}^2}{2}. \tag{4.71}$$

So far, the approach is the same as the one in Section 4.2.7, apart from the use of the mass-conservative equation. However, this fact alone does not guarantee conservation; as we saw in that section, the problem stems from the application of quadrature to the aggregation birth term.

Decomposition of the Aggregation Birth Term

We now employ the transformation introduced by Thompson (1968) in the context of moment methods, already encountered in Section 4.2.4. This involves applying the change of variable $u = v - w$ to the aggregation birth integral and, when applied to Eq. 4.70 results in[6]:

$$\hat{v}_i \frac{dn_i(t)}{dt} = \int_{v_{i-1}-w}^{v_i-w} \int_{v_{min}}^{\frac{u+w}{2}} (u+w')\beta(w',u)n(w',t)n(u,t)dw'du$$
$$- n_i(t) \int_{v_{i-1}}^{v_i} \int_{v_{min}}^{v_{max}} v\beta(v,w)n(w,t)dwdv, \tag{4.72}$$

where the dummy variable w' has been used in the integrand to be distinguished from w in the integration limits. Splitting the aggregation birth term into two terms, we have:

$$\hat{v}_i \frac{dn_i(t)}{dt} = \int_{v_{i-1}-w}^{v_i-w} \int_{v_{min}}^{\frac{u+w}{2}} u\beta(w',u)n(w',t)n(u,t)dw'du$$
$$+ \int_{v_{i-1}-w}^{v_i-w} \int_{v_{min}}^{\frac{u+w}{2}} w'\beta(w',u)n(w',t)n(u,t)dw'du$$
$$- n_i(t) \int_{v_{i-1}}^{v_i} \int_{v_{min}}^{v_{max}} v\beta(v,w)n(w,t)dwdv. \tag{4.73}$$

[6] Note that w in the outer integral is not a variable; instead, the w-space is partitioned into sub-intervals whose limits are determined with a geometrical procedure, as shown in Appendix B.

We note that the convolution-like term has given its place to two aggregation birth terms, each of which has the same form as the death term. In other words, *the aggregation birth term has been decomposed into its constituent aggregation death terms.* Each of the three terms now *corresponds to a flux of particles.* The two birth fluxes in interval i correspond to death fluxes from intervals j and k. It is thus possible to calculate, for example, only the death fluxes and subsequently obtain the birth fluxes by summing up the appropriate pairs. Therefore, conservation of the first moment is ensured by construction.

Evaluation of Integrals

In order to be evaluated, the integrals in Eq. 4.73 must be converted into sums. Since the distribution has been approximated with pairwise constant functions, it is possible to partition the domain of integration in such a way that the number density product is constant in each subdomain. For example, the death term integral can be written as follows:

$$n_i(t)\int_{v_{i-1}}^{v_i}\int_{v_{min}}^{v_{max}} v\beta(v,w)n(w,t)dwdv = n_i(t)\sum_{j=1}^{N}\left(n_j(t)\int_{v_{i-1}}^{v_i}\int_{w_{j-1}}^{w_j} v\beta(v,w)dwdv\right), \tag{4.74}$$

where w_{j-1} and w_j are the bounds of each subdomain and N is the number of intervals. This way, quadrature needs to be applied only to integrals of the following type:

$$I = \int_{v_{i-1}}^{v_i}\int_{w_{j-1}}^{w_j} v\beta(v,w)dwdv. \tag{4.75}$$

These integrals can be evaluated once and tabulated for a given grid, so that their computation does not affect the CPU needed at run-time. Liu and Rigopoulos (2019) evaluated them at a single point, while O'Sullivan and Rigopoulos (2022) showed a procedure using two- or three-point Gaussian quadrature. In their tests, the one-point evaluation method yielded an error of less than 1% in a grid comprising 40 intervals, while the three-point quadrature had an error smaller than 0.3% with a grid as coarse as 10 nodes and across all kernels. The quadrature error is thus negligible and CPU cost of the three-point method can be overcome with tabulation, while the one-point method is still quite accurate.

The Aggregation Map

The final step in the method is to identify and map the subdomains on which the double integrals are to be evaluated. As mentioned, the domain of integration must be partitioned such that the number densities are constant within each sub-domain. Furthermore, the subdomains of the birth term integral must be mapped to the corresponding subdomains in the death term integral. This way, a single flux is calculated for each transfer of particle mass.

The partitioning of the domain and the mapping between birth and death terms is called the *aggregation map*. The first description of its construction was shown by Liu and Rigopoulos (2019), while O'Sullivan and Rigopoulos (2022) showed a more comprehensive account that includes also fragmentation. The latter exposition is summarised in Appendix B.

The construction of the aggregation map can be carried out once and tabulated, together with the evaluation of the kernel integrals. As a result, the operations to be carried out during the temporal integration of the discretised PBE are very simple and consist of updating the birth/death terms and advancing the discretised number densities in time. The last operation can be carried out with explicit methods as the ODEs resulting from the PBE discretisation are usually not stiff; in O'Sullivan and Rigopoulos (2022), the explicit Euler method was used.

An alternative formulation of the conservative finite volume discretisation approach for aggregation has been recently proposed by Sewerin (2023). In that work, a linear bivariate coordinate transformation from (u, w) to (ξ, η) is applied, and the computation of the integrals is carried out in the transformed space. Furthermore, the procedure for locating the parent–daughter interval combinations is different from the one described in Appendix B. As a result, the partitioning of the integration domain and the subsequent evaluations of integrals are considerably simplified. In that method, the kernel is evaluated once for every pair of parent intervals. It is also shown how the integration can be carried out with the number density being approximated with either a piecewise constant or a piecewise linear approximation. The details of that method can be found in the aforementioned paper.

Conservative Finite Volume Discretisation of the Fragmentation PBE

Following the same process as with aggregation, the starting point here is the fragmentation equation multiplied by volume and truncated to lie within the range (v_{min}, v_{max}). The lower truncation point implies that fragmentation cannot proceed to infinitely small particles. The equation is:

$$\frac{\partial}{\partial t}(vn(v,t)) = v \int_{v}^{v_{max}} \gamma(z) b(v,z) n(z,t) dw - v\gamma(v) n(v,t). \tag{4.76}$$

This equation is now integrated over an interval to yield:

$$\hat{v}_i \frac{dn_i}{dt} = \int_{v_{i-1}}^{v_i} v \int_{v}^{v_{max}} \gamma(z) b(v,z) n(z,t) dz dv - n_i \int_{v_{i-1}}^{v_i} v\gamma(v) dv. \tag{4.77}$$

In a similar vein to the procedure followed in aggregation, only the fragmentation birth terms are calculated. Similar to the aggregation map, a *fragmentation map* ensures that the volume subtracted from the parent intervals is assigned to the appropriate daughter intervals. However, fragmentation is considerably simpler than aggregation, in that no decomposition of the double integral in Eq. 4.77 is required; rather, it can be readily used in its current form to construct the fragmentation map. The construction of the map is shown in Appendix B.

All operations involved in the construction of the aggregation map, that is, the location and classification of subdomains and the evaluation of integrals, are carried out only once and the results are tabulated. Thus, at each time step the only operations to be carried out consist of the time advancement of the discretised number densities.

Validation of the Conservative Finite Volume Method

Since the conservative finite volume method will be employed in the case studies in Chapter 6, it is worthwhile to look at some of the validation cases shown in O'Sullivan and Rigopoulos (2022). Fig. 4.3 shows four representative cases including aggregation, fragmentation and combined aggregation–fragmentation. The top-left figure shows a solution of the aggregation PBE with the sum kernel (Eq. 4.21, $\beta_0 = 1$) and an exponential initial distribution (Eq. 4.18), and comparison is made with the analytical solution of Scott (1968), shown in the form of Gelbard and Seinfeld (1978a) in Eq. 4.22, for two different times. The top-right figure shows a calculation of the self-preserving distribution for Brownian aggregation. As discussed in Section 4.2.3, there is no analytical solution to compare with (apart from asymptotic solutions), but the numerical solution of the equivalent Smoluchowski equation with time marching (Eq. 4.16) is very accurate and can be regarded as a substitute for an analytical solution. The bottom-left figure shows a comparison with the solution of Ziff and McGrady (1985) for uniform binary fragmentation (Eq. 4.25), while the bottom right shows the combined aggregation–fragmentation case for which a similarity solution was obtained by Patil and Andrews (1998) and corrected by Lage (2002). In all cases, the solution of the discretised PBE is in excellent agreement with the reference solution. More validation cases can be found in O'Sullivan and Rigopoulos (2022). Results for the evolution of the moments were also shown in Table 2 of O'Sullivan and Rigopoulos (2022); compared with the reference solution, the error in m_1 was always zero, thus confirming that the method is conservative with respect to this moment. Results for m_0 and $m_{2/3}$ were also shown; the latter is proportional to surface area and thus important for many problems. The error in these moments was very small and in almost all cases less than 1%.

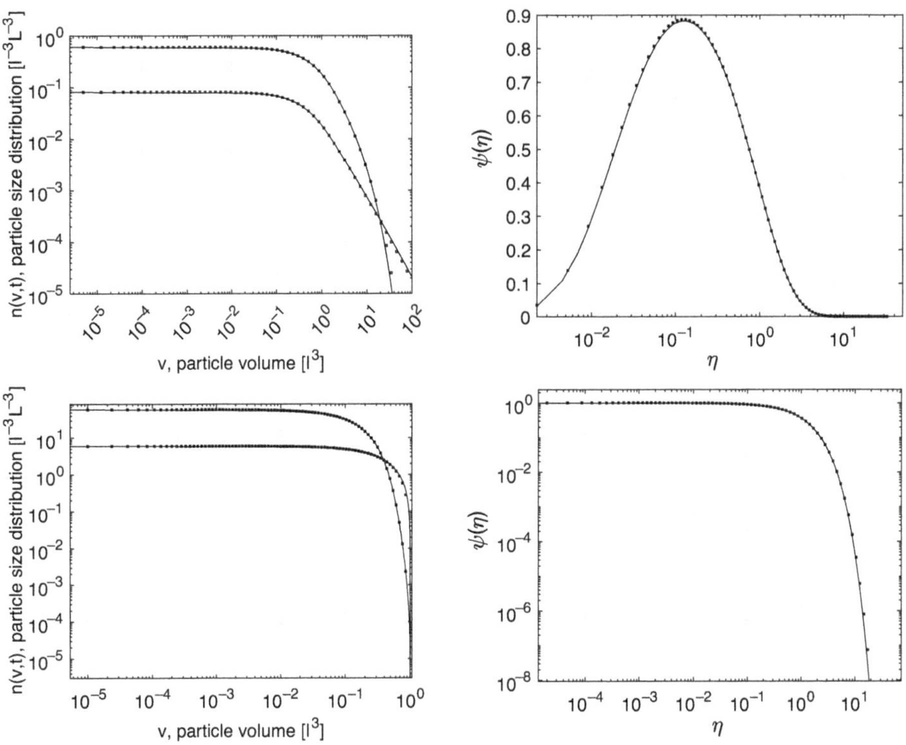

Figure 4.3 Comparisons of solutions of the PBE obtained with the conservative finite volume method with analytical and discrete PBE solutions. Top left: aggregation with the sum kernel (cf. Section 4.2.2), analytical (line) and numerical (symbols) solutions for $t = 0.5$ s and $t = 2.5$ s. Top right: self-preserving distribution for Brownian aggregation (cf. Section 4.2.3), numerical solutions of the discrete PBE (lines) and discretised PBE (symbols). Bottom left: uniform binary fragmentation (cf. Section 4.2.2), analytical (line) and numerical (symbols) solutions for $t = 1$ s and $t = 10$ s. Bottom right: steady state solution of the joint aggregation–fragmentation case (Patil and Andrews, 1998; Lage, 2002), analytical (line) and numerical (symbols) solutions. In all cases, 80 intervals were used in the discretised PBE. l and L indicate particle and physical size dimensions, respectively. Reproduced from O'Sullivan and Rigopoulos (2022) under CC-BY 4.0 license.

Discussion of Two-Dimensional Quadrature Methods

As will be shown, in two-dimensional quadrature methods where the distribution is approximated with a stepwise constant function, the integration of the birth and death integrals is exact and quadrature is applied only for the aggregation kernels. It has been shown (O'Sullivan and Rigopoulos, 2022) that this quadrature can be carried out with very high accuracy, and therefore the only remaining discrepancy stems from the discrete representation of the distribution via a stepwise constant function, which can be controlled by refining the grid. Furthermore, the kernel integrations can be pre-tabulated and thus not incur a computational overhead during run-time. The methods in this category can thus attain accuracy in the prediction of the distribution, conservation of one moment and computational efficiency. Pointwise approximation methods can attain the conservation of more than one moment by construction, although this can be at the expense of accuracy in the distribution. In terms of implementation, however, two-dimensional quadrature methods are more complex due to the need to map all fluxes; some individual methods in this category are simpler by focussing on the geometric grid. The conservative finite volume method described in some detail in this section can accommodate both aggregation and fragmentation on an arbitrary grid by means of an algorithm for constructing the map.

4.2.9 Monte Carlo Methods

Basic Concepts

Monte Carlo methods are different from the solution methods discussed so far in that the PBE is not solved directly, but rather its solution is obtained indirectly via a computational simulation of an ensemble of particles that interact in a stochastic manner. The stochastic interaction is designed so as to emulate the kinetic processes involved and approximate the solution, in the sense that the distribution of the properties of interest among the stochastic particles is an approximation of the one in the actual system. Many variations of Monte Carlo methods exist that differ in terms of the formulation of the stochastic process or the computational implementation, with the aim to improve convergence and efficiency.

One of the earliest applications of Monte Carlo methods to a process that resembles aggregation (coalescence and breakup of a dispersed phase) is due to Spielman and Levenspiel (1965). Gillespie (1975) proposed one of the first practical Monte Carlo algorithms for a population balance problem, the coalescence of droplets in a cloud. Ramkrishna (1981) provided a theoretical analysis of the Monte Carlo approach for the population balance. A detailed review of early works and of the statistical foundations of the method can be found in Ramkrishna (2000). Further theoretical developments were carried out by Eibeck and Wagner (2000a,b, 2001, 2003). The following discussion will outline the basic concepts involved, the main variations and the overall structure of a Monte Carlo algorithm for the aggregation PBE. The reader may refer to the references that will be provided for the details of the methods or their many applications.

While Monte Carlo methods are particularly effective for multidimensional problems, the main concepts underlying the design of a Monte Carlo algorithm can be laid bare by considering a one-dimensional aggregation process, as will be shown below, and extended in a straightforward way to include more kinetic processes or multiple dimensions.

Types of PBE Monte Carlo Algorithms

In a Monte Carlo algorithm, the evolution of a notional ensemble of particles is controlled by a random variable. There are two main choices for this variable, yielding two types of algorithms:

- *Event-driven*: The random variable is the time between two events (in this case, the event is an aggregation between two particles), which is assumed to be exponentially distributed. Early works on event-driven methods includes the papers by Gillespie (1975) and Shah et al. (1977), based on earlier concepts put forward by Kendall (1950).
- *Time-driven*: The random variable is number of events occurring within a given time step. This approach was put forward by Rod and Míšek (1982) and Rajamani et al. (1986)

A second feature with respect to which the Monte Carlo algorithms for aggregation may be classified is the treatment of the number of particles in the ensemble, which is an issue because aggregation events reduce the number of particles. Regarding this aspect, methods can be distinguished into:

- *Constant volume*: The number of particles is allowed to be reduced by aggregation events, maintaining the total volume constant. This approach is employed in the earlier works such as Gillespie (1975).
- *Constant number*: The number of particles is maintained constant by inserting new particles to compensate for the number reduction in each aggregation event. Matsoukas and Friedlander (1991) introduced this feature into an event-driven Monte Carlo simulation, and the idea was later formalised by Smith and Matsoukas (1998).

A third feature pertains to the computation of the kernel, which is the most expensive part of the algorithm because it has to be computed for all particle pairs and is usually nonlinear. Eibeck and Wagner (2000b) introduced an algorithm where the aggregation kernel is replaced by a simpler function, the *majorant kernel*, $\hat{\beta}(v,w)$, that obeys the property $\hat{\beta}(v,w) \geq \beta(v,w)$, while additional *fictitious jumps* are introduced. The use of the majorant kernel speeds up computations, but a suitable function must be found for every aggregation kernel; Eibeck and Wagner (2000b) employed the sum kernel to substitute for the free molecule regime kernel. A more efficient majorant kernel for the same regime was proposed by Goodson and Kraft (2002).

Main Steps of a PBE Monte Carlo Algorithm

A Monte Carlo method is best described in the form of an algorithm. In recent years, the event-based constant-number approach is used more often and the main steps of the basic algorithm will be outlined below. First, an initial ensemble of particles is generated. Subsequently, the following procedure is repeated for the desired simulation time:

- A time step is randomly selected, following an exponential distribution.
- The aggregation process is emulated by generating a number of aggregation events between randomly selected particles. The probabilities of these events must be computed by comparing the kernel for the selected pair with all possible pairs. This part of the algorithm is computationally expensive and can be speeded up with the use of a majorant kernel.
- The statistics of the particles are collected to estimate the distribution of the desired properties at the new time instant.

This procedure can be augmented with steps to emulate other kinetic processes such as fragmentation, growth and nucleation. The initial methods (such as that of Gillespie, 1975) were direct, in the sense that the kernel was computed for all particle pairs. For the implementation of the majorant kernel and fictitious jumps algorithm, the reader is referred to Eibeck and Wagner (2000b) and Goodson and Kraft (2002).

Applications of Monte Carlo Methods

Monte Carlo methods are considerably more expensive than deterministic methods when it comes to solving the one-dimensional aggregation PBE. They offer a computational advantage, however, for the treatment of multidimensional problems (cf. Section 4.2.10), due to the fact that the computational effort scales linearly with dimensionality (rather than exponentially, as is the case with discretisation methods). They are also useful for some physico-chemical processes where the formulation of a deterministic PBE may be difficult and for small populations where fluctuations are significant.

Van Peborgh Gooch and Hounslow (1996) applied the Monte Carlo method to crystallisation and considered nucleation and growth as well as aggregation. After testing the algorithm with a one-dimensional PBE, they applied it to problems with up to four dimensions. Monte Carlo methods have also been employed for aerosols (Kruis et al., 2000; Kostoglou and Konstandopoulos, 2001; Kostoglou et al., 2006), particularly to address the problem of aggregation and coalescence. In soot formation, the method has also been used to study the fundamentals of aggregation (Kaplan and Gentry, 1988) and aggregation-growth (Mitchell and Frenklach, 1998,

2003). An algorithm for simulating nucleation, surface processes and coagulation in one-dimensional premixed flames was proposed by Balthasar and Kraft (2003). Later works addressed soot morphology (Balthasar and Frenklach, 2005b; Patterson and Kraft, 2007) and PAH growth (Raj et al., 2009).

4.2.10 Multidimensional PBEs

Multidimensional and coupled PBEs for aggregation and fragmentation are solved for with extensions of the methods already presented. Not all methods, however, are as readily adaptable to multiple dimensions. Some analytical solutions for aggregation (and also growth) in multicomponent aerosols have been derived by Gelbard and Seinfeld (1978b). Below, we briefly discuss the extensions of moment, discretisation and Monte Carlo methods.

Moment Methods

The extension of moment methods to multidimensional problems does not pose problems from the computational point of view, as the number of variables involved is still small, and the issues lie in formulating and obtaining closure of the multidimensional moment equations.

The MOMIC method has been extended to two dimensions in order to model the formation of fractal aggregates by Kazakov and Frenklach (1998). It is assumed that particles either coalesce into spherical ones or combine into aggregates with constant fractal dimension (cf. Eq. 3.38). The independent variables in the distribution are particle mass and number of primary particles per aggregate. The equations are summarised by Frenklach (2002a). Furthermore, Balthasar and Frenklach (2005a) and Mueller et al. (2009b) put forward MOMIC formulations that employ the particle surface area as the second independent variable.

The QMOM approach has also been applied to two-dimensional problems. A bivariate formulation of QMOM to address coagulation and sintering of particles can be found in Wright et al. (2001). Marchisio and Fox (2005), in their presentation of the DQMOM approach, showed also the multivariate formulation of the equations. DQMOM and other QMOM variants are described in detail by Marchisio and Fox (2013). Blanquart and Pitsch (2009) formulated a trivariate PBE for soot formation in terms of volume, surface area and number of active sites and solved it with a DQMOM method. Finally, Mueller et al. (2009a) developed the Hybrid Method of Moments (HMOM), which combines concepts from both MOMIC and DQMOM and is based on a bivariate PBE in terms of particle volume and surface area.

Discretisation Methods

Multidimensional discretisation methods for aggregation and fragmentation can be designed upon the same principles as the one-dimensional ones. However, the CPU time required by discretisation methods increases exponentially with the number of dimensions. As a result, there have been relatively few works adopting this approach. A two-dimensional discretised PBE for aggregation and sintering has been developed by Xiong and Pratsinis (1993), with particle volume and surface area being the independent variables of the distribution. The method is based on the principles of two-dimensional quadrature methods. Another study employing discretisation in multiple dimensions for an aggregation problem is that of Campos and Lage (2003), which is based on an extension of the moving grid approach by Kumar and Ramkrishna (1997) (cf. Section 4.3.5).

Monte Carlo Methods

Monte Carlo methods are naturally adapted to multidimensional problems; in fact, this is one of their main advantages due to the fact that the computational requirements scale only linearly with dimensionality. The main algorithm, outlined in Section 4.2.9, is extended to multiple dimensions in a straightforward manner, while the majority of applications mentioned in that section involve multidimensional problems.

4.2.11 Coupled PBEs

The coupled PBE approach was introduced in Section 2.4.2 as an alternative means of modelling aggregate morphology. It introduces additional equations, rather than dimensions, and is thus amenable to solution via discretisation methods and coupling with CFD. We will now show how a discretisation method can be adapted to the coupled PBE formulation. The combination of the two-PBE formulation and the conservative finite volume method will be employed in the case studies in Chapter 6, in particular for soot (Section 6.2) and for silica nanoparticle synthesis (Section 6.1). However, as we will see below, the present methodology can be applied to any discretisation method in a straightforward manner.

In Section 2.4.2, we derived two equations for the coupled PBE model, Eqs. 2.33 and 2.34. Of these, the former is a standard aggregation PBE and can be addressed with any of the discretisation methods previously described. The latter is an equation for the number of primary particles of volume v, denoted by $n_p(v,t)$. The particle volume domain is discretised into a number of intervals, with each of them (denoted by i) covering a volume range (v_{i-1}, v_i) and featuring uniform number densities n_i and $n_{p,i}$ of aggregates and primary particles, respectively. Also, let $n_{av,i}$ denote the average number of primary particles per aggregate in interval i, that is:

$$n_{av,i} = \frac{n_{p,i}}{n_i}. \tag{4.78}$$

Consider now a particle at interval j colliding with a particle at interval k to form a new particle at interval i. The change in the number of aggregates dn_i and the change in the number of primary particles $dn_{p,i}$ are then related as follows:

$$dn_i = -dn_j = -dn_k, \tag{4.79}$$

$$dn_{p,i} = -n_{av,j}dn_j + -n_{av,k}dn_k. \tag{4.80}$$

Thus, once the change in the number density of aggregates, dn_i, has been calculated by means of any discretisation method, the change in primary particle number density, $dn_{p,i}$, can be readily calculated. The coupled PBE approach for aggregation, therefore, does not require substantially more computational effort than solving a single PBE.

Finally, the discretised form of the term corresponding to the fusing of primary particles in the PBE for the primary particles can be written as follows:

$$S_{f,i} = \left.\frac{dn_{av,i}}{dt}\right|_f n_i = -\frac{3}{\tau_f}\left[\frac{n_{p,i}}{n_i} - \left(\frac{n_{p,i}}{n_i}\right)^{2/3}\right] n_i, \tag{4.81}$$

where the models for the characteristic timescale τ_f have been discussed in Section 3.2.3.

4.3 Population Balance with a Growth Term

Processes such as surface growth, soot oxidation and droplet evaporation are referred to collectively as growth in the present book (in some of them, the growth is negative, i.e. the particles are shrinking). All of these phenomena are represented in the PBE via a term that denotes a continuous change in the size of the particle.

The presence of a growth term gives to the PBE the character of a hyperbolic partial differential equation, in which the other kinetic processes appear as source terms. In the present section, we will deal with PBEs that include a growth term, either on its own or combined with other kinetic processes. In the latter case, the approach undertaken is usually an augmentation of the methods for aggregation and fragmentation that were discussed in the previous section with those for treating the growth term.

4.3.1 General Features

Forms of the Growth PBE

In some cases, such as aerosols and soot, surface phenomena may involve the addition or removal of a monomer, such as a molecule. In those cases, growth can be described with the discrete PBE, but the range of sizes involved can be enormous. In other problems, such as the evaporation of droplets and the dissolution of bubbles, a monomer cannot even be defined. Therefore, in the present section we will consider only the continuous PBE with a growth term (with the exception of Monte Carlo methods, which can be based on the discrete PBE).

In many problems such as crystallisation and aerosol formation, growth is accompanied by nucleation, as both processes are brought about by supersaturation. While nucleation is sometimes approached as growth at the zero-volume boundary, in the present book it will be considered as a source term at a given nuclei volume v_0.

With the above considerations, the form of the PBE that will serve as the basis for the present section is:

$$\frac{\partial n(v,t)}{\partial t} + \frac{\partial (G(v)n(v,t))}{\partial v} = B\delta(v - v_0) + S(v,n), \tag{4.82}$$

where the source term, $S(v,n)$, may include aggregation and/or fragmentation.

In Chapter 2 it was noted that, when aggregation and fragmentation are not present, it is advantageous to formulate the PBE in terms of a linear size dimension, such as particle diameter or radius. The PBE for nucleation and growth in terms of a linear dimension, ℓ, is Eq. 2.24, shown also here for reference:

$$\frac{\partial n(\ell,t)}{\partial t} + \frac{\partial (G_\ell(\ell)n(\ell,t))}{\partial \ell} = B\delta(\ell - \ell_0), \tag{4.83}$$

where ℓ_0 is the size of the nucleus and G_ℓ is the growth along the linear dimension.

Solution Methods

Without the presence of aggregation and fragmentation, the growth PBE is a hyperbolic first-order partial differential equation that has the form of an advection equation. Nucleation, deposition and inflow/outflow (in the case of a continuous perfectly stirred reactor) appear as sources or sinks. Analytical or semi-analytical solutions can be sought using the *method of characteristics*, which converts a PDE into a set of ODEs.

Numerical solution is often required, however, when other kinetic processes are present or when a growth problem is coupled with mass and energy balances that may feature nonlinear reaction source terms. For problems with aggregation and/or fragmentation, the solution methods are largely based on the techniques of Section 4.2, which are extended to accommodate the movement of the distribution due to the advection-type growth term. The numerical solution of advection equations poses several computational challenges, which have been investigated in an extensive literature, particularly in the context of gas dynamics. Some of the methods that will be shown here have their origin in that literature.

4.3.2 Analytical Solutions

Growth and Nucleation-Growth

The analytical method for solving first-order hyperbolic equations is the method of characteristics, a detailed account of which can be found, for example, in Leveque (2002). To apply this method, we first rewrite Eq. 4.82 as follows:

$$\frac{\partial n(v,t)}{\partial t} + G(v)\frac{\partial n(v,t)}{\partial v} = -n(v)\frac{dG(v)}{dv} + S(v,n), \tag{4.84}$$

which is called the *non-conservation form* of the equation (as opposed to the *conservation form* where the product $G(v)n(v,t)$ is inside the derivative[7]). The first term on the right-hand side would be zero in the case of size-independent growth. Physically, the equation describes the advection of a population of particles with distribution $n(v,t)$ alongside the particle size domain with velocity $G(v)$, while experiencing a source term that may include the effect of size-dependent growth. We now formulate the problem in a Lagrangian perspective, that is following the particles as they 'move' along the size domain due to growth. The trajectory of the particles is called a *characteristic curve*, $s(v,t)$, and is defined by:

$$\frac{dv}{ds} = G(v), \tag{4.85a}$$

$$\frac{dt}{ds} = 1. \tag{4.85b}$$

By applying the chain rule and using Eqs. 4.85, we have:

$$\frac{dn(s)}{ds} = \frac{\partial n(v,t)}{\partial t} + G(v)\frac{\partial n(v,t)}{\partial v}. \tag{4.86}$$

Eq. 4.84, therefore, becomes:

$$\frac{dn(s)}{ds} = -n(v)\frac{dG(v)}{dv} + S(v,n). \tag{4.87}$$

Thus the change in the distribution along the characteristic curve is described by an ODE rather than a PDE. Solution of this ODE subject to appropriate initial and boundary conditions yields the evolution of the distribution in this moving frame, from which the evolution in the stationary frame (i.e. the original v, t coordinates) can be constructed.

For population balance problems without aggregation and fragmentation, where the PBE is typically formulated in terms of a linear size dimension (Eq. 4.83), a considerable simplification arises when the linear growth rate is size-independent.

[7] This term can be interpreted as the *growth flux* (cf. Section 4.3.4). See also Section A.2 about the conservation and non-conservation forms of the Navier–Stokes equations.

This situation arises often, as it implies that the volumetric growth rate is proportional to surface area (cf. Section 2.3.2). In crystallisation, the independence of the linear growth rate from the particle size is called McGabe's ΔL law (Randolph and Larson, 1988) and is observed in several problems. The simplest case is for size-independent growth and no source term, which can describe the growth of crystals from seeds:

$$\frac{\partial n(\ell,t)}{\partial t} + G_\ell \frac{\partial n(\ell,t)}{\partial \ell} = 0, \tag{4.88}$$

with the initial condition $n(\ell,0) = n_0(\ell)$. To solve this equation, we apply the following change of coordinates:

$$\frac{dv}{ds} = G_\ell, \tag{4.89a}$$

$$\frac{dt}{ds} = 1. \tag{4.89b}$$

The equation is readily transformed into:

$$\frac{dn(s)}{ds} = 0, \tag{4.90}$$

which indicates that the initial distribution is unaltered while it moves along the size domain with a constant velocity G_ℓ. The solution, therefore, is:

$$n(\ell,t) = n_0(\ell - G_\ell t). \tag{4.91}$$

Fig. 4.4 shows the initial condition and an instance of the solution.

Another important case is that of nucleation and size-independent growth, for which the equation is:

$$\frac{\partial n(\ell,t)}{\partial t} + G_\ell \frac{\partial n(\ell,t)}{\partial \ell} = B\delta(\ell - \ell_0), \tag{4.92}$$

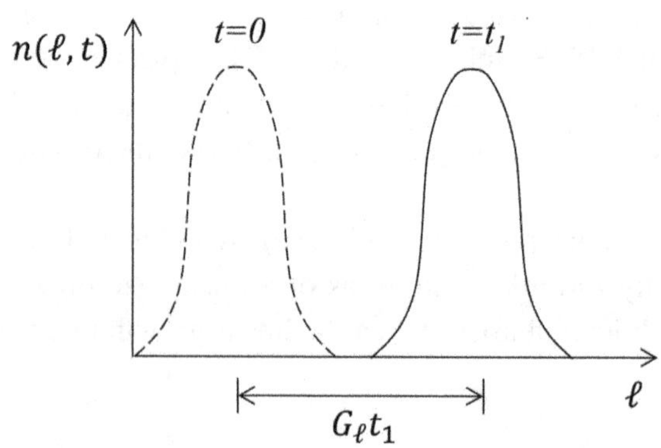

Figure 4.4 Solution of the PBE for size-independent growth.

where ℓ_0 is the size of the nucleus. Together with the following initial condition:

$$n(v, 0) = 0, \quad v > 0. \tag{4.93}$$

The solution is:

$$n(\ell, t) = \frac{n_0}{G_\ell} H(t - \frac{\ell}{G_\ell}), \tag{4.94}$$

where H stands for the Heaviside step function.

An important case is that of steady-state nucleation and size-independent growth. In the crystallisation literature, this formulation is called the mixed-suspension mixed-product removal (MSMPR) model (Randolph and Larson, 1988). It is a model that may be conveniently used for experiments on kinetics extraction (cf. Section 3.5.2). Assuming no particles present at the inlet, the equation is Eq. 3.79. Nucleation may be conveniently be represented as a boundary condition, as in Eq. 3.80. The solution is:

$$n(\ell) = n_0 \exp\left(-\frac{\ell}{G_\ell \tau}\right), \tag{4.95}$$

where τ is the residence time.

Randolph and Larson (1988) derived a solution for steady state nucleation and size-dependent growth, the latter being described by the following function employed in crystallisation:

$$G_\ell = G_0(1 + \gamma\ell)^b, \qquad b < 1. \tag{4.96}$$

This function involves two empirical parameters, γ and b. The equation to be solved, therefore, is:

$$\frac{d}{d\ell}\left[G_0(1 + \gamma\ell)^b n(\ell)\right] + \frac{n(\ell)}{\tau} = 0, \tag{4.97}$$

with the boundary condition given again by Eq. 3.80. The solution is:

$$n(\ell) = K n_0 (1 + \gamma\ell)^{-b} \exp\left[-\frac{(1 + \gamma\ell)^{1-b}}{G_0 \tau \gamma (1 - b)}\right], \tag{4.98}$$

where

$$K = \exp\left[\frac{1}{G_0 \tau \gamma (1 - b)}\right]. \tag{4.99}$$

As mentioned in Section 4.3.1, while the one-dimensional nucleation and growth equation can be solved analytically in many cases, more complex problems with further kinetic processes and coupling with species and energy transport usually require numerical methods. Nevertheless, the concept of the method of characteristics is underpinning some of the numerical methods that will be shown in the following sections.

Aggregation-Growth

Analytical solutions for combined aggregation-growth problems have been derived by Ramabhadran et al. (1976) and Peterson et al. (1978) with the method of Laplace transforms. We show here, without derivation, one of the solutions of Ramabhadran et al. (1976) that has often been employed for validating numerical methods. The solution is for the following PBE with a linear size-dependent growth rate and aggregation with the constant kernel, together with an exponential initial condition:

$$\frac{\partial n(v,t)}{\partial t} + \frac{\partial (k_g v n(v,t))}{\partial v} = \frac{1}{2}\int_0^v \beta_0 n(w,t) n(v-w,t) dw - n(v,t)\int_0^\infty \beta_0 n(w,t) dw, \tag{4.100a}$$

$$n(v,0) = \frac{N_0}{v_0}\exp\left(-\frac{v}{v_0}\right), \tag{4.100b}$$

where N_0 is the initial total number of particles and v_0 is the initial average particle volume.

The solution is:

$$n(v,t) = \frac{m_0^2(t)}{m_1(t)}\exp\left(-\frac{m_0(t)}{m_1(t)}v\right), \tag{4.101}$$

where the moments are given in terms of the following expressions:

$$m_0(t) = \frac{2N_0}{2+\beta_0 N_0 t}, \tag{4.102a}$$

$$m_1(t) = N_0 v_0 \exp(\sigma_1 t). \tag{4.102b}$$

Fig. 4.5 shows this analytical solution for $\beta_0 = 1$ and $k_g = 0.5$, as well as a comparison with the results of the conservative finite volume method discussed in Section 4.2.8 with 80 intervals (the discretisation of the growth term in the context of that method will be presented in Section 4.3.4). This case is good for testing numerical methods because it combines rapid movement of the distribution due to aggregation with a sharp front that develops owing to the growth term.

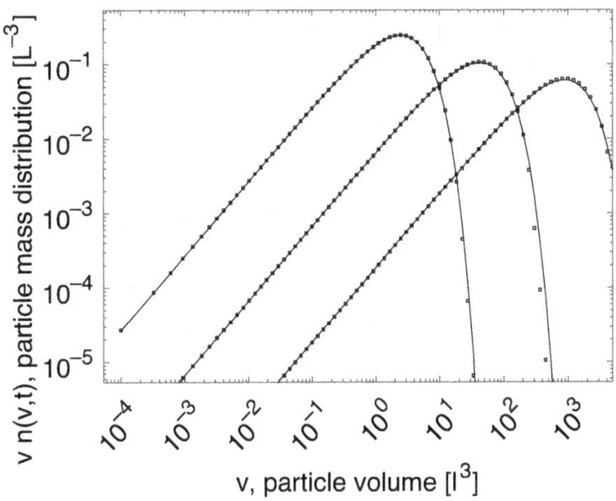

Figure 4.5 The analytical solution of Ramabhadran et al. (1976), Eq. 4.101 compared with a numerical solution with the conservative finite volume method (cf. Section 4.2.8) at different time instants. Reproduced from O'Sullivan and Rigopoulos (2022) under CC-BY 4.0 license.

4.3.3 Moment Methods

Moment Equations for the PBE with a Growth Term

Moment methods were among the first approaches used for solving the nucleation-growth PBE (Hulburt and Katz, 1964). Following the same procedure as in Section 4.2.4, the moment equations for the PBE with a growth term are obtained by multiplying Eq. 4.82 by v^k and integrating:

$$\frac{dm_k}{dt} + \int_0^\infty v^k \frac{\partial (G(v)n(v,t))}{\partial v} dv = \int_0^\infty v^k B\delta(v - v_0)dv + \int_0^\infty v^k S(v,n)dv, \tag{4.103}$$

or:

$$\frac{dm_k}{dt} - k\int_0^\infty v^{k-1} G(v)n(v,t)dv = v_0^k B + \int_0^\infty v^k S(v,n)dv. \tag{4.104}$$

We thus see that, as in the case of aggregation and fragmentation, the moment transformation of the growth term results in an unclosed, in general, system of equations. The ability to bring it to a closed form depends on the growth function. Below, we look first at cases where a closed form can be obtained.

Closed Forms

A closed form can be obtained when the PBE is expressed in terms of a linear size dimension, aggregation and fragmentation are not present and growth has a linear dependence on size. By multiplying Eq. 2.23 by ℓ^k and integrating, we obtain the following moment-transformed PBE:

$$\frac{\partial m_k(t)}{\partial t} + \int_0^\infty \ell^k G_\ell(\ell) \frac{\partial n(\ell, t)}{\partial \ell} d\ell = \ell_0^k B, \tag{4.105}$$

where ℓ_0 is the size of the nuclei. For a linear size-dependent growth function, $G_\ell(\ell) = G_0(1 + \alpha\ell)$, we now have:

$$\frac{dm_k(t)}{dt} - kG_0\alpha m_k(t) - kG_0 m_{k-1}(t) = \ell_0^k B, \qquad k \geq 1, \tag{4.106a}$$

$$\frac{dm_0(t)}{dt} = B, \tag{4.106b}$$

where it is assumed that growth is positive (i.e. there is no evaporation of nuclei). Typically, the first four moments (zero to third) are solved for, standing for the total number, length, surface area and volume of the particles.

For other size-dependent growth functions, however, closure assumptions are, in general, needed to arrive at a closed set of equations. The closure methods are based on the same approaches that were presented in Section 4.2.5 in the context of aggregation. Their extensions to growth will be briefly described below.

Closure Methods

Closure is readily achieved if the shape of the distribution can be presumed. The use of the log-normal distribution for aerosols was discussed in Section 4.2.5. Williams (1986) and Pratsinis (1988) discuss moment methods for condensational growth of aerosols, in addition to other kinetic processes such as aggregation and nucleation. Hulburt and Katz (1964) proposed closure using a series expansion.

The MOMIC method can be applied to obtain closure of the growth term with a straightforward application of the interpolation concept on which it is based (cf. Section 4.2.5); the respective equations can be found in Frenklach (2002a).

The QMOM method was originally proposed for the treatment of the growth term (McGraw, 1997). The fundamentals of this method were explained in Section 4.2.5; in brief, the unclosed integral is approximated with Gaussian quadrature, with the number density being regarded as the weight function. Hence, the approximation for the unclosed growth term is:

$$k \int_0^\infty v^{k-1} G(v) n(v, t) dv \approx k \sum_{i=1}^{n} v_i^{k-1}(t) W_i(t) G(v_i), \tag{4.107}$$

where the abscissas, v_i, and weights, W_i, both change with time and are to be determined with a matrix inversion process. McGraw (1997) shows an example of an application to diffusion controlled growth. As mentioned in Section 4.2.5, a later method based on the QMOM concept is DQMOM, where differential equations are solved directly for the quadrature parameters, thus eliminating the need for the matrix inversion. A case of multidimensional growth solved with DQMOM was shown in the paper with the first presentation of the method (Marchisio and Fox, 2005). DQMOM and other QMOM variants are comprehensively discussed by Marchisio and Fox (2013).

4.3.4 Discretisation Methods: Fixed Grid

Overview

The discretisation of the growth term poses a different computational problem from that of aggregation and fragmentation: as it is a first order derivative, it can lead to sharp fronts. The most extreme case is that of size-independent growth with no aggregation or fragmentation. In Section 4.3.2, we saw that the analytical solution of the growth equation in that case is one where the initial distribution propagates unaltered along the size domain. The use of low-order schemes can result in considerable *numerical diffusion*, which results in a smearing of the shape of the distribution. On the other hand, straightforward high-order schemes result in *numerical dispersion* which manifests itself in spurious oscillations and can even be unstable.

In the present section, we consider fixed grid methods. Pointwise methods employ custom schemes that ensure conservation of the moments, in accordance with the overall spirit of these methods. In finite element and finite volume methods, high-resolution schemes have been proposed to counter numerical diffusion. The following sections will consider schemes that combine discretisation with the method of characteristics (cf. Section 4.3.2) and adaptive grids.

Discretisation of the Growth Term in Pointwise Methods

Pointwise approximation methods were introduced in Section 4.2.6 as methods for solving aggregation and fragmentation problems. It was shown that these methods, which approximate the distribution in a semi-discrete manner as a sum of delta functions, place emphasis on the conservation of a number of selected moments rather than on the accurate prediction of the distribution itself. The treatment of the growth term in the context of these methods is in the same spirit: equations are solved that ensure that selected moments are conserved.

In the method of Hounslow et al. (1988), an approximate expression is proposed that relates the growth term at a given node with the number densities at the nodes before and after. This expression introduces three unknown parameters, each associated with one of these nodes. By taking the moment transformation of the PBE and introducing in it this approximate expression, an equation is obtained for the conservation of an arbitrary moment that involves these unknown parameters. Thus, three chosen moments can be conserved by solving three of these equations to obtain expressions for the parameters. Hounslow et al. (1988) chose to conserve the zeroth, first and second moment with respect to particle length and the resulting expressions can be found in that work. In accordance with the overall objective of the method, this scheme ensures the conservation of the selected moments; however, the predicted distribution can exhibit considerable numerical diffusion and oscillations. A discussion of this and other early methods can be found in Kostoglou and Karabelas (1995).

In the original work of Kumar and Ramkrishna (1996a), growth was not considered, while later work by the same authors (Kumar and Ramkrishna, 1997) proposed a method belonging to the category of characteristics-based approaches that will be described in Section 4.3.5. Park and Rogak (2004) combined the original Kumar and Ramkrishna method with a three-point, three-parameter scheme derived using the principle of the Hounslow et al. (1988) approach, for the PBE expressed in terms of particle volume. However, rather than conserving three moments, they chose to conserve two (the zeroth and the first moment) and leave one parameter free. This free parameter is adjusted according to the local slope of the distribution, in order to obtain a balance between stability and mitigation of numerical diffusion. With respect to this feature, the method is similar to the high-resolution schemes that will be discussed below in the context of the finite volume method.

Finite Volume Discretisation of the PBE with a Growth Term

We now consider fixed grid schemes for the growth term that can be implemented within the framework of finite element or finite volume methods and combined with the methods discussed in Sections 4.2.7 and 4.2.8 for aggregation and fragmentation. In the following discussion, the finite volume method will be employed as a generic framework for implementing these schemes.[8]

As a starting point, we consider the discretisation of Eq. 4.82 with the finite volume method. Integrating Eq. 4.82 over an interval, we obtain:

$$\int_{v_{i-1}}^{v_i} \frac{\partial n(v,t)}{\partial t} dv + \int_{v_{i-1}}^{v_i} \frac{\partial (G(v)n(v,t))}{\partial v} dv = \int_{v_{i-1}}^{v_i} B\delta(v - v_0) dv + \int_{v_{i-1}}^{v_i} S(v,n) dv. \tag{4.108}$$

The treatment of the aggregation and fragmentation source term has been covered in Sections 4.2.6–4.2.8, while nucleation is straightforward: if the nucleation rate

[8] For problems with growth only, the schemes discussed below can also be implemented in finite difference methods, as in Ma et al. (2002).

is given by $B\delta(v - v_{nuc})$ and the nuclei volume is located within the first interval, (v_{min}, v_1), we have:

$$\int_{v_{min}}^{v_1} B\delta(v - v_{nuc})dv = B. \tag{4.109}$$

The following discussion will thus concentrate on the discretisation of the growth term.

Upwind Schemes and Basic Concepts

The product $G(v)n(v,t)$ can be interpreted as a flux of particles along the particle volume domain and will be called the *growth flux*. The integral of the growth term can then be written as follows:

$$\int_{v_{i-1}}^{v_i} \frac{\partial(G(v)n(v,t))}{\partial v} dv = F_{g,i} - F_{g,i-1}, \tag{4.110}$$

where $F_{g,i-1}$ and $F_{g,i}$ are approximations of the growth flux at v_{i-1} and v_i, respectively. This expression is identical to the one obtained by the discretisation of the convection term in the scalar transport equation in a finite volume CFD code.

To proceed, we need to provide expressions for approximating the growth fluxes. If a zeroth-order approximation is employed, that is the distribution is represented as a histogram with a single value, n_i, over each interval (which is the case with most finite volume methods), then the number density changes discontinuously at interval boundaries and an expression must be sought for the growth fluxes in terms of the neighbouring interval values.

The PBE with a growth term is a first-order advection equation without a diffusion term. The features of such equations with respect to numerical solution are extensively discussed by Leveque (2002). One of the most important aspects is that, to maintain stability, *upwind* schemes are required. Such schemes use information from upstream nodes for the evaluation of the growth term, where 'upstream' refers to the direction of the advection velocity – in case of the PBE, the growth rate. Physically, this depicts the fact that the transported quantity moves in a single direction – in the case of positive growth, for example, particles grow only from smaller to larger sizes. On the other hand, the use of the central differencing scheme (which uses information from both sides and is typical for diffusion processes) results in instability.

In order to describe the following numerical schemes, we consider a zeroth-order stepwise discretisation of the number density, such that interval i has a range (v_{i-1}, v_i) and a constant value of the number density, n_i. This discretisation and the associated growth fluxes are shown in Fig. 4.6.

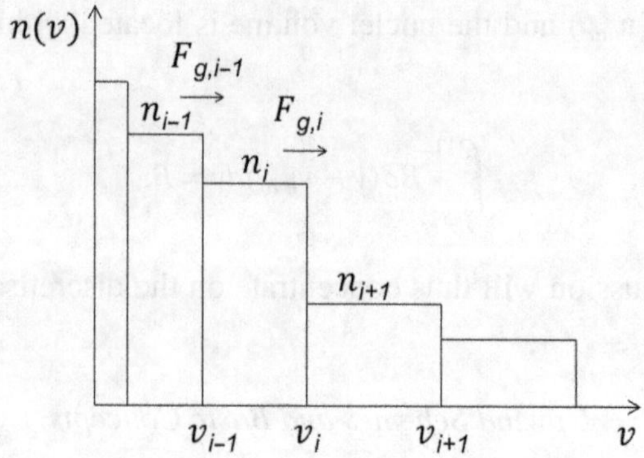

Figure 4.6 Zeroth-order approximation of the distribution and notation for the derivation of growth discretisation schemes.

The simplest upwind scheme is the *first-order upwind scheme*, which evaluates the growth flux using the upstream value:

$$F_{g,i} = G(v_i)n_i. \tag{4.111}$$

The application of this scheme yields the following approximation of the growth-term integral:

$$\int_{v_{i-1}}^{v_i} \frac{\partial(G(v)n(v,t))}{\partial v} dv = G(v_i)n_i - G(v_{i-1})n_{i-1}. \tag{4.112}$$

While this scheme guarantees stability, it causes considerable numerical diffusion. This can be seen in Fig. 4.7, which compares several methods applied to the solution of an equation expressing advection of a unit step.[9] This is the most stringent test for a numerical method for advection problems, as it consists of a discontinuity propagating along the domain. Such a discontinuity can occur in size-independent growth with no aggregation or fragmentation and an initial monodisperse distribution or a nucleation source (e.g. as in Eq. 4.94). The solutions in Fig. 4.7 utilise a very small number of nodes (20) and so the smearing of the front with fixed grid methods is inevitable, but it can be seen that the first-order upwind scheme features the highest level of smearing of the front among the methods compared.

An important necessary (but not sufficient) condition for stability of numerical schemes for advection problems is the *Courant–Friedrichs–Lewy (CFL) condition*, a detailed discussion of which can be found in Leveque (2002). In CFD codes, the CFL condition is employed in the form of the *Courant number* that relates velocity,

[9] To obtain the results shown in this figure, the first-order scheme was implemented with a collocation rather than a finite volume method, but this has no impact on the point made here about numerical diffusion.

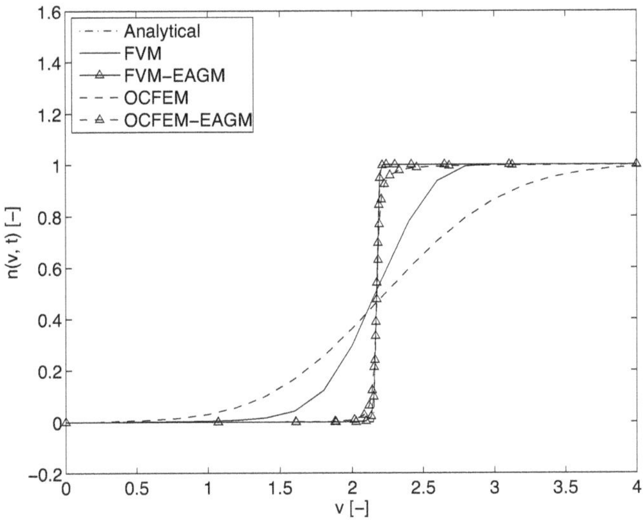

Figure 4.7 Comparison of analytical solution for the advection of a unit step with numerical solutions obtained with fixed grid schemes and with EAGM. The methods are: orthogonal collocation on finite elements with a first-order upwind scheme (OCFEM), finite volume with the TVD scheme of Koren (1993) and Qamar et al. (2006) (FVM), and both of the above with EAGM (OCFEM–EAGM and FVM–EAGM). 20 finite volume cells/finite elements are employed. Reprinted from Sewerin and Rigopoulos (2017a) with permission from Elsevier.

cell width and time step. Since the growth rate assumes the role of velocity in the PBE, an equivalent Courant number can be defined for the PBE as follows:

$$C_{PBE} = \frac{G\delta t}{\delta v}. \tag{4.113}$$

The PBE CFL condition is that $C_{PBE} \leq 1$, if explicit time integration is employed, and should be borne in mind with respect to the choice of time step. It has been used in the context of PBE discretisation by Ma et al. (2002) and Gunawan et al. (2004), who proposed some high-resolution schemes that will be discussed below.

High-Resolution Schemes

The reason for numerical diffusion in the first-order upwind scheme is the fact that the first-order upwind scheme represents, in fact, the solution to an advection-diffusion problem, as can be shown with a Taylor series expansion. High-resolution schemes counter this by adding an anti-diffusion term. They have been developed primarily for use in gas dynamics, and more details on their origins and derivation can be found in Leveque (2002). Their use in the context of the PBE was pioneered by Ma et al. (2002) and Gunawan et al. (2004). Qamar et al. (2006) compared several such schemes and adopted for the PBE a scheme based on the flux limiter of Koren (1993) that has been used by several subsequent works, including the case studies that will be discussed in Sections 6.2 and 6.3. For this reason, the equations for this scheme are summarised below.

The basic equation for the flux is:

$$F_{g,i} = G(v_i)\left[n_{u(v_i)} + \frac{\phi(r_i)}{2}\left(n_{u(v_i)} - n_{uu(v_i)}\right)\right], \tag{4.114}$$

where $n_{u(v_i)}$ and $n_{uu(v_i)}$ are the number densities of the two intervals upstream of node v_i, while $\phi(r_i)$ is the *flux limiter* at node v_i given by:

$$\phi(r_i) = \max\left[0, \min\left[2r_i, \min\left(\frac{1}{3}\frac{v_i - v_{i-1}}{v_{m,i} - v_{m,i-1}} + \frac{2r_i}{3}\frac{v_i - v_{i-1}}{v_{m,i+1} - v_{m,i}}, 2\right)\right]\right], \tag{4.115}$$

where $v_{m,i} = 0.5(v_i + v_{i-1})$ and r_i is the upwind ratio of two consecutive number density gradients (Koren, 1993):

$$r_i = \frac{n_{i+1} - n_i + \epsilon}{n_i - n_{i-1} + \epsilon}, \tag{4.116}$$

where ϵ is a small number employed to avoid division by zero.

Fig. 4.7 shows that this scheme considerably reduces the numerical diffusion caused by the first-order upwind scheme. While some smearing of the front is still present, it is inevitable due to the small number of nodes (20) and, furthermore, the unit step is the most stringent test for an advection discretisation scheme. Fig. 4.5 shows the simulation of an aggregation-growth case and a comparison with the analytical solution of Ramabhadran et al. (1976) (constant aggregation kernel with $\beta(v,w) = 1$, linear size-dependent growth kernel with $G(v) = 0.5v$). The simulation employs the conservative finite volume of Liu and Rigopoulos (2019) and O'Sullivan and Rigopoulos (2022) described in Section 4.2.8 for the aggregation terms and the Qamar et al. (2006) scheme for growth. 80 intervals are employed but relatively few are placed at the front at each time instant, yet the front is still reasonably well captured while the distribution evolves over 8 orders of magnitude.

Growth Discretisation in Mass-Conservative Schemes

The inclusion of growth in the conservative finite volume method for aggregation and fragmentation described in Section 4.2.8 requires some special attention because that method involves multiplying the PBE by v. If a growth term is included, the integration over an interval results in:

$$\frac{d}{dt}\int_{v_{i-1}}^{v_i} vn(v,t)dv + \int_{v_{i-1}}^{v_i} v\frac{\partial(G(v)n(v,t))}{\partial v}dv = \int_{v_{i-1}}^{v_i} vSdv, \tag{4.117}$$

where S is a source term accounting for nucleation, aggregation and fragmentation. The growth term can now be further elaborated as follows:

$$\int_{v_{i-1}}^{v_i} v\frac{\partial(G(v)n(v,t))}{\partial v}dv = v_iF_{g,i} - v_{i-1}F_{g,i-1} - \int_{v_{i-1}}^{v_i} n(v,t)G(v)dv. \quad (4.118)$$

It can be seen that, apart from the two growth fluxes, a new term appears, which can be regarded as a growth source term and is due to the change of particle volume due to growth within interval Δv_i.

An alternative way of dealing with growth in mass-conservative schemes is to employ a fractional step method (cf. Section 4.4.4) to split the PBE and treat each term in its own step, similar to the splitting in the Vervisch et al. (2023) approach shown in the next section. The growth term can then be treated in its original form (i.e. without multiplying by v). In many cases, this approach is preferable because the form of the growth term may not permit a straightforward treatment of Eq. 4.118; this is particularly the case when dealing with fractal aggregates. The mass-conservative discretisation of the growth term can also cause numerical problems in the case of size-independent growth, although in these cases the PBE is usually formulated in terms of particle length and aggregation is not present, hence there is no need to multiply by v.

4.3.5 Discretisation Methods: Characteristics-Based

In Section 4.3.2, the method of characteristics was shown to provide analytical solutions to growth PBEs. The elements of this approach can be combined with discretisation to build methods that overcome the numerical diffusion problem.

Gelbard (1990) introduced the *moving sectional method*. The main concept of this method is that the boundaries of the discretisation intervals (sections) are moving according to the method of characteristics:

$$\frac{dv_i}{dt} = G(v_i). \quad (4.119)$$

As a result, the distribution moves smoothly without the effect of numerical diffusion. Kim and Seinfeld (1990) extended this approach to multicomponent aerosol dynamics (cf. Section 4.3.8). Kumar and Ramkrishna (1997) combined this concept with their earlier approach to aggregation and fragmentation (Kumar and Ramkrishna, 1996a,b) and also proposed a method for treating nucleation, which is an issue in this approach due to the fact that the sections move away from the nucleation range. Another group of works (Tsang and Brock, 1983; Tsang and Rao, 1990; Mahoney and Ramkrishna, 2002) have utilised a moving finite element method, also using the concept of the method of characteristics.

All of the aforementioned methods involve a moving grid, in which they resemble the adaptive grid methods to be discussed in the next section. They differ from those methods, however, in that the change of the grid is dictated by the growth kinetics rather than by the shape of the solution. As such, the grid changes in the characteristics-based methods are specific to growth-dominated problems. Both classes of methods face a problem when it comes to coupling with CFD: if different PBE grids are used at different spatial locations, the treatment of the transport terms is not straightforward.

Vervisch et al. (2023) proposed a characteristics-based method that does not involve a moving grid. The non-conservation form of the growth equation, Eq. 4.84, is taken as the starting point and split into two equations. This splitting is based on the concept of fractional steps that will be described in Section 4.4.4 in the context of the coupling of PBE with CFD. The equations are:

$$\frac{\partial n(v,t)}{\partial t} + G(v)\frac{\partial n(v,t)}{\partial v} = 0, \tag{4.120a}$$

$$\frac{\partial n(v,t)}{\partial t} = -n(v)\frac{dG(v)}{dv} + S(v,n). \tag{4.120b}$$

The first of these equations can be readily solved as shown in Section 4.3.2, to yield the following:

$$n^*(v) = n(v - G\delta t, t). \tag{4.121}$$

Here, $n^*(v)$ is an intermediate value to be used in the next fractional step, which is the integration of Eq. 4.120b:

$$n(v, t+\delta t) = \int_t^{t+\delta t}\left(-n^*(v)\frac{dG(v)}{dv} + S(v,n^*)\right)dt. \tag{4.122}$$

For an aggregation-growth problem, the second fractional step (Eq. 4.120b) can be split into two further fractional steps. Vervisch et al. (2023) showed a way for treating the aggregation death term analytically in the context of this approach. Owing to its linearity, this term be combined with the ndG/dv term and treated in the same fractional step as follows:

$$\frac{\partial n(v,t)}{\partial t} = -n(v)\frac{dG(v)}{dv} - n(v,t)\int_0^\infty \beta(v,w)n(w,t)dw, \tag{4.123}$$

the solution to which is:

$$n^{**}(v) = n^*(v)\exp\left[-\left(\frac{dG(v)}{dv} + \int_0^\infty \beta(v,w)n^*(w)dw\right)\delta t\right]. \tag{4.124}$$

The result, $n^{**}(v)$, can be used for the solution of the next fractional step:

$$\frac{\partial n(v,t)}{\partial t} = S(v,n), \tag{4.125}$$

where the source term includes the aggregation birth as well as nucleation, if present. The integration of this equation yields $n(v, t + \delta t)$.

This procedure is a fractional step method for integrating in time which is first-order accurate. Its applications, however, rests on the ability to employ Eq. 4.121, which is not straightforward if a zeroth-order approximation of the distribution is employed. Therefore, the distribution is approximated within each interval with high-order polynomials (fifth-order were used by Vervisch et al., 2023). When it comes to coupling with CFD, this approach does not share the problem of different PBE grids used throughout the domain. However, the high-order polynomials increase considerably the number of dependent variables, thus rendering the approach potentially expensive. In Vervisch et al., 2023, it is recommended that a data-driven method (artificial neural networks, cf. Section 4.4.7) is used to reduce it back to one variable per interval and attain coupling with CFD. Another possibility is the local reconstruction of the interpolation points from a zeroth-order approximation at every time step.[10]

4.3.6 Discretisation Methods: Adaptive Grids

Adaptive grid methods overcome the numerical diffusion problem by refining the grid in regions where higher resolution is needed. They differ from characteristics-based moving grid methods in that the grid changes are governed by the shape of the distribution rather than the kinetics of the growth term. As such, they constitute a more general approach that can be employed for PBEs with any combination of kinetic processes, although growth-dominated problems are the ones with the greatest need for resolving local gradients with a highly refined grid. Adaptive grid methods for the PBE have been proposed by several authors, including Lee et al. (1999, 2001) and Qamar et al. (2007). As mentioned in the previous section, however, the main issue with methods that involve a changing grid is their coupling with CFD, as different PBE grids at different spatial locations will have to be somehow matched in order to address spatial transport.

We now describe an adaptive grid method due to Sewerin and Rigopoulos (2017a) that has been developed with prospect for incorporation into CFD codes. Only the basic method will be described here, while the treatment of spatial transport is discussed in Section 4.4.3. In this approach, the grid adaptivity is based on

[10] Vervisch, L., private communication.

the gradient of the distribution. The method is based on a coordinate transformation between a uniform grid in a reference space and the actual grid in the particle size domain, and ensures that the nodes in the latter are distributed in an optimal way so as to capture accurately the steep gradients that arise from the presence of the growth term. The coordinate transformation is marched in time in an explicit manner, in that the evolution of the transformation over a time step is prescribed on the basis of the present transformation and distribution, hence the approach is referred to as an explicit adaptive grid method (EAGM). As will be discussed in Section 4.4.3, this coordinate transformation is key for the coupling of the method with CFD, because it allows the derivation of a transport equation for the transformed number density, in which the grid is uniform everywhere in the domain, while certain new terms arising from the coordinate transformation can be treated as source terms.

Fig. 4.7 shows results obtained with this method for the advection of a unit step, which is the most stringent test case for an advection problem. EAGM has been implemented within two different discretisation approaches – an orthogonal collocation on finite elements with a first-order upwind scheme and a finite volume method with the high-resolution scheme of Koren (1993) and Qamar et al. (2006) and is compared with the same methods without EAGM. All solutions use 20 nodes, which is a very small number for a problem with a discontinuity, so the failure of the fixed grid schemes to provide a sharp gradient is expected. However, by redistributing most of the nodes on the steep front, EAGM manages to follow very closely the steep front of the analytical solution and avoid the numerical diffusion marring the fixed grid schemes.

4.3.7 Monte Carlo Method

In Monte Carlo methods, the PBE is usually formulated in a discrete manner. Growth, positive or negative, can then be accommodated by the addition or removal of a single monomer, as can be seen in the algorithm of Balthasar and Kraft (2003) for modelling soot formation.

A different use of the Monte Carlo concept, in the context of the continuous PBE, was proposed by Bouaniche et al. (2019). In that approach, the aggregation terms in the PBE are treated with a discretisation method, while the Monte Carlo method is used on the growth term to resolve the problem of numerical diffusion. The derivation is based on recasting the PBE into an equation for the probability (rather than number density) and an additional equation for zeroth moment. Subsequently, the probability distribution is represented by an ensemble of stochastic particles which are moved to simulate the effect of the growth term. The aggregation terms are accounted for by distributing the particles to adjacent sections, in accordance with the Kumar and Ramkrishna (1996a) method. The movement of particles is not subject to numerical diffusion, hence the method can reproduce even extreme cases of particle growth such as discontinuous fronts. Like all Monte Carlo methods, convergence of the statistics must be achieved with a sufficiently large number of particles. An application of the method to soot formation in laminar premixed flames can be found in Bouaniche et al. (2020). The coupling of this method with a general CFD code is not straightforward, however, and the authors have put forward a data-driven method (artificial neural networks) for that purpose that will be discussed in Section 4.4.7.

4.3.8 Multidimensional and Coupled PBEs

Multidimensional PBE

The solution of a multidimensional growth PBE can be accomplished with extensions of the approaches discussed so far, and the main concern is that of computational requirements. We first note that some analytical solutions for growth and aggregation in multicomponent aerosols have been derived by Gelbard and Seinfeld (1978b).

Moment methods employ a small number of variables and Monte Carlo methods scale only linearly with dimensionality, hence these methods are not severely affected by the presence of multiple dimensions. With discretisation methods, however, the CPU requirements scale exponentially with dimensionality. The use of an adaptive grid is one way of addressing this issue.

High-dimensional growth problems appear often in crystallisation, as discussed in Section 2.4.1. High-resolution discretisation schemes for nucleation-growth problems have been presented and applied to crystallisation by Ma et al. (2002) and Gunawan et al. (2004). Qamar et al. (2007) proposed an adaptive grid together with a high-resolution scheme and applied it also to crystallisation. Kim and Seinfeld (1990) also employed a moving grid method for the solution of the PBE for a multicomponent aerosol. Simultaneous growth and aggregation with multiple components has been approached by Campos and Lage (2003) with a moving grid approach based on the method of Kumar and Ramkrishna (1997). Furthermore, Mantzaris et al. (2001a,b,c) have explored finite difference, spectral and finite element methods for a biochemical problem of multivariable cell growth.

Coupled PBEs

One of the aims of the coupled PBE approach described in Section 2.4.2 is to provide a more consistent account of surface growth in fractal aggregates, as opposed to the one-PBE approach. In the latter, the volume of the primary particles composing the aggregates is assumed to be 'frozen', even though the entire aggregate is changing its volume due to surface growth. This inconsistency is resolved with the two-PBE approach, where the volume of the primary particles within the aggregates is dynamic.

The numerical treatment of the growth term in the two-PBE approach is based on the principle that the growth of primary particles is related to that of aggregates, and thus an identical growth rate must be employed. For example, in the context of a finite volume method, when aggregates enter interval $i + 1$ due to surface growth, all of the associated primary particles also enter this section. If surface growth is the only kinetic process, then the aggregate volume increases while the number of primary particles maintains the same, leading to an increased primary particle size. The finite volume discretisation of the growth term is thus identical in both PBEs (for aggregates and primary particles) and needs to be computed only once.

4.4 Coupling of PBE with CFD: Non-Inertial and Low-Inertia Particles

The formulation of a spatially dependent population balance problem requires coupling of the PBE with fluid dynamics. The objective of the present section is to show how the solution methods for the PBE presented so far can be combined with the methods of CFD. The section is focussed on the case of non-inertial particles, where the particle velocity is equal to the fluid velocity. The concepts involved are extended in a straightforward manner to the case of low Stokes number inertial particles, where the particle velocity can be obtained as a function of the fluid velocity. Particles of arbitrary inertia require a fundamentally different treatment that will be only briefly discussed in Section 4.5, as it is comprehensively treated in the multiphase flow literature.

The discussion here is limited to the numerical issues related to the coupling of PBE and CFD and does not include the modelling issues associated with the presence of turbulence and its interaction with the population balance, as the latter will be discussed in Chapter 5. Finally, Chapter 6 will show applications of the coupled CFD-PBE approach for soot formation, flame synthesis of nanoparticles and crystallisation.

4.4.1 Overview

The most straightforward way to couple the PBE with a CFD code is via an Eulerian framework, where every cell has a local particle size distribution. One issue with this formulation is numerical diffusion in the spatial domain and the low diffusivity associated with particle transport. Another issue is associated with the use of moving or adaptive grids for the PBE discretisation in growth-dominated problems, which aim to counter numerical diffusion in the particle size domain. While it is easy to change the grid in the numerical solution of a spatially homogeneous

PBE, in CFD one has to account for spatial transport of the distribution, and this is not straightforward if different grids are employed to represent the distribution in different spatial locations.

An alternative is to solve the PBE along a sample of Lagrangian trajectories. This is effectively a stochastic approach, as only a sample of trajectories can be simulated and this should be large enough to ensure convergence. A more difficult problem to overcome in this approach is the coupling of the PBE with the continuous phase. In many problems, the kinetic processes in the PBE bring associated changes in transported variables such as concentrations, temperature and density which should appear as source terms in the corresponding transport equations. If coupling is not a problem, then the PBE may be solved as a post-processing step along the Lagrangian trajectories. This approach is typically used in conjunction with Monte Carlo methods because their coupling with CFD in an Eulerian fashion is too computationally expensive.

Moment methods were the first ones to be used for coupling of PBE with CFD due to the reduced memory requirements that they offer. Discretisation methods are increasingly being used in later years, owing to the increased memory capacity of modern computers. Monte Carlo methods are expensive to use in the context of Eulerian CFD simulations because each spatial cell needs to have a sufficiently large number of particles to approximate the local distribution. However, some attempts have been made to use them in conjunction with Lagrangian methods (Section 4.4.6).

In the Eulerian approach, once a moment or discretisation method has been applied to the PBE, the moments or discretised number densities can be treated as additional scalars in a CFD code, since they are convected by the fluid velocity (in the non-inertial case). A convenient method of implementing the population balance source terms is the *method of fractional steps*, introduced in Section 4.4.4. Moment methods pose some further closure issues that will be discussed in Section 4.4.2. With discretisation methods, coupling is straightforward with fixed grids but further issues arise in the case of adaptive grids and coupled PBEs, which will be discussed in Section 4.4.3.

The solution of the overall problem requires numerical solution of the continuity and Navier–Stokes equations. Furthermore, it may be required to solve transport equations for chemical species and energy, as the particle formation may have feedback effects on these variables. In the case of variable density flows, such effects may also result in density changes and thus have feedback on the flow field. In these cases, the momentum, energy and mass transport equations have to be solved together with the PBE in a coupled manner. Appendix A reviews the main formulations of these equations that can be encountered in the context of coupled CFD-PBE problems.

The numerical solution of the continuity, Navier–Stokes, energy and species transport equations must be accomplished using CFD methods, which will not be covered in the present book, as there are many treatises available (e.g. Hirsch, 2007; Versteeg and Malalasekera, 2007; Ferziger et al., 2020). Furthermore, many population balance problems involve reacting flows whose source terms require special treatment, especially when turbulence and fast reactions are involved. References that discuss methods for this purpose include Bałdyga and Bourne (1999), Peters (2000), Veynante and Vervisch (2002), Fox (2003), Poinsot and Veynante (2005) and Cant and Mastorakos (2008).

As a starting point, we take the spatially dependent PBE for non-inertial particles, Eq. 2.21. For the purpose of the present section, it is convenient to rewrite it with the growth term on the right-hand side:

$$\frac{\partial n}{\partial t} + \frac{\partial (u_i n)}{\partial x_i} = \frac{\partial}{\partial x_i}\left(D_p \frac{\partial n}{\partial x_i}\right) - \frac{\partial (Gn)}{\partial v} + S(n). \tag{4.126}$$

In this form, the equation is compatible with the generic convection-diffusion-source transport equation that is being solved in CFD codes, with growth viewed as part of the source term. For brevity, the dependence of variables on space, time and particle volume will be omitted in this section.

The formulation of the PBE for low Stokes number inertial particles was shown in Section 2.7.3. Written in a similar form as Eq. 4.126, it is:

$$\frac{\partial n}{\partial t} + \frac{\partial (u_{p,i} n)}{\partial x_i} = \frac{\partial}{\partial x_i}\left(D_p \frac{\partial n}{\partial x_i}\right) - \frac{\partial (Gn)}{\partial v} + S(n). \tag{4.127}$$

The difference of this equation from Eq. 4.126 is that the particle velocity in the convection term is not equal to the fluid velocity but can be computed from that, i.e, $u_{p,i} = f(u_i)$, according to Eq. 3.145. This can be implemented in the CFD-PBE code in the form of a correction to the particle convection velocity. Hence, the treatment of this case is based on the same methods as for the non-inertial case.

The following discussion will be based on the one-dimensional PBE, as the extension to the multidimensional PBE with respect to the issues treated here is straightforward, apart from the associated increase in computational time that may be a factor in the choice of method.

4.4.2 Moment Methods

The application of the moment transformation to Eq. 4.126 results in the following:

$$\frac{\partial m_k}{\partial t} + \int_0^\infty v^k \frac{\partial (u_i n)}{\partial x_i} dv = \int_0^\infty v^k \frac{\partial}{\partial x_i}\left(D_p \frac{\partial n}{\partial x_i}\right) dv - \int_0^\infty v^k \frac{\partial (Gn)}{\partial v} dv + \int_0^\infty v^k S dv. \tag{4.128}$$

We have seen (Sections 4.2.4 and 4.3.3) that the terms related to the kinetic processes (growth and source term) are, in general, unclosed, and a closure method such as a presumed distribution approach or QMOM must be applied, except for special cases of kinetic models. In the present section, we focus on the treatment of the extra terms arising from the coupling of the moment-transformed PBE with fluid dynamics.

The convection terms does not pose any further issues, as the derivative and integral can be interchanged:

$$\int_0^\infty v^k \frac{\partial (u_i n)}{\partial x_i} dv = \frac{\partial (u_i m_k)}{\partial x_i}. \tag{4.129}$$

The diffusion term, however, poses an issue due to the dependence of the particle diffusivity on both space and size. Eq. 3.128 yields a d_p^{-1} dependency in the continuum regime, while in the free molecule regime the friction coefficient results in a d_p^{-2} dependency (cf. Eqs. 3.127 and 3.115). In the case of a volume-based distribution and the continuum regime, by writing the diffusivity as $D_p(v, \mathbf{x}, t) = A(\mathbf{x}, t)/v^{1/3}$, the moment transformation results in the following expression:

$$\int_0^\infty v^k \frac{\partial}{\partial x_i} \left(D_p \frac{\partial n}{\partial x_i} \right) dv = \int_0^\infty \frac{\partial}{\partial x_i} \left(v^k \frac{A}{v^{1/3}} \frac{\partial n}{\partial x_i} \right) dv,$$

which can be written as:

$$\int_0^\infty v^k \frac{\partial}{\partial x_i} \left(D_p \frac{\partial n}{\partial x_i} \right) dv = \frac{\partial}{\partial x_i} \left(A \frac{\partial m_{k-1/3}}{\partial x_i} \right). \tag{4.130}$$

We can see that a new fractional moment appears in the diffusion term, which would be $m_{k-2/3}$ in the case of the free molecule regime. In a formulation where the particle diameter is used as the independent variable in the distribution, the new moment appearing would be a lower-order integer moment. Settumba and Garrick (2003, 2004) studied the aggregation of nanoparticles in a mixing layer with a moment method and assumed a log-normal distribution to obtain closure of all unclosed terms. In the second of these works, a comparison was made between a model that accounted for the fractional moments in the diffusion term and a simplified one where the diffusion coefficient was based on the geometric mean volume. The latter allows writing the diffusion term for a given moment, m_k, in the following way:

$$\int_0^\infty v^k \frac{\partial}{\partial x_i} \left(D_p \frac{\partial n}{\partial x_i} \right) dv \approx \frac{\partial}{\partial x_i} \left(D_p' \frac{\partial m_k}{\partial x_i} \right), \tag{4.131}$$

where D_p' is the modified diffusion coefficient based on the geometric mean volume. In their test cases, the differences between the two models were negligible as far as the moments and mean particle size were concerned, although a small but noticeable difference was observed in the geometric standard deviation of the particle size distribution.

If the approximation of Eq. 4.131 is adopted, the moment transport equation can be written as follows:

$$\frac{\partial m_k}{\partial t} + \frac{\partial (u_i m_k)}{\partial x_i} = \frac{\partial}{\partial x_i}\left(D_p' \frac{\partial m_k}{\partial x_i}\right) - \int_0^\infty v^k \frac{\partial (Gn)}{\partial v} dv + \int_0^\infty v^k S dv, \quad (4.132)$$

where closure for the last two terms must be obtained using one of the methods described in Sections 4.2.5 and 4.3.3. When turbulence modelling is employed, the diffusion term is usually neglected as it is much smaller than turbulent dispersion.

Once a closed form for the moment transport equation has been obtained, it can be introduced into a CFD code as an additional scalar transport equation with a source term using the methods that will be described in Sections 4.4.4 and 4.4.5. Therefore, the introduction of moment methods in CFD is straightforward and the main issue is the choice of a closure method for the kinetic terms. Despite the closure issues, moment methods offer an economic way of coupling PBE with CFD, as the number of extra variables are kept to a minimum.

4.4.3 Discretisation Methods

Transport Equations for the Discretised Population Balance

The application of a PBE discretisation method to Eq. 4.126 results into a system of transport equations for the set of discretised number densities, $(n_1, n_2, \ldots, n_K)$. With the introduction of the spatial convection and diffusion terms, the transport equations for the discretised population balance have the following form:

$$\frac{\partial n_i}{\partial t} + \frac{\partial (u_j n_i)}{\partial x_j} = \frac{\partial}{\partial x_j}\left(D_{p,i} \frac{\partial n_i}{\partial x_j}\right) - \frac{\partial (G n_i)}{\partial v} + S_i(n_1, n_2, \ldots, n_K), \quad (4.133)$$

where $D_{p,i}$ is the diffusion coefficient for particles in interval i while the source term, in general, depends on all discretised number densities, since it may include aggregation and fragmentation.

Eq. 4.133 have the form of a generic transport equation and can be readily integrated into a CFD code. In principle, any of the discretisation methods discussed so far can be employed. The main issue is the large number of variables and equations that result. In Sections 4.3.5 and 4.3.6, we saw that moving and adaptive grids can be employed to minimise numerical diffusion in growth-dominated problems, but such methods pose an extra difficulty in coupling with CFD. We now discuss some implementations based on fixed and adaptive grids.

Fixed PBE Grids

With a fixed PBE grid, the integration of the discretised PBE with CFD is straightforward. Eq. 4.133 can be treated as a set of scalar transport equations with source terms. Further details on coupling it with CFD will be given in Sections 4.4.4 and 4.4.5.

Since discretisation methods entail a considerable number of additional variables added to the CFD code, the issue of computational efficiency is brought to the fore. A detailed breakdown of the CPU time in the context of a CFD-PBE simulation of a turbulent sooting flame with the conservative finite volume method described in Section 4.2.8 and a grid of 60 intervals will be shown in Section 6.2 (Table 6.5). In that case, the solution of the PBE requires about 18% of the total CPU time and is thus not a major overload in the total CPU time. The time required for the spatial transport of the discretised number densities should also be considered; in that case, the time taken for convection/diffusion of scalars is 12.25%, about 29% of which is associated with the PBE (since the chemical mechanism involved 149 scalars). Of course, this proportion depends on the other components of the simulation; in that problem, the integration of the chemistry is the main contribution to the CPU requirements. In general, the computational requirements of the PBE solver should be considered in the context of the whole simulation. An analysis of the scaling of the CPU requirements of the conservative finite volume method in the context of parallel computer architectures was provided by O'Sullivan and Rigopoulos (2022).

Moving and Adaptive Grids

PBE discretisation methods with moving or adaptive grids pose an additional problem when it comes to coupling with CFD. As we have seen in Sections 4.3.5 and 4.3.6, such methods can be used to minimise numerical diffusion in the particle volume domain, especially when a growth process is present. In a spatially dependent formulation, however, the adaptation of the grid in the particle volume domain has to be carried out locally, thus resulting in different grids in different parts of the domain. This is a difficult issue to resolve, as it means that the discretised distributions in neighbouring cells will effectively be incompatible with respect to the spatial transport terms.

The majority of CFD-PBE simulations with discretisation methods have been performed with fixed grids. In problems featuring aggregation, such as aerosols and soot, the resulting distributions are usually smooth and do not feature sharp fronts that would require grid adaptivity. In crystallisation problems, the distribution may move to greater particle volumes once the kinetics shift from nucleation-dominated to growth-dominated. One example is the DNS simulation of turbulent precipitation by Tang et al. (2020), where this change was accommodated by a composite grid (cf. Section 6.3.5).

Campos and Lage (2003) developed a method for solution of the spatially-dependent PBE that employs a moving grid based on the method of characteristics (cf. Section 4.3.2). They proposed two possibilities for the spatial discretisation: either a fixed grid is maintained throughout the domain or a new grid, also uniform for the whole domain, is determined at each time step. The local distribution at each location is then mapped to the uniform grid. This approach manages to reconcile the moving grid with the spatial transport of the distribution. However, the use of a single grid means that, if different features of the distribution develop at different locations, the grid will not accommodate them in an optimal way; rather, it represents a compromise throughout the whole domain.

Sewerin and Rigopoulos (2017a) proposed an approach that allows for different PBE grids to be present in different parts of the domain. The basics of their explicit adaptive grid method (EAGM) were discussed in Section 4.3.6, so here we will concentrate on how it reconciles the locally adapted grid with the spatial transport operators. The main concept is that the particle volume domain is mapped invertibly to a reference space that is both space and timedependent, that is, $v = \bar{v}(\tau, \mathbf{x}, t)$, where τ is the parameter corresponding to particle volume in the reference space. We can thus write:

$$n(v, \mathbf{x}, t) \equiv f(\bar{\tau}(v, \mathbf{x}, t), \mathbf{x}, t), \tag{4.134}$$

where f is the number density in the transformed space and $\bar{\tau}$ is the inverse of the map. By applying the chain rule to the PBE, the following equation is obtained:

$$\begin{aligned}
\frac{\partial f}{\partial t} + \sum_{i=1}^{3} \frac{\partial (u_i f)}{\partial x_i} &= \sum_{i=1}^{3} \frac{\partial}{\partial x_i}\left(D_p \frac{\partial f}{\partial x_i}\right) - f\left.\frac{\partial G}{\partial v}\right|_{\bar{v}} \\
&\quad - \frac{\partial f}{\partial \tau}\left(G\left.\frac{\partial \bar{\tau}}{\partial v}\right|_{\bar{v}} + \left.\frac{\partial \bar{\tau}}{\partial t}\right|_{\bar{v}} + \sum_{i=1}^{3}\left(u_i - \frac{\partial D_p}{\partial x_i}\right)\left.\frac{\partial \bar{\tau}}{\partial x_i}\right|_{\bar{v}} - \sum_{i=1}^{3} D_p \left.\frac{\partial^2 \bar{\tau}}{\partial x_i^2}\right|_{\bar{v}}\right) \\
&\quad + D_p \sum_{i=1}^{3}\left(\frac{\partial^2 f}{\partial \tau^2}\left.\frac{\partial \bar{\tau}}{\partial x_i}\right|_{\bar{v}}^2 + 2\frac{\partial^2 f}{\partial \tau \partial x_i}\left.\frac{\partial \bar{\tau}}{\partial x_i}\right|_{\bar{v}}\right) + S.
\end{aligned} \tag{4.135}$$

The first three terms can be readily identified as the accumulation, convection and diffusion terms. A number of new terms have arisen from the coordinate transformation, but they can all be collected together with the PBE source term and treated with the method of fractional steps (to be discussed in Section 4.4.4).

To demonstrate the grid adaptation in a spatially dependent problem, Sewerin and Rigopoulos (2017a) considered the precipitation of $BaSO_4$ in an unsteady plug-flow reactor, accounting for nucleation, growth and spatial transport in one

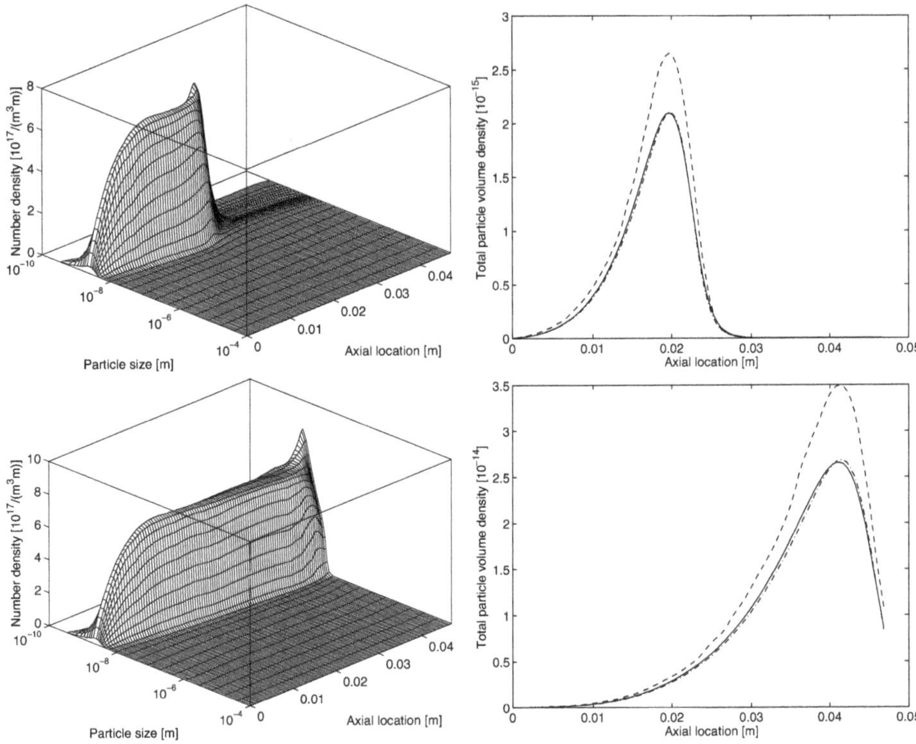

Figure 4.8 Simulation of unsteady precipitation of $BaSO_4$ in a plug flow reactor with the explicit adaptive grid method and 30 nodes in the particle size domain at two different time instants. Left – evolution of particle size distribution along the reactor length; right – comparison of volume density computed with the adaptive grid (dash-dotted line) with a simulation without grid adaptivity and the same number of nodes (dashed line), as well as with a reference solution with 1,000 nodes (solid line). Reprinted from Sewerin and Rigopoulos (2017a) with permission from Elsevier.

coordinate. The PBE was also coupled with the species mass transport and reaction equations, which yielded the local supersaturation on which the nucleation and growth rates depend. Fig. 4.8 shows the particle size distribution at all locations at two different times. Also shown is the total particle volume density (i.e. the concentration of the third moment) as a function of the axial location, where comparison is made with a computation without grid adaptation and the same number of nodes (30) and a reference solution with 1,000 grid nodes. In spite of the coarse grid, the solution with the adaptive grid is almost indistinguishable from the reference one. This method has been coupled with CFD and applied to simulations of aerosol condensation (Sewerin and Rigopoulos, 2017b) and sooting formation in turbulent flames (Sewerin and Rigopoulos, 2018, 2019).

Coupled PBE

The implementation of the two-PBE approach for primary particles and aggregates in the context of a CFD code requires some special attention, as it must be ensured that the flux of the primary particle number density is consistent with that of the aggregate number density. Using Eq. 4.78, the convective and diffusive fluxes of $n_{p,i}$ can then be obtained by multiplying those for n_i with $n_{av,i}$. The transport equation for $n_{p,i}$ can thus be written as follows:

$$\frac{\partial n_{p,i}}{\partial t}+\frac{\partial(n_{av,i}u_jn_i)}{\partial x_j}=\frac{\partial}{\partial x_j}\left(n_{av,i}D_{p,i}\frac{\partial n_i}{\partial x_j}\right)+S_{p,i}(n_1,n_2,\ldots,n_K,n_{p,i}),\tag{4.136}$$

where the source term, $S_{p,i}$, is related to the distribution of aggregates (due to aggregation) and to $n_{p,i}$ (due to fusing) as shown in Section 4.2.11. We now substitute $n_i=n_{p,i}/n_{av,i}$ and use the chain rule to obtain:

$$\begin{aligned}\frac{\partial n_{p,i}}{\partial t}+\frac{\partial u_jn_{p,i}}{\partial x_j}&=\frac{\partial}{\partial x_j}\left(D_{p,i}\frac{\partial n_{p,i}}{\partial x_j}\right)-\frac{\partial}{\partial x_j}\left(D_{p,i}\frac{n_{p,i}}{n_{av,i}}\frac{\partial n_{av,i}}{\partial x_j}\right)\\&\quad+S_{p,i}(n_1,n_2,\ldots,n_K,n_{p,i}).\end{aligned}\tag{4.137}$$

Note that the second term in the right-hand side accounts for the connection between the transport of soot aggregates and primary particles.

4.4.4 Introduction of the PBE Source Term in CFD

Most CFD codes provide for the transport of additional scalar variables, possibly with source terms, via the following generic equation:

$$\frac{\partial\phi}{\partial t}+\frac{\partial(u_i\phi)}{\partial x_i}=\frac{\partial}{\partial x_i}\left(D_\phi\frac{\partial\phi}{\partial x_i}\right)+S_\phi,\tag{4.138}$$

where ϕ is a scalar intensive variable, that is a quantity per unit volume. For variable-density flows, the following equation may be used instead:

$$\frac{\partial\rho\phi'}{\partial t}+\frac{\partial(\rho u_i\phi')}{\partial x_i}=\frac{\partial}{\partial x_i}\left[\rho D_\phi\frac{\partial\phi'}{\partial x_i}\right]+S_\phi,\tag{4.139}$$

where ϕ' represents, in this case, a quantity per unit mass.

We have seen that, once a moment or discretisation equation has been applied to the PBE, the resulting equations have the form of Eq. 4.138 or 4.139, with the scalars being the moments or discretised number densities. From here on, two alternative ways may be sought. The first one is to introduce the PBE source term into the scalar transport equation and solve the latter with the methods available in the CFD code. The second one is to employ the method of *fractional steps* or *operator splitting* to decouple the convection-diffusion and source terms, which offers the advantage that the solver for the spatially homogeneous PBE can be introduced directly into the CFD code as a plug-in, without the need for modifications in either code. This approach has been used for the case studies in Chapter 6, so it will be briefly described below, while the reader may refer to Leveque (2002) for more details.

The main concept of the fractional step will be demonstrated with the equation for the array of discretised number densities, Eq. 4.133. For simplicity, we consider flow in one dimension, size-independent growth and no particle diffusion. The equation to be approximated, therefore, is:

$$\frac{\partial n_i}{\partial t} + \frac{\partial (u n_i)}{\partial x_j} = -G\frac{\partial n_i}{\partial v} + S_i(n_1, n_2, \ldots, n_K). \tag{4.140}$$

This equation is now split into two equations as follows:

$$\frac{\partial n_i}{\partial t} + \frac{\partial (u n_i)}{\partial x_j} = 0, \tag{4.141a}$$

$$\frac{\partial n_i}{\partial t} = -G\frac{\partial n_i}{\partial v} + S_i(n_1, n_2, \ldots, n_K). \tag{4.141b}$$

Note that the second of these equations is the one that results from the application of the discretisation method to the spatially homogeneous population balance equation. This is the main advantage of the fractional step method: *the existing PBE solver can be used straight away.*

The simplest fractional step method would be to use the explicit Euler method for the temporal discretisation and any approach for the spatial discretisation. For brevity, this will be demonstrated here using the upwind method. Let $n_{i,j}(t_n)$ denote the discretised number density at the jth spatial node and at time instant t_n, while space, time and particle volume are discretised with uniform steps δx, δt and δv, respectively. We first solve Eq. 4.141a to obtain a set of intermediate values, $n^*_{i,j}$:

$$n^*_{i,j} = n_{i,j}(t_n) - \frac{u\delta t}{\delta x}(n_{i,j}(t_n) - n_{i,j-1}(t_n)). \tag{4.142}$$

The values at time instant t_{n+1} are then given by solving Eq. 4.141b as follows:

$$n_{i,j}(t_{n+1}) = n^*_{i,j} + \delta t\left[-\frac{G}{\delta v}(n^*_{i,j} - n^*_{i-1,j}) + S_i(n^*_{1,j}, n^*_{2,j}, \ldots, n^*_{K,j})\right]. \tag{4.143}$$

This fractional step scheme results in a first-order accurate approximation of Eq. 4.140. More elaborate schemes, such as the second-order accurate *Strang splitting*, are described in Leveque (2002).

4.4.5 Convection and Diffusion of Particles

In the coupling of moment and discretisation methods with CFD in an Eulerian fashion, the convection and diffusion terms for particle transport conform to the generic scalar transport equation Eq. 4.138 and can thus be treated with the standard methods available in the CFD code. Two issues require further attention and will be discussed here.

For non-inertial particles, convection can be treated in a standard manner, with the particle velocity being the local fluid velocity. For low Stokes number inertial particles, the two velocities deviate. However, by applying Eq. 3.145, we can see that the particle velocity can be computed as a function of the local fluid velocity. Thus, the implementation involves simply adding a correction term to the latter.

Regarding the diffusion term, a note should be made about the particle diffusivity, which is typically very low and decreases with particle size according to Eq. 3.128. The ratio of kinematic viscosity to diffusivity is called the *Schmidt number*:

$$Sc = \frac{\nu}{D}. \tag{4.144}$$

While gas-phase species have Sc of the order of one, the Sc of particles can be easily in the range of thousands due to their low diffusivity. Resolving the scalar field at high Sc can be very computationally expensive and even not feasible in the case of turbulent flows. In that case, the flow field exhibits a range of scales (cf. Section 5.2.3), the smallest of which are the *Kolmogorov scales*. At high Sc, the spatial distribution of a scalar exhibits even smaller scales, down to the *Batchelor scales*, the resolution of which is not possible at present for most problems. In particular, the length scale of the Batchelor scales, η_c, is related to that of the Kolmogorov scales, η, according to $\eta_c = \eta / \sqrt{Sc}$ (Davidson, 2004). This means that, for $Sc = 1{,}000$, η_c, is $\sqrt{1{,}000}$ or about 31 times smaller than the Kolmogorov scale. The use of a computational grid finer by this factor in each spatial dimension means using $31^3 = 29{,}791$ times more cells; thus, if 10 million cells are required to resolve η, about 300 billion cells are needed for η_c.

While studies of transport at high Schmidt numbers have appeared in the literature (Donzis et al., 2010, 2014; Donzis and Yeung, 2010; Derksen, 2012; Buaria et al., 2021; Ranjan and Menon, 2021), the have been mainly on passive and inert scalars. The effect of the Batchelor scales on the kinetic processes in the population balance has been explored in very few studies. Garrick and Khakpour (2004) studied the effect of particle diffusivity on aggregation in a two-dimensional mixing layer, while Cifuentes et al. (2020) simulated nanoparticle formation in a two-dimensional flame-vortex interaction including nucleation and aggregation. A hybrid Euler–Lagrange method for introducing high Sc number in CFD has recently been proposed by Leer et al. (2022) and demonstrated for the transport of a passive scalar in a mixing layer.

A recent study by Tang et al. (2023) investigated particle nucleation and growth at $Sc = 1$ and $Sc = 10$ and results from that work will be shown in Section 6.3.7. It was found that, while the zones of particle formation were thinner at higher Sc, the precipitation rates were also more intense, and the competition between these two effects resulted in a very small effect on the PSD. These considerations, however, refer to direct numerical simulations (DNS). In Reynolds-Averaged Navier–Stokes (RANS) or Large–Eddy Simulation (LES) (cf. Section 5.2), even η_c is not resolved. Furthermore, the uncertainties in kinetic models are usually of a much higher order than the issue of particle diffusivity.

4.4.6 Lagrangian Monte Carlo Methods

An alternative method of introducing the population balance into CFD is via simulating the PBE alongside Lagrangian trajectories of fluid elements containing particles. A sample of such trajectories is extracted, making this a form of Monte Carlo simulation. The PBE can then be solved as a post-processing step along these trajectories. The form of the PBE to be solved within a fluid element is a spatially homogeneous one (since the element operates as an ideal reactor). Therefore, any of the methods for the solution of the PBE described in Sections 4.2–4.3 can be employed in a straightforward manner.

The Lagrangian approach is well suited to Monte Carlo solution methods for the PBE (as well as being a Monte Carlo simulation in itself). In these methods, as we saw in Section 4.2.9, an ensemble of stochastic particles is constructed so that its statistics represent the PSD. In an Eulerian CFD-PBE simulation, such an ensemble must be present at each spatial location, thus making the CPU requirements prohibitively expensive. By contrast, the use of a Lagrangian approach allows performing the PBE Monte Carlo only along the selected trajectories.

An example of a Lagrangian approach for coupling a Monte Carlo method for the PBE with a fluid flow problem can be found in the work of Balthasar and Kraft (2003). In that work, a one-dimensional laminar premixed sooting flame is simulated. The steady-state solution of the flame is first obtained and then used for the Monte Carlo PBE solution in a Lagrangian manner. Other works with this approach include Schwarzer et al. (2006), Attili and Bisetti (2013) and Lucchesi et al. (2017).

One advantage offered by Lagrangian simulations is that they overcome the issue of the high Schmidt number of the particles, since there is no need to introduce a diffusion term in the PBE. Apart from the issue of statistical convergence, another difficulty is related to the coupling between CFD and PBE. In many cases, the PBE is solved as a post-processing step after the CFD has been completed. If there is no feedback from the kinetic processes in the PBE to the flow and transport processes, this approach can be directly applied. However, if the kinetics of the particulate phase alter the concentration and temperature environment, for example via species consumption or particle radiation, this coupling cannot be addressed in a straightforward manner. One solution to this problem, applied by Balthasar and Kraft (2003) in their laminar sooting flame simulation, is to perform first an Eulerian CFD-PBE simulation with a simplified PBE solution method – in that case, a method of moments – so that the feedback effect of the particles can be accounted for in the steady-state solution. The Monte Carlo PBE solution can then be applied to obtain more details about the particle population (i.e. the PSD) than could be produced by the moment-based approach. A similar approach was adopted by Lucchesi et al. (2017), where the HMOM method (cf. Section 4.2.10) was employed.

4.4.7 Data-Driven Methods

Data-driven methods are approaches to problem solving that are guided by data collected from previous solutions, as opposed to using directly equations derived from physical principles. In recent years, data-driven methods have witnessed a rapid growth, precipitated by the increasing availability of computer power and storage, and are finding application in an increasing number of scientific fields as diverse as fluid dynamics and drug development (as well as a multitude of non-scientific fields). At the time of writing, there have been very few attempts to bring data-driven methods to population balance modelling, but more can be expected in the future. In order to provide an outlook to future developments, we will provide here an introduction to two such methods.

Data-driven methods are, by their nature, confined to a range of problems determined by the training dataset. However, they can be very computationally efficient because they condense the information contained in the that dataset and replace the solution of integro-differential equations with much simpler models. As such, their potential uses are for applications where a very large number of PBEs must be solved, such as in coupled CFD-PBE or optimisation problems. The two approaches that will be described below are motivated by CFD-PBE applications.

In general, the deployment of a data-driven method for the solution of a scientific problem that involves prediction of certain quantities from given data can be broken into the following steps. The first one is *data generation*, in which a dataset of solutions is generated to 'drive' the method. These solutions can be obtained with any of the conventional methods that can be used for the problem in question – in the case of the PBE, for example, it could be any of the methods described so far. This step is crucial, as it determines the range of validity of the final model; a data-driven method usually has limited capacity for extrapolating far from the conditions covered by the dataset employed. The second step is the *formulation of the data-driven model*, in which a model is postulated to replace the original problem (the PBE in our case). This model typically involves free parameters that will be determined during the next step, based on the dataset provided. The next step is *training*, in which the parameters of the model are determined from the dataset provided. The final step is *application*, where the resulting model is used to replace the original one and make the desired predictions. Testing with unseen data (i.e. not included in the training dataset) is also required before the model can be applied with confidence.

Seltz et al. (2021) put forward a method based on artificial neural networks (ANNs) for solving the PBE for soot formation in flames. ANNs are a class of machine learning models that were initially inspired by the biological processes that take place in the brain and can be trained to perform tasks such as regression or pattern recognition. Technically, they are nonlinear algebraic models with coefficients to be determined during the training process. Owing to the intended use for predicting soot formation, the training problems used in Seltz et al. (2021) are laminar one-dimensional premixed sooting flames. The PBE for the evolution of the soot PSD is solved using the method of Bouaniche et al. (2019) described in Section 4.3.7, which combines a Monte Carlo approach for the growth term with the pointwise discretisation of Kumar and Ramkrishna (1996a) for aggregation. The model used to represent the PBE is the convolutional neural network (CNN), a particular type of ANN, while an additional ANN is used to replace the solution of the species and energy transport equations. ANNs are nonlinear models, hence their training is via a nonlinear optimisation procedure. The test results show that the method can produce accurate results at reduced computational cost, but they were performed on the same configuration as the one used for the training. The aim of future work in this direction is to explore the ability of ANNs to perform well on a range of realistic problems while being treated with canonical ones (such as the one-dimensional premixed flame).

Sewerin (2022) proposed a method for solving the PBE based on proper orthogonal decomposition (POD), which is an approach for decomposing a function into a set of orthogonal modes. This decomposition involves a set of coefficients from which the original function (in this case, the number density) can be uniquely reconstructed during the application stage. The PBE-POD method is oriented towards growth-dominated problems, and the PBE is thus reformulated with the aid of the method of characteristics. The datasets employed by Sewerin (2022) are obtained by solutions of the PBE in ideal flow configurations, such as perfectly stirred and plug flow reactors. Two adaptive grid methods are used to obtain the solutions of the training problems: EAGM, which was described in Section 4.3.6, and a new one based on the method of characteristics; the latter is found to be more amenable to model reduction. In POD, the training process is a linear procedure (as opposed to a nonlinear optimisation in the case of ANNs). The test results show that the PSD can be predicted with moderate accuracy with a very small number of degrees of freedom (i.e. dependent variables), while coping with the presence of sharp fronts in growth-dominated problems. For a given accuracy, the PBE-POD method reduces the number of degrees of freedom involved in adaptive grid methods by over four times, although it is computationally more expensive per degree of freedom. The method also possesses some robustness towards inadequacies of the training dataset; in particular, for any given training dataset, the method will always return the best possible solution in the least squares sense. As with all data-driven methods, the predictive capacity depends on the design of the training problems, which should reflect the physical effects that shape the PSD in the target application.

4.5 Coupling of PBE with CFD: Inertial Particles

For particles of arbitrary inertia, where the low Stokes number approximation cannot be invoked, the formulations discussed in Section 2.7.4 must be used. The most complete one is the kinetic approach, which allows for a velocity distribution at each particle size to be considered. The formulation and theoretical framework of this approach has been reviewed by Reeks et al. (2017) and Reeks (2021).

The direct solution of this approach in an Eulerian framework requires discretisation along all velocity components as well as particle size, physical space and time; it has been attempted by a few studies but, in general, is prohibitively expensive. Therefore, various simplified approaches have been developed for CFD calculations, which amount to various forms of moment methods. Alternatively, Lagrangian methods can be employed where a sample of particle trajectories is simulated. Since these methods are described in multiphase flow treatises and reviews such as Crowe et al. (1998), Drew and Passman (1998), Mashayek and Pandya (2003) Ishii and Hibiki (2005), Sirignano (2010), Marchisio and Fox (2013) and Michaelides and Feng (2017), they will be discussed here only briefly.

4.5.1 PBE with Discretised Velocity Coordinates

The numerical solution of a PBE with a velocity distribution, such as the equation proposed by Williams (1958, 1961) for spray combustion, by means of discretisation in all coordinates poses formidable difficulties, due to the number of independent variables present. Works that attempted to provide numerical solutions in this manner include Sutton et al. (1972), Gany et al. (1976) and Westbrook

(1977). The last of these works employed 10 intervals for each spatial coordinate, 5 for each velocity coordinate and 14 for the droplet radius, while no coagulation, break-up or source terms were included. Gupta and Bracco (1978) compared the discretisation method with an approach where the spray is simulated as a discrete number of sprays, each injected at a different time with an initial distribution. In general, this approach has not been widely adopted because of the computational demands involved in the discretisation of a state space of at least 7 dimensions (for three-dimensional flows and one size variable).

4.5.2 Moment Methods and Multi-Fluid Models

In the multi-fluid model, the number density is discretised into size intervals, or classes. A moment transform is then performed over the velocity domain for each size class to derive a mean momentum equation for the particles in that size class. This results in a coupled system of equations for the volume fraction and momentum of the particles in each class. The concept is similar to the coupled system of equations used to describe particle morphology. If an integration is performed over the size domain as well, a system of two equations results, one for the total particle volume (or volume fraction) and one for the total momentum of the dispersed phase, to be coupled with the momentum equation for the continuous phase. This approach is referred to as the two-fluid model. Two-fluid and multi-fluid models are called Eulerian–Eulerian models in the multiphase flow literature (Crowe et al., 1998). The procedure outlined above shows their link to the PBE as moments equations in the velocity domain. The derivation of the multi-fluid model as a discretised population balance is evident in the work of Greenberg et al. (1993). Furthermore, Laurent and Massot (2001) show the link of the multi-fluid approach with the PBE (in the form of the Williams spray equation, cf. Section 2.7.4), as well as with the Lagrangian approach that will be discussed in the next section.

Like all moment formulations, multi-fluid models are unclosed. Closure can be obtained in various ways, such as via extensions of the QMOM approach. This allows for incorporation of two-way momentum exchange between the continuous and dispersed phases in dense flows. The derivation of multi-fluid models and the closure methods have been shown in detail by Fox (2007). Jakobsen et al. (2005), Massot (2007) and Hjertager (2007) have discussed multi-fluid models for bubble column reactors, sprays and chemical reactors, respectively. A wide range of formulations and closure hypotheses for multi-fluid and two-fluid models is presently available; recent and detailed reviews of them from a population balance perspective can be found in Marchisio and Fox (2013) and Yeoh et al. (2014).

4.5.3 Lagrangian Methods

Lagrangian methods for inertial particles are stochastic (Monte Carlo) methods where trajectories of particles are being simulated. This approach is similar to the one described in Section 4.4.6 for non-inertial particles, although in this case the inertia of particles is taken into account.

These methods are often employed for sprays. The spray is described by an ensemble of 'parcels of droplets', with each parcel containing droplets of the same size. The velocity of the parcels includes a deterministic part, which is obtained from the equation of motion for a particle (cf. Section 3.6.5), and a stochastic part that accounts for the effect of turbulent fluctuations. This approach has its origins in the works of Dukowicz (1980) and Gosman and Ioannides (1983) and is also the basis of the KIVA code (Amsden et al., 1989). Jones and Sheen (2000) employed it for the modelling of a turbulent spray via a PDF transport equation solving for the PDF of size, position, temperature and number of droplets. The approach was extended to LES by Bini and Jones (2008), while the stochastic models employed in that work have been described in Bini and Jones (2007). Other LES-based works with this approach include Pesmazoglou et al. (2014, 2017).

A discussion of this viewpoint, its assumptions and its link with the PBE has been provided by Subramaniam (2001). Another type of stochastic models is based on considering the properties of both particle and fluid element 'viewed' by the particle; more information on this can be found in the review of Peirano et al. (2006). A Lagrangian approach that draws a close link with the population balance is that of Salehi et al. (2017).

The Lagrangian approach provides a natural way for modelling the motion and dispersion of inertial particles, since it is straightforward to introduce the forces acting on the particles of a given size via the equation of motion. In turbulent flows, modelling is required to account for the effect of stochastic fluctuations on particle motion. Finally, it is more difficult to account for aggregation with this viewpoint, as the collisions between the particles must be tracked in the whole spatial domain.

4.6 Summary and Concluding Remarks

A very wide range of methods has been developed for solving various forms of population balance problems. In the present chapter, the methods were discussed progressively based on the kinetic and transport processes involved.

In terms of methodology, the solution methods for the kinetic part of the PBE were classified into several families: analytical/similarity, moment, discretisation and Monte Carlo methods. The transport part can be treated in an Eulerian or Lagrangian fashion. The coupling with CFD was shown for the case of non-inertial and low Stokes number inertial particles. Inertial particles, which feature a distribution of velocities, are treated with methods described in the multiphase flow literature. For each method, the advantages and disadvantages that determine its suitability for certain classes of problems were discussed. A few final remarks will be made here.

The presence of several processes of a different nature in the most general form of the PBE is the main reason why it is difficult for a single method to fit all circumstances. However, most population balance problems involve only a selection of these processes. The choice of method, therefore, has to be partly dictated by the problem formulation. The data available for the kinetic parameters, as well as the accuracy of these data, should also be taken into account. Finally, one should consider the level of information that is desired from the model. Depending on the application, the distribution may be important, or it may be sufficient to predict certain moments of it.

Analytical and similarity methods are available for few cases but are useful for model validation and in a limited range of practical problems. Moment methods greatly reduce the number of variables needed to describe the problem and are thus well suited for coupling for CFD. However, they involve assumptions that limit their applicability or require further closure models, while not predicting directly the distribution. Discretisation methods predict the distribution without closure assumptions but require more computational time and memory. Specialised schemes may need to be employed to overcome issues such as conservativeness, stability and lack of numerical diffusion. They also become prohibitively expensive at higher dimensionality, for which coupled PBE formulations can often be employed instead. Coupling with CFD can be attained in various ways. Monte Carlo methods are simple to implement but require a large number of stochastic particles to attain convergence, which makes them less suited to CFD applications. On the other hand, they are easily amenable to multidimensional PBE formulations.

Aggregation and fragmentation processes involve the solution of integral equations and require more specialised methods, particularly for aggregation. For this reason, they were discussed first in this chapter. If a problem is dominated by these processes, considerations related to them must take precedence in the choice of method. Pointwise methods place emphasis on the prediction of the moments, while finite element or volume methods prioritise the distribution. For the latter, conservation of moments in aggregation is an important issue and can be overcome by ensuring the balancing of birth and death terms, as in two-dimensional quadrature methods. Moment methods face the problem of closure for physically realistic kernels, while Monte Carlo methods must ensure convergence of the statistics. Further considerations of the physical problem provide more clues into method selection;

for example, if there are reasons to expect that the distribution will assume a standard form such as the log-normal or a self-preserving one, moment or similarity methods are sufficient.

Growth processes render the PBE into a first-order hyperbolic equation. This raises the problems of numerical diffusion and dispersion in discretisation methods. Several methods have been proposed for this problem, from flux limiters that mitigate numerical diffusion to characteristics-based methods and adaptive grids that eliminate it. The appropriate option to a particular problem has again to be guided by the kinetics and the extent to which the shape of the distribution is dominated by growth. Moving or adaptive grids also pose issues for coupling with fluid dynamics, which is the next consideration. Moment and Monte Carlo methods do not share these issues, although a closure model may be required for the former.

For spatially dependent problems, several ways were discussed for treating the transport terms in CFD codes, including Eulerian, Lagrangian and, more recently, data-driven formulations. At this point, it must be acknowledged that even the more computationally expensive methods (or those requiring finer grids to attain the accuracy demanded) are easily affordable with modern computers if solution of a single PBE is required. The issue of computational efficiency comes to the fore when a large number of PBEs must be solved. This occurs mainly when the PBE is to be coupled with CFD or to be used for optimisation purposes. Coupled CFD-PBE problems, in particular, may require the solution of millions of PBEs over a large number of time steps. In such cases, both the number of variables and the CPU time required for the PBE solver must be kept to a manageable level. Regarding the number of variables, we must consider the memory capacity of the computer as well as the CPU time required for the spatial transport terms. The CPU time spent in the PBE solver should be considered as a proportion of the CPU time required by the overall simulation.

It should be emphasised that the formulation of the PBE should be considered together with the choice of solution method. More complex formulations, such as multidimensional PBEs, apart from requiring more time-consuming solution methods, involve also more complex kinetic models with more parameters. This raises the issue of the validity and accuracy of the kinetic models, as well as their sensitivity to the parameters. In a complex problem, it may be beneficial to focus on identifying the key steps and formulate a simpler model, than resort to a complex formulation and associated solution method.

Chapter 6 will show case studies featuring several of the solution methods encountered in the present chapter and discuss the selection and implementation in each case.

5

Population Balance in Turbulent Flow

5.1 Overview

The population balance problems encountered in applications often take place within turbulent flows. Crystallisation reactors such as stirred tanks and other mixers operate under turbulent flow conditions. The flow in turbines where soot is emitted and in industrial-scale reactors for flame synthesis of nanoparticles is turbulent. Particle-laden plumes emitted from stacks, volcanic ash clouds and other atmospheric flows with aerosols are turbulent. The presence of turbulence introduces new theoretical and computational problems into population balance modelling, which will be the subject of the present chapter.

Turbulence is a chaotic and unsteady state of fluid motion. While the deterministic equations underpinning fluid mechanics and transport phenomena still apply, the chaotic nature of turbulent flow means that solutions are extremely sensitive to the initial and boundary conditions and any individual realisation of a turbulent flow is not reproducible. The direct numerical simulation (DNS) of turbulent flows is feasible only for simple flows and a limited range of Reynolds numbers, and is thus relevant only for fundamental studies. The prediction of turbulent flows for practical purposes is based on either time-averaged equations – the Reynolds-averaged Navier Stokes (RANS) approach – or on space-filtered equations – the large eddy simulation (LES) approach. In the former, the focus is on the mean values of the variables of interest – velocity, temperature, pressure, species composition and particle number density in population balance problems – while the effect of fluctuations is taken into account via closure models. In the latter, the larger scales of the flow that contain most of the energy and are more directly affected by the boundary conditions are being resolved, while models are employed for the smaller scales whose effect is more amenable to general-purpose models. LES is considerably more expensive than RANS but also offers significant advantages: apart from more detail on the flow structures, the mean quantities are predicted more reliably and the method is applicable to a wider range of flows with less reliance on problem specific model selection and tuning. This is due to the diminished impact of the small-scale models involved in LES on the prediction of the mean quantities.

The effect of turbulence on the kinetic and transport processes that appear in the population balance may be manifested in three ways. First, fluctuations in concentrations of precursor species, temperature and particle number density result in fluctuating rates of kinetic processes, manifested via nonlinear correlations that are unclosed under turbulence modelling approaches. Second, the mechanisms of particle collision and fragmentation are different in turbulent flow. Third, the fluctuations in fluid velocity give rise to particle dispersion.

Of these issues, the second one was briefly considered in Chapter 3 where the kernels for turbulent aggregation and fragmentation were presented. The issue of turbulent dispersion can be treated with standard gradient models for non-inertial particles. It is much more complex for inertial particles, however, where a population balance formulation requires a consideration of the velocity distribution. The theoretical framework for this approach was discussed in Section 2.7.4, while the solution approaches (discretisation of velocity space, multi-fluid and Lagrangian models) were briefly discussed in Section 4.5. The modelling of turbulent dispersion in the context of these approaches is treated extensively in multiphase flow treatises including the references provided in that section, hence it will not be discussed any further in this chapter.

The present chapter, therefore, is devoted to the effect of turbulent fluctuations in the kinetic processes in the population balance and the methods for taking such effects into account. It commences with a brief summary of the features of turbulent flow that are most relevant to the following discussion and a description of the categories of modelling approaches. The subject of turbulent flow modelling and simulation is vast and has an extensive literature of its own, hence only a few rudimentary aspects of it will be summarised here, to the extent that is needed to set the stage and establish the notation for the discussion of population balance in turbulent flows that follows. More information on turbulent flows can be found in Tennekes and Lumley (1974), Pope (2000), Mathieu and Scott (2000) or Bernard and Wallace (2002), among many other excellent books. We then consider the closure problem for the kinetic processes in the PBE, the unclosed terms that result and their impact. Subsequently, we examine the methods that have been proposed for obtaining closure. Finally, the effect of turbulence on transport processes is discussed. Applications in the context of specific problems such as soot formation and crystallisation will be shown in Chapter 6.

5.2 Fundamentals of Turbulent Flow Modelling and Simulation

5.2.1 Turbulent Flow, Averages and PDFs

Turbulent flow is described by the same set of deterministic equations that govern the dynamics of any flow, that is, the continuity and Navier–Stokes equations together with any further transport equations required for a particular problem. However, obtaining a direct numerical solution to these equations is not practically possible in most cases for two reasons. First, the chaotic nature of turbulence means that any small disturbance in the initial and boundary conditions is sufficient to cause significant alterations in the flow field. It is impossible to know and control the boundary conditions (e.g. the microstructure of a rough surface) with sufficient accuracy to render a turbulent flow experiment repeatable with respect to instantaneous measurements. As a result, only the statistics of a turbulent flow can be reproducible. Second, turbulence gives rise to space and time scales in the motion of the fluid that range over several orders of magnitude. In order to resolve a turbulent flow via a numerical method, the discretisation employed must be fine enough to resolve the smallest scales, which implies using a large number of spatial nodes and time steps (estimates of which will be shown in Section 5.2.4). While such DNS studies have indeed been performed for simplified flows, this is not an approach that can be employed for practical problems in the present or foreseeable future. Therefore, models have been constructed for predicting the average values of variables (rather the instantaneous ones), which are often all that is needed in applications.

Averaging in turbulence can be defined and interpreted in three ways: *time, space and ensemble average*. The last one is the most general and can be interpreted as the average value of a property over a very large number of realisations – thought experiments with nominally identical conditions – of the turbulent flow. For a scalar field, $\phi(\mathbf{x},t)$, the ensemble average is defined as follows:

$$\langle\phi(\mathbf{x},t)\rangle = \frac{1}{N}\sum_{j=1}^{N}\phi(\mathbf{x},t). \tag{5.1}$$

Since the ensemble average is not usually feasible to practically carry out in experiments and DNS, it is often replaced by the time average, which is defined for statistically stationary flows in the following way:

$$\langle\phi(\mathbf{x},t)\rangle = \frac{1}{T}\int_{t}^{t+T}\phi(\mathbf{x},\tau)d\tau. \tag{5.2}$$

A spatial averaging operation can also be defined in a similar way. In the following, the notation for averaging will imply the ensemble average.

A probability density function (PDF) may be associated with every quantity in a turbulent flow. The *one-time, one-point PDF* of ϕ, $f(\psi;\mathbf{x},t)$, is the PDF of ϕ at a single point in space and at a single time instant,[1] while ψ is a variable that may assume the same range of values as ϕ (it is called a *sample-space variable*). The mean of ϕ is related to its PDF as follows (using the ensemble average definition):

$$\langle\phi(\mathbf{x},t)\rangle = \int_{-\infty}^{\infty} \psi f(\psi;\mathbf{x},t)d\psi. \tag{5.3}$$

A PDF of n scalar variables would thus be denoted as $f(\psi_1,\ldots,\psi_n;\mathbf{x},t)$. Multi-time and multi-point PDFs can also be defined. The PDF will play an important role in the models that will be discussed in the present chapter.

We will now consider briefly the modelling and simulation approaches that are employed to predict flow and other quantities in turbulent flow. These approaches can be combined with the PBE to formulate models of population balance problems in turbulent flows.

5.2.2 Reynolds-Averaged Navier–Stokes (RANS) Models

A family of models are based on the Reynolds-Averaged Navier–Stokes (RANS) equations, which are obtained by averaging the Navier–Stokes equations. To arrive at the RANS equations, each variable is first decomposed into a mean and a fluctuating component – for example, for a scalar ϕ, we have:

$$\phi(\mathbf{x},t) = \langle\phi(\mathbf{x},t)\rangle + \phi'(\mathbf{x},t). \tag{5.4}$$

A set of equations can then be derived for the mean values, which are often sufficient for applications. However, these equations are unclosed, as will be shown below.

The averaging of the continuity and Navier–Stokes equations yields:

$$\frac{\partial\langle u_i\rangle}{\partial x_i} = 0, \tag{5.5}$$

$$\frac{\partial\langle u_i\rangle}{\partial t} + \frac{\partial\left(\langle u_j\rangle\langle u_i\rangle\right)}{\partial x_j} + \frac{\partial\langle u_j'u_i'\rangle}{\partial x_j} = -\frac{1}{\rho}\frac{\partial\langle p\rangle}{\partial x_i} + \nu\frac{\partial^2\langle u_i\rangle}{\partial x_j\partial x_j} + g_i, \tag{5.6}$$

where the constant density and viscosity form of the Navier–Stokes equations (Eqs. A.2 and A.25) has been employed; the variable density case will be considered in Section 5.2.6. To derive these equations, we used the facts that averaging is a linear

[1] The symbol ';' is used to indicate that f is a PDF with respect to ψ only; see Section 1.5.

operation and commutes with differentiation, and also the identities $\langle\langle \mathbf{u} \rangle\rangle = \langle \mathbf{u} \rangle$ and $\langle \mathbf{u}' \rangle = 0$.

The RANS equations are unclosed because they contain the tensor $\langle u_j' u_i' \rangle$ involving the nonlinear product of velocity fluctuations, and therefore a model is required to obtain a closed set of equations. The quantities $\rho \langle u_j' u_i' \rangle$ represent momentum fluxes due to turbulent fluctuations and are also called *Reynolds stresses*, as they have the same dimensions as stress. They can also be regarded as the joint second moments of the velocity components; in this point of view, the RANS equations are moment equations with a closure problem similar to that of the moment equations for the PBE. In both cases, the equations cannot be brought into a form where the number of unknowns is the same as the number of equations.

In the case of a passive scalar subject to convection and diffusion, a similar averaging of the relevant transport equation (akin to Eq. A.13 without the source terms) results in the following:

$$\frac{\partial \langle \phi \rangle}{\partial t} + \frac{\partial \left(\langle u_i \rangle \langle \phi \rangle \right)}{\partial x_i} + \frac{\partial \langle u_i' \phi' \rangle}{\partial x_i} = -\frac{\partial \langle J_{\phi,i} \rangle}{\partial x_i}. \tag{5.7}$$

If the scalar flux is given by Fick's law (Eq. A.14) and the diffusivity is constant, we have:

$$\frac{\partial \langle \phi \rangle}{\partial t} + \frac{\partial \left(\langle u_i \rangle \langle \phi \rangle \right)}{\partial x_i} + \frac{\partial \langle u_i' \phi' \rangle}{\partial x_i} = D \frac{\partial^2 \langle \phi \rangle}{\partial x_i^2}. \tag{5.8}$$

In this equation, the term $\langle u_i' \phi' \rangle$ represents the scalar transport due to turbulent fluctuations and the equation is unclosed. The case of reacting scalars presents an additional closure problem that will be discussed in Section 5.2.6.

A wide range of RANS models have been developed for obtaining the Reynolds stresses and turbulent scalar transport terms from the mean values, thus providing closure to Eqs. 5.6 and 5.8. One class of models relates the Reynolds stresses to the mean velocity gradients via an *eddy kinematic viscosity*, ν_t, as follows:

$$\langle u_j' u_i' \rangle = -\nu_t \left(\frac{\partial \langle u_i \rangle}{\partial x_j} + \frac{\partial \langle u_j \rangle}{\partial x_i} \right) + \frac{2}{3} k \delta_{ij}, \tag{5.9}$$

where k is the *turbulent kinetic energy*, that is, the kinetic energy of the fluctuating motion:

$$k = \frac{1}{2} u_i'^2. \tag{5.10}$$

Similarly, the *eddy diffusivity* can be defined in terms of the eddy kinematic viscosity:

$$\Gamma_t = \frac{\nu_t}{Sc_t} \tag{5.11}$$

where Sc_t is the turbulent Schmidt number. The averaged scalar transport equation, Eq. 5.8, can then be closed as follows:

$$\langle u_i' \phi' \rangle = -\Gamma_t \frac{\partial \langle \phi \rangle}{\partial x_i}. \tag{5.12}$$

The eddy kinematic viscosity is spatially dependent and can be related to the mean quantities in various ways. A common way is to relate it to the turbulent kinetic energy and its dissipation rate, ϵ. The latter is the rate at which energy is dissipated at the small scales via viscous action (see the next section). The following relationship is used to relate these three quantities:

$$\nu_t = C_\mu \frac{k^2}{\epsilon}, \tag{5.13}$$

where C_μ is a proportionality constant. The widely used k-ϵ model (Jones and Launder, 1972) requires the solution of two additional differential equations for k and ϵ. Other models do not involve the eddy kinematic viscosity but rather solve equations for the Reynolds stresses directly. The details of these models are out of the scope of the present text and can be found, for example, in the books of Hanjalić and Launder (2022) or Leschziner (2016).

Equations can be derived also for other averaged quantities, such as the average kinetic energy, dissipation rate and Reynolds stresses. These equations will not be shown here, as they are used in the derivation of turbulence models and can be found in the aforementioned books. One quantity that will be important in this chapter, however, is the variance of the scalar fluctuations. In population balance problems in reacting flows, such as precipitation or aerosol synthesis, the reactants must mix in the molecular scale before reacting, and the variance of concentration fluctuations is an indication of the state of mixing. The equation for the scalar variance is:

$$\frac{\partial \langle \phi'^2 \rangle}{\partial t} + \overline{u}_i \frac{\partial \langle \phi'^2 \rangle}{\partial x_i} = D \frac{\partial^2 \langle \phi'^2 \rangle}{\partial x_i^2} - \frac{\partial \langle u_i' \phi'^2 \rangle}{\partial x_i} - 2 \langle u_i' \phi' \rangle \frac{\partial \langle \phi \rangle}{\partial x_i} - 2D \left\langle \frac{\partial \phi'}{\partial x_i} \frac{\partial \phi'}{\partial x_i} \right\rangle. \tag{5.14}$$

For the derivation of this equation, see Poinsot and Veynante (2005), for example. On the right-hand side, the first term is the transport of scalar variance by molecular diffusion, which is usually negligible, while the second term is the turbulent transport and the third term is the production due to mean gradient. The most important term is the last one, the *scalar dissipation*,[2] which indicates the rate at

[2] Various related expressions have been associated with this term in the literature. The factor 2 may or may not be included. Furthermore, the scalar dissipation can be associated with the instantaneous value of the scalar and resolved into dissipation due to mean and fluctuations, but the former can be neglected. See Poinsot and Veynante (2005) for more details.

which inhomogeneities in concentration are disappearing. This term plays a role in models for turbulent mixing.

RANS has been the basis of several CFD-PBE studies on soot formation, aerosol synthesis and crystallisation.[3] The case studies in Chapter 6 that involve turbulent flow are LES- or DNS-based, but the concepts of RANS models will play an important role in the rest of the chapter.

5.2.3 Scales of Turbulent Motion

Turbulence consists of a range of coherent fluid motions, or *eddies*, which exhibit a wide range of length and time scales. RANS models approximate this process with a single mean flow and aim to account for the effects of eddies at all scales on the mean values. This is a formidable task, given the variety of eddy structures encountered in turbulent flows. DNS and LES address this issue by computing directly the flow either at all scales or at a proportion of them, respectively. Before we proceed to describe these approaches, a few notes should be made about how the magnitude of the scales of turbulence motion can be estimated.

The following considerations are based on Kolmogorov's phenomenological theory (for an account of it see, e.g., Tennekes and Lumley, 1974). According to this theory, the smallest scales of turbulent motion, known as *Kolmogorov scales*, have universal properties that depend only on the kinematic viscosity, ν, and the rate of energy dissipation, ϵ. Length and time scales for this motion can be defined, and are denoted by η and τ_η, respectively. On the basis of a scaling argument, they are given by the following expressions:

$$\eta \equiv \left(\frac{\nu^3}{\epsilon}\right)^{1/4}, \tag{5.15}$$

$$\tau_\eta \equiv \left(\frac{\nu}{\epsilon}\right)^{1/2}. \tag{5.16}$$

If a Reynolds number is defined on the basis of a characteristic scale of the flow domain, L, and the root-mean-square (rms) velocity, $u_{rms} = \sqrt{\langle u^2 \rangle}$, as $Re = Lu_{rms}/\nu$, then it can be shown (see e.g. Bernard and Wallace, 2002) that:

$$\frac{L}{\eta} \sim Re^{3/4}. \tag{5.17}$$

This relationship will prove to be of fundamental importance for estimating the computational needs of DNS.

[3] Reviews of related papers can be found, for example, in Rigopoulos (2019) for soot formation, Tsagkaridis et al. (2022) for aerosol synthesis and Tang et al. (2023) for crystallisation.

In turbulent flows, most of the energy is contained in the larger eddies, which have features determined by the boundary conditions of a particular flow. These eddies gradually break down and generate a spectrum of eddy sizes, down to the Kolmogorov scales. As the eddies break down, their statistics gradually become more and more universal. This feature implies that the smaller scales are more amenable to approximation by a universal model and is exploited by LES, where modelling is employed for the smaller scales, as opposed to RANS, where the model accounts for all scales.

5.2.4 Direct Numerical Simulation (DNS)

DNS consists of the numerical solution of the Navier–Stokes equations, which do not require any closure models in their original form. However, DNS must employ a spatial and temporal discretisation such that all scales of turbulent motion are resolved. Therefore, while the conduct of a DNS does not involve any modelling questions, several issues regarding the numerical method must be considered to ensure that all scales are indeed resolved; further details on these matters can be found in Moin and Mahesh (1998). The need for resolving all scales renders DNS prohibitively expensive with increasing *Re*, as will be shown below.

The computational cost of DNS can be estimated as follows (Bernard and Wallace, 2002). If the size of the grid cells is proportional to η, then the grid should comprise approximately L/η cells along a single coordinate, which is proportional to $Re^{3/4}$ (Eq. 5.17). Therefore, the number of cells in three dimensions is proportional to $Re^{9/4}$, which means that the computation of a flow with a *Re* of 10^4 requires a grid of the order of 10^9 cells. Furthermore, if η/u_{rms} is an estimate of the shortest time scale to be resolved and L/u_{rms} is a characteristic time of the large-scale motions, then the ratio of these two timescales, L/η, is indicative of the number of time steps that must be performed in order to collect statistics for the average quantities, hence of the order of $Re^{3/4}$. Accordingly, the total number of operations required by a DNS scales with Re^3, hence an increase of *Re* by a factor of 10 results in an increase in the number of operations by a factor of 1000. For this reason, the range of *Re* that can be investigated by DNS is limited.

Due to the aforementioned *Re* constraint, the application of DNS is restricted to fundamental studies where the objective is to understand certain features of a fluid flow by examining the results of a simulation free of phenomenological models. In such studies, the simulation is treated as a substitute for a physical experiment.

When DNS is combined with other models, such as chemical reactions or the PBE, it must be acknowledged that the overall simulation does not constitute a DNS in the sense of being a truly model-free numerical solution of the problem as a whole. This is because the additional models, such as the kinetics employed in the PBE, involve assumptions of their own. A DNS-PBE simulation, therefore, must be interpreted as a DNS with respect to fluid dynamics only, where the role of DNS is to exclude the effects of turbulence and turbulence–PBE interaction modelling and allow us to study a population balance problem in a fluid flow, bearing in mind the remaining uncertainties inherent in the kinetic models.

While a DNS-PBE simulation is computationally expensive due to the need to solve the PBE in addition to the flow at all scales, it is nevertheless feasible for suitably simplified cases and allows the detailed study of the interaction of turbulence with kinetic processes. Several DNS-PBE studies have been conducted in recent years; for example, Tsagkaridis et al. (2022) conducted a DNS-PBE simulation of coagulation in a turbulent flow in order to study the effect of turbulent fluctuations on coagulation while Tang et al. (2020, 2022, 2023) studied crystallisation with DNS-PBE in order to provide insight into the interaction of turbulence with nucleation and growth and the effect of turbulent fluctuations on modelling approaches. The latter work will be shown in more detail in Section 6.3, in the form of a case study.

5.2.5 Large Eddy Simulation (LES)

The essence of large eddy simulation (LES) is to compute directly the larger flow structures and employ a model for the smaller ones. LES has its origins in the pioneering work of Smagorinsky (1963) and, while more computationally expensive than RANS, it is finding increasing use with the rise in processing power and memory of contemporary computers.

LES is motivated by certain features of turbulence that have been observed in a wide range of flows and are expressed succinctly via Kolmogorov's phenomenological theory. As mentioned in Section 5.2.3, while the larger scales have a structure that is particular to a given flow, the smaller eddies have increasingly universal features and are more amenable to the application of a general-purpose model. The simulation of the large scales in different flows is not very sensitive to the choice of model for the small scales, a fact that endows LES with substantial predictive capability across different flows. By contrast, a RANS model has much more impact on the prediction of the mean values and should be carefully selected for flows with particular features. It should be noted, however, that the concept of scale separation on which LES relies is more prominent in high Reynolds number flows and away from walls. It should also be emphasised that, while RANS yields a single outcome for the mean values, a DNS or LES corresponds to an individual realisation of a turbulent flow (among an ensemble of many possible ones with the same nominal boundary conditions). Since ensemble averaging is not practically possible, time averaging is typically applied to DNS or LES to obtain mean values.

The flow structures that are resolved in an LES are those greater than the filter size, which is often taken to be the grid cell size. The effect of the smaller ones, or *subgrid scales*, is approximated via a model. The finer the grid, the more flow scales are resolved. This means that, in LES, it is not possible to obtain a grid independent solution in the sense of this concept in conventional numerical methods. The choice of grid in LES is driven by considerations about the proportion of the turbulent kinetic energy that is being resolved – in general, at least 80–90% of it must be resolved. In the limit of all energy being resolved, an LES becomes a DNS.

While the basic concept of LES is simple, its successful application relies on a number of choices such as the grid, subgrid model and wall treatment. The details of these matters are out of the scope of the present text and can be found in many good references such as Rogallo and Moin (1984), Piomelli and Chasnov (1996), Fröhlich and Rodi (2002) and Sagaut (2006). In the following, we will summarise only the basic equations of LES, in order to establish the notation for the LES-based population balance models to be presented in the following sections.

In LES, spatial filtering (rather than average, as in RANS) is applied to the flow equations. For a given flow variable, $f(\mathbf{x})$, this operation can be defined via a convolution integral as follows[4]:

$$\overline{f}(\mathbf{x}) = \int_{\Omega} f(\mathbf{x}')G(\mathbf{x}-\mathbf{x}')d\mathbf{x}', \tag{5.18}$$

where the function G is the *filter* and should be a non-negative function such that $\int G(\mathbf{x}-\mathbf{x}')d\mathbf{x}' = 1$. It should be noted that, in general, $\overline{\overline{f}}(x) \neq \overline{f}(x)$, unlike Reynolds averaging where $\langle\langle f\rangle\rangle = \langle f\rangle$. One of the most commonly used filters, and the one that appears in the case studies in the present book, is the *box filter*:

$$G(\mathbf{x}-\mathbf{x}') = \begin{cases} 1/\Delta^3 & \text{if } |\mathbf{x}-\mathbf{x}'| \leq \Delta/2, \\ 0 & \text{otherwise,} \end{cases} \tag{5.19}$$

where Δ is filter width. The latter can vary with position and is often chosen to coincide with the grid cell size. In that case, all flow and scalar structures at scales smaller than the cell are filtered out and must be accounted for by subgrid models.

With the aid of the above definitions, the following filtered equations of motion and passive scalar transport can be derived:

$$\frac{\partial \overline{u}_i}{\partial x_i} = 0, \tag{5.20}$$

[4] Note that, in the present text, angle brackets denote averaging while overlines denote spatial filtering.

$$\frac{\partial \overline{u}_i}{\partial t} + \frac{\partial \left(\overline{u}_j \overline{u}_i\right)}{\partial x_j} = -\frac{1}{\rho}\frac{\partial \overline{p}}{\partial x_i} + \nu \frac{\partial^2 \overline{u}_i}{\partial x_j \partial x_j} - \frac{\partial \tau_{ij}^{sgs}}{\partial x_j} + g_i, \tag{5.21}$$

$$\frac{\partial \overline{\phi}}{\partial t} + \frac{\partial \left(\overline{u}_i \overline{\phi}\right)}{\partial x_i} = D\frac{\partial^2 \overline{\phi}}{\partial x_i^2} - \frac{\partial J_{\phi,i}^{sgs}}{\partial x_i}, \tag{5.22}$$

where, as in Section 5.2.2, the constant density and viscosity form of the Navier–Stokes equations (Eqs. A.2 and A.25) have been employed for simplicity, as well as constant diffusivity in Eq. 5.22. Like the RANS equations, the filtered equations are unclosed. In the momentum equation, the unclosed term is τ_{ij}^{sgs}, a tensor whose components have the dimensions of stress (like the Reynolds stresses) and represents the contribution of subgrid motions on the filtered velocity field. It is given by:

$$\tau_{ij}^{sgs} = \overline{u_i u_j} - \overline{u}_i \overline{u}_j. \tag{5.23}$$

In the scalar transport equation, the unclosed term is $J_{\phi,i}^{sgs}$, which is a vector that represents the effect of sub-grid scales on scalar transport:

$$J_{\phi,i}^{sgs} = \overline{u_i \phi} - \overline{u}_i \overline{\phi}. \tag{5.24}$$

Subgrid models must be provided to obtain closure of the LES equations. However, as mentioned before, the success of an LES simulation relies mainly on resolving the majority of the energetic motions rather than on the subgrid models.[5] The oldest subgrid model, and still a commonly employed one, is the Smagorinsky model, which relates the subgrid stress tensor to the resolved rate of strain tensor with the aid of an eddy viscosity:

$$\tau_{ij}^{sgs} - \frac{1}{3}\delta_{ij}\tau_{kk} = -2\nu_t \overline{e}_{ij}, \tag{5.25}$$

where the eddy viscosity, ν_t, is defined as:

$$\nu_t = (C_S \Delta)^2 \left| |\overline{e}_{ij}| \right|, \tag{5.26}$$

where C_S is the Smagorinsky constant, Δ is the filter width (usually taken as the cubic root of the local cell volume) and $||\overline{e}_{ij}|| = \sqrt{2\overline{e}_{ij}\overline{e}_{ij}}$ where $\overline{e}_{ij}$ is the resolved rate of strain tensor given by:

$$\overline{e}_{ij} = \frac{1}{2}\left(\frac{\partial \overline{u}_i}{\partial x_j} + \frac{\partial \overline{u}_j}{\partial x_i}\right). \tag{5.27}$$

A commonly used value of C_S is 0.1 (Fröhlich and Rodi, 2002). The scalar transport equation is closed in a similar way, with an eddy diffusivity defined as per Eq.

[5] In fact, one approach known as Implicit LES employs no subgrid models and relies on the dissipation in the numerical scheme (Grinstein et al., 2007).

5.11. Other models compute C_S dynamically or involve different concepts, and further models are also needed near walls; the details of those can be found in the aforementioned references.

5.2.6 Reacting Flows

Reacting flows pose an additional closure problem associated with the source term. This can be seen by considering Eq. 5.8 with the addition of a source term, S_ϕ, which is a nonlinear function of ϕ:

$$\frac{\partial\langle\phi\rangle}{\partial t}+\frac{\partial\left(\langle u_i\rangle\langle\phi\rangle\right)}{\partial x_i}+\frac{\partial\langle u_i'\phi'\rangle}{\partial x_i}=D\frac{\partial^2\langle\phi\rangle}{\partial x_i^2}+\langle S_\phi(\phi)\rangle. \tag{5.28}$$

A new unclosed term appears, $\langle S(\phi)\rangle$. This term is unclosed because the average value of a nonlinear function is not equal to the function of the average values, that is:

$$\langle S_\phi(\phi)\rangle \neq S_\phi(\langle\phi\rangle). \tag{5.29}$$

The general form of the reaction source term is shown in Section A.1.4. In an isothermal flow, the nonlinearity arises from the product of concentrations involved in the reaction kinetics. For a second-order chemical reaction between two species A and B, we have:

$$\langle S_\phi\rangle = k\langle C_A C_B\rangle, \tag{5.30}$$

which is unclosed. In a non-isothermal flow, further nonlinearity arises from the Arrhenius temperature dependence:

$$\langle S_\phi\rangle = A\langle e^{-\frac{E_0}{RT}} C_A C_B\rangle, \tag{5.31}$$

where A is the pre-exponential factor and E_0 is the activation energy.

It is also worth considering the physical implications of the reaction source term closure problem. In a very slow reaction, there is enough time for complete mixing to occur before reaction takes place and, therefore, the effect of such fluctuations can be negligible. When the reaction is extremely fast, it can be assumed to take place instantaneously and reach complete conversion or chemical equilibrium. In this case, fluctuations are important, but the problem can be described with a reduced model in terms of a single variable describing the extent of mixing (cf. Section 5.4). When the reaction timescale is comparable to the mixing one, such a reduction is not possible and the effect of fluctuations on finite rate kinetics must be considered. The above can be quantified with the aid of the Damköhler number, which is defined as follows:

$$Da = \frac{\tau_f}{\tau_c}, \tag{5.32}$$

where τ_f and τ_c are the characteristic time scales associated with flow and chemistry, respectively. High Da values imply that the problem is mixing controlled, while low values indicates the opposite. In the case of fast reactions (as compared with the time for mixing), such as those occurring in a flame in soot formation or aerosol synthesis, incomplete mixing can have a profound effect on the reaction rate as it can control phenomena like ignition and extinction. In Section 6.3.6, as part of the case study on precipitation, two separate Damköhler numbers will be defined in order to characterise the interaction between turbulent mixing, nucleation and growth.

In LES, a similar closure problem arises, as can be seen from adding a source term to Eq. 5.22:

$$\frac{\partial \overline{\phi}}{\partial t} + \frac{\partial \left(\overline{u}_i \overline{\phi}\right)}{\partial x_i} = D \frac{\partial^2 \overline{\phi}}{\partial x_i^2} - \frac{\partial J^{sgs}_{\phi,i}}{\partial x_i} + \overline{S_\phi(\phi)}, \tag{5.33}$$

where the unclosed term is the filtered nonlinear source term, $\overline{S_\phi(\phi)}$.

5.2.7 Variable Density Flows

While compressible flows will not be considered in the present book, a category of variable density flows is relevant to a number of population balance problems. In these flows, the Mach number is small and they can thus be treated as incompressible (i.e. the density does not change significantly with pressure), but the density is nevertheless variable due to temperature and composition changes. In soot formation and aerosol synthesis, for example, the density changes considerably as a result of the chemical reaction.

When dealing with the effect of turbulence on variable-density flows, it is convenient to employ *density-weighted* or *Favre averaging* (Favre, 1969). In this approach, all variables except pressure are averaged, as shown below for the velocity in the case of RANS:

$$\widetilde{u}_i = \frac{\langle \rho u_i \rangle}{\langle \rho \rangle}, \tag{5.34}$$

where a tilde ($\widetilde{\cdot}$) over any quantity denotes density-weighted averaging. In the above, the velocity was decomposed into its density-weighted average and fluctuation about that as follows:

$$u_i = \widetilde{u}_i + u_i''. \tag{5.35}$$

It should be noted that $\langle \mathbf{u}'' \rangle \neq 0$ but rather $\langle \rho \mathbf{u}'' \rangle = 0$. Based on this, Favre-averaged equations can be derived; for example, the equivalent of Eq. 5.8 is:

$$\frac{\partial \overline{\rho}\widetilde{\phi}}{\partial t} + \frac{\partial \left(\overline{\rho}\widetilde{u}_i\widetilde{\phi}\right)}{\partial x_i} + \frac{\partial \langle \rho u_i'' \phi'' \rangle}{\partial x_i} = \frac{\partial}{\partial x_i}\left\langle \rho D \frac{\partial \phi}{\partial x_i} \right\rangle, \tag{5.36}$$

where we have considered the case of varying diffusivity. The full set of Favre-averaged equations can be found in Libby and Williams (1980).

In the case of LES, we first define the filtered density field:

$$\overline{\rho}(\mathbf{x},t) = \int_{\Omega} \rho(\mathbf{x}',t) G(\mathbf{x}-\mathbf{x}') d\mathbf{x}'. \tag{5.37}$$

With the aid of this, density-weighted filtered values can be defined in a way analogous to Eq. 5.34:

$$\overset{\approx}{u}_i = \frac{\overline{\rho u_i}}{\overline{\rho}}, \tag{5.38}$$

where the double tilde $(\overset{\approx}{\cdot})$ is used to denote density-weighted filtering.

5.3 Closure Problem for Population Balance in Turbulent Flow

We now focus on the effect of turbulence on kinetic processes, namely nucleation, growth, aggregation and fragmentation. First, the averaged PBE is shown and the unclosed terms are identified. The same analysis is carried afterwards for the moment equations. The types of unclosed terms are then classified, and this is followed by a first look into the closure methods. The following sections will discuss these methods in more detail.

5.3.1 Averaging of the PBE

For the purpose of the following discussion, it is instructive to write the PBE for non-inertial particles with the transport terms collected on the left-hand side and the kinetic ones on the right-hand side (in the following, arguments will be omitted except when necessary):

$$\begin{aligned}\frac{\partial n}{\partial t} + \frac{\partial (u_i n)}{\partial x_i} - D_p(v)\frac{\partial^2 n}{\partial x_i \partial x_i} &= B(Y_\alpha)\delta(v - v_0) - \frac{\partial (G(Y_\alpha, v)n)}{\partial v} \\ &+ \frac{1}{2}\int_0^v \beta(w, v-w) n(w,\mathbf{x},t) n(v-w,\mathbf{x},t) dw - n(v,\mathbf{x},t)\int_0^\infty \beta(v,w) n(w,\mathbf{x},t) dw \\ &+ \int_v^\infty \gamma(w) b(v,w) n(w,\mathbf{x},t) dw - \gamma(v) n(v,\mathbf{x},t),\end{aligned} \tag{5.39}$$

where the nucleation and growth terms are functions of scalars, Y_α, which typically include precursor concentrations and/or temperature and must be computed from the solution of the corresponding transport equation. In order to be employed in the context of a RANS or LES approach, this equation must be averaged or filtered, respectively. In this section we consider RANS, but it should be noted that the same unclosed terms appear in the case of LES.

In order to perform the averaging of the PBE, we first decompose the velocity, transported scalars and number density into mean and fluctuating values:

$$u_i = \langle u_i \rangle + u_i', \tag{5.40}$$

$$Y_\alpha = \langle Y_\alpha \rangle + Y_\alpha', \tag{5.41}$$

$$n = \langle n \rangle + n'. \tag{5.42}$$

We now take the average of each term in Eq. 5.39. The averaging of the accumulation and diffusion terms is straightforward, as they involve linear operators:

$$\left\langle \frac{\partial n}{\partial t} \right\rangle = \frac{\partial \langle n \rangle}{\partial t}, \tag{5.43}$$

$$\left\langle D_p(v) \frac{\partial^2 n}{\partial x_i \partial x_i} \right\rangle = D_p(v) \frac{\partial^2 \langle n \rangle}{\partial x_i \partial x_i}. \tag{5.44}$$

The average of the convective transport term can be decomposed into two terms:

$$\left\langle \frac{\partial (u_i n)}{\partial x_i} \right\rangle = \frac{\partial (\langle u_i \rangle \langle n \rangle)}{\partial x_i} + \frac{\partial \langle u_i' n' \rangle}{\partial x_i}. \tag{5.45}$$

The second term on the right-hand side is unclosed and represents the transport of the number density fluctuation by the velocity fluctuation. For non-inertial particles, the methods used for modelling for this term are the same methods as those used for turbulent transport of scalars.

The averaging of the nucleation term yields the following expression:

$$\langle B(Y_\alpha) \delta(v - v_0) \rangle = \langle B(Y_\alpha) \rangle \, \delta(v - v_0). \tag{5.46}$$

Further manipulation of this term is not possible because nucleation is, in general, a nonlinear function of Y_α. This term poses a closure issue akin to that of the source term in reacting flows (Section 5.2.6).

The growth term can be averaged in the following way:

$$\left\langle \frac{\partial (G(Y_\alpha, v) n)}{\partial v} \right\rangle = \frac{\partial}{\partial v} \left(\langle G(Y_\alpha, v) \rangle \langle n \rangle \right) + \frac{\partial}{\partial v} \langle G(Y_\alpha', v) n' \rangle. \tag{5.47}$$

Since growth is, like nucleation, usually a nonlinear function of the scalars Y_α, both terms resulting from this averaging are unclosed. The first of these terms is unclosed due to the presence of $\langle G(Y_\alpha, v) \rangle$ and thus poses a closure problem akin to that of the nucleation term, while the second one involves a correlation between fluctuating growth rate and number density.

We now consider the averaging of the aggregation terms. The average is elaborated as follows for the birth term:

$$\left\langle \frac{1}{2}\int_0^v \beta(w,v-w)n(w,\mathbf{x},t)n(v-w,\mathbf{x},t)dw \right\rangle$$
$$= \frac{1}{2}\int_0^v \beta(w,v-w)\langle n(w,\mathbf{x},t)\rangle\langle n(v-w,\mathbf{x},t)\rangle dw$$
$$+ \frac{1}{2}\int_0^v \beta(w,v-w)\langle n'(w,\mathbf{x},t)n'(v-w,\mathbf{x},t)\rangle dw, \tag{5.48}$$

while for the death term:

$$\left\langle n(v,\mathbf{x},t)\int_0^\infty \beta(v,w)n(w,\mathbf{x},t)dw \right\rangle$$
$$= \langle n(v,\mathbf{x},t)\rangle \int_0^\infty \beta(v,w)\langle n(w,\mathbf{x},t)\rangle dw$$
$$+ \int_0^\infty \beta(v,w)\langle n'(v,\mathbf{x},t)n'(w,\mathbf{x},t)\rangle dw. \tag{5.49}$$

In both cases, the integral involves a product of number densities and is thus unclosed.

Finally, for fragmentation, we have:

$$\left\langle \int_v^\infty \gamma(w)b(v,w)n(w,\mathbf{x},t)dw \right\rangle = \int_v^\infty \gamma(w)b(v,w)\langle n(w,\mathbf{x},t)\rangle dw. \tag{5.50}$$

$$\langle \gamma(v)n(v,\mathbf{x},t)\rangle = \gamma(v)\langle n(v,\mathbf{x},t)\rangle. \tag{5.51}$$

Both terms are closed, as can be expected from their linearity.

5.3.2 Averaging of the Moment Equations

Moment equations pose two kinds of closure problems in turbulent flows. In Sections 4.2.4 and 4.3.3, it was shown that moment equations are, in general, unclosed, except for special cases such as size independent growth and aggregation with the constant kernel. The presence of turbulence introduces an additional closure problem, due to the nonlinear terms involving fluctuations.

The issue can be shown initially with the aggregation equation. In Section 4.2.5, we saw that the moment transformation results in the following:

$$\frac{dm_k}{dt} = \frac{1}{2}\int_0^\infty \int_0^\infty \left[(v+w)^k - v^k - w^k\right] \beta(v,w)n(v,t)n(w,t)dwdv, \tag{5.52}$$

which cannot be further reduced into a closed system involving a finite set of m_k except for special cases of the kernel $\beta(v, w)$. Now, introducing transport and averaging this equation for the case of turbulent flow yields the following result:

$$\frac{\partial \langle m_k \rangle}{\partial t} + \frac{\partial (\langle u_i \rangle \langle m_k \rangle)}{\partial x_i} + \frac{\partial \langle u_i' m_k' \rangle}{\partial x_i} = D_p(v) \frac{\partial^2 \langle m_k \rangle}{\partial x_i \partial x_i}$$
$$+ \frac{1}{2} \int_0^\infty \int_0^\infty \left[(v+w)^k - v^k - w^k \right] \beta(v,w) \langle n(v,\mathbf{x},t) \rangle \langle n(w,\mathbf{x},t) \rangle dw dv$$
$$+ \frac{1}{2} \int_0^\infty \int_0^\infty \left[(v+w)^k - v^k - w^k \right] \beta(v,w) \langle n'(v,\mathbf{x},t) n'(w,\mathbf{x},t) \rangle dw dv, \tag{5.53}$$

which includes an extra unclosed term in the right-hand side akin to the one encountered in Eqs. 5.48 and 5.49.

In the case of the constant kernel, we saw that a closed form for the moment equations can be derived in the form of Eqs. 4.48 and 4.49. For the spatially dependent PBE in turbulent flow, the averaging of these equations yields:

$$\frac{\partial \langle m_0 \rangle}{\partial t} + \frac{\partial (\langle u_i \rangle \langle m_0 \rangle)}{\partial x_i} + \frac{\partial \langle u_i' m_0' \rangle}{\partial x_i} = D_p(v) \frac{\partial^2 \langle m_0 \rangle}{\partial x_i \partial x_i} - \frac{1}{2} \beta_0 \langle m_0 \rangle^2$$
$$- \frac{1}{2} \beta_0 \int_0^\infty \int_0^\infty \langle n'(v,\mathbf{x},t) n'(w,\mathbf{x},t) \rangle dw dv \tag{5.54}$$

and

$$\frac{\partial \langle m_k \rangle}{\partial t} + \frac{\partial (\langle u_i \rangle \langle m_k \rangle)}{\partial x_i} + \frac{\partial \langle u_i' m_k' \rangle}{\partial x_i} = D_p(v) \frac{\partial^2 \langle m_k \rangle}{\partial x_i \partial x_i} + \frac{1}{2} \beta_0 \sum_{i=1}^{k-1} \binom{k}{i} \langle m_i \rangle \langle m_{k-i} \rangle$$
$$+ \frac{1}{2} \int_0^\infty \int_0^\infty \left[(v+w)^k - v^k - w^k \right] \beta(v,w) \langle n'(v,\mathbf{x},t) n'(w,\mathbf{x},t) \rangle dw dv, \quad \forall k > 1, \tag{5.55}$$

that is, the equations are now unclosed due to the extra term representing the effect of number density fluctuations.

In Chapter 4, we saw that the fragmentation and growth terms under the moment transformation can pose a closure problem for arbitrary fragmentation kernels and growth rate functions. However, these terms are linear in terms of the number density, hence the presence of turbulence does not present an additional closure problem. For example, the averaging of the growth term integral (the one in the second term on the left-hand side of Eq. 4.104) yields:

$$\left\langle k \int_0^\infty v^{k-1} G(v) n(v,\mathbf{x},t) dv \right\rangle = k \int_0^\infty v^{k-1} G(v) \langle n(v,\mathbf{x},t) \rangle dv. \tag{5.56}$$

Therefore, the closure problem associated with the growth integral remains, but there is no additional closure problem due to turbulence since the integral does not involve a product of number densities (as in the case of aggregation).

If, however, the nucleation and growth rates are functions of the composition, an additional closure problem results as discussed in Section 5.3.1. For example, the equivalent of Eq. 4.104 for the case of a spatially dependent problem and turbulent flow is:

$$\frac{\partial\langle m_k\rangle}{\partial t}+\frac{\partial(\langle u_i\rangle\langle m_k\rangle)}{\partial x_i}+\frac{\partial\langle u_i'm_k'\rangle}{\partial x_i}=D_p(v)\frac{\partial^2\langle m_k\rangle}{\partial x_i\partial x_i}$$
$$+k\int_0^\infty v^{k-1}\langle G(Y_\alpha,v)\rangle\langle n(v,\mathbf{x},t)\rangle dv+k\int_0^\infty v^{k-1}\langle G(Y_\alpha',v)n'(v,\mathbf{x},t)\rangle dv$$
$$+v_0^k\langle B(Y_\alpha)\rangle+\left\langle\int_0^\infty v^kS(v,n)dv\right\rangle, \tag{5.57}$$

where $\langle G(Y_\alpha',v)n'(v,\mathbf{x},t)\rangle$ and $\langle B(Y_\alpha)\rangle$ are also unclosed as the nucleation and growth rate functions are, in general, nonlinear.

As in the previous section, all of the considerations in this section apply also to LES with the averaging operation replaced by filtering.

5.3.3 Classification of Unclosed Terms

To summarise, the averaging of the PBE for non-inertial particles results in unclosed terms of the following types:

1. Correlation between fluctuating velocity and number density.
2. nonlinear dependency of nucleation and growth kinetics on fluctuating composition and/or temperature.
3. Correlation between fluctuating growth rate and number density.
4. Correlations between number density fluctuations.

These terms appear both in the averaged PBE and the averaged moment equations. Unclosed terms of the first type are typically addressed with models for turbulent transport of scalars. In the following sections, we will concentrate on methods for obtaining closure for the other three terms that are associated with kinetic processes.

The physical meaning and importance of the unclosed terms in the PBE for turbulent flow depends on the case. An evaluation of the importance of these unclosed terms can be carried out by DNS, in the context of particular problems. Tsagkaridis et al. (2022) studied the correlations between number density fluctuations in the coagulation of particles in a turbulent jet. Tang et al. (2023) studied the effect of fluctuations in a crystallisation problem with nucleation and growth kinetics. The second of these works will be discussed further as part of a case study in Section 6.3.1. It will be shown there that nucleation, in particular, has a strong nonlinear dependence on supersaturation, which implies that small fluctuations in supersaturation can result in large fluctuations in the nucleation rate. The interaction between turbulent mixing and precipitation will be quantified with the aid of suitably defined Damköhler numbers for nucleation and growth (Section 6.3.6).

Finally, a note should be made about LES. As mentioned, the unclosed terms in the filtered transport equations are the same as in the averaged ones. However, as with all subgrid models employed in LES, the impact of these terms is smaller than in RANS and diminishes with the use of increasingly fine grids. Still, the closure models described in Sections 5.3.4 and 5.5 can be applied also to LES.

5.3.4 Overview of Closure Methods

We now consider models that have been proposed for closure of the unclosed terms discussed in the previous sections. The number of studies of population balance in turbulent flows is relatively limited. However, a similar closure problem occurs in turbulent combustion, where the closure of the nonlinear reaction term is vital for combustion models. As a result, several modelling approaches have been developed in that field. The methods for the PBE in turbulent flow are based on the same concepts.

As mentioned in Section 5.2.6, the closure problem stems from the fact that the average of a nonlinear function, $F(\phi)$, is not equal to the function of the mean value of ϕ, that is, $\langle F(\phi)\rangle \neq F(\langle\phi\rangle)$. Therefore, such a term remains unclosed upon averaging and the resulting equation cannot be reduced into a form where the only dependent variable is $\langle\phi\rangle$. If, however, the PDF of ϕ, $f(\psi)$, is known, then the average of F can be evaluated as follows:

$$\langle F(\phi)\rangle = \int_{-\infty}^{\infty} F(\psi) f(\psi) d\psi. \tag{5.58}$$

From this point on, two types of approach can be adopted. In the first one, an assumption is made that allows us to presume the functional form of the PDF, $f(\psi; t)$. Once this is available, the average in Eq. 5.58 can be carried out. This type of closure is called a *presumed PDF method.* In the second approach, no assumption is made on the PDF; rather, it is computed as part of the solution. This is called a *transported PDF method* because a transport equation is derived and solved for the PDF itself.

Presumed PDF methods can be paralleled to presumed distribution moment methods for the solution of the PBE (cf. Section 4.2.5). In both cases, closure is obtained via an assumption on the shape of the distribution. On the other hand, transported PDF methods are the equivalent of discretisation methods for the PBE. Other similarities can also be drawn with the solution methods: transported PDF methods are more memory and computationally intensive but involve fewer assumptions. Unlike the case of the PBE, however, the closure problem cannot be avoided completely in the case of turbulence and some assumptions are essential even in transported PDF methods. Furthermore, stochastic numerical methods need to be employed in the latter.

In the following sections, we will consider briefly the basics of each approach and describe some representative works. A more extensive review of related works can be found in Rigopoulos (2019) in the context of soot formation, or in Rigopoulos (2010) in a more general context. The transported PDF approach will then be described in some detail, as it is employed in the case study on soot formation in Section 6.2. It should also be mentioned that most of these methods, while initially developed in a RANS context, have been adapted also to LES.

5.4 Presumed PDF Methods

The presumed PDF approach to modelling turbulent flows with nonlinear source terms originates in the combustion literature. As such, the earliest forms of population balance models adopting this approach can be found in the studies of soot formation. A similar approach has also been applied to turbulent precipitation.

Most presumed PDF methods are applied to two-stream mixing problems, where reactants from the two streams form a mixture with concentrations that induce kinetic processes such as nucleation and growth. The presumed PDF is then used to address the effect of fluctuations in rates of these processes.

The concept of a *mixture fraction* is usually employed in presumed PDF methods. In a two-stream mixing problem, the mixture fraction, Z, can be defined as the mass fraction originating in a certain specified inlet. By definition, Z is equal to 1 in that inlet (Fig. 5.1). This approach has its origins in the field of combustion, where Z is defined as the mass fraction of the mixture that originated in the fuel inlet stream, and most problems where the fuel and oxidant enter via separate streams can be reduced to a binary mixing model like this. For example, in a non-premixed

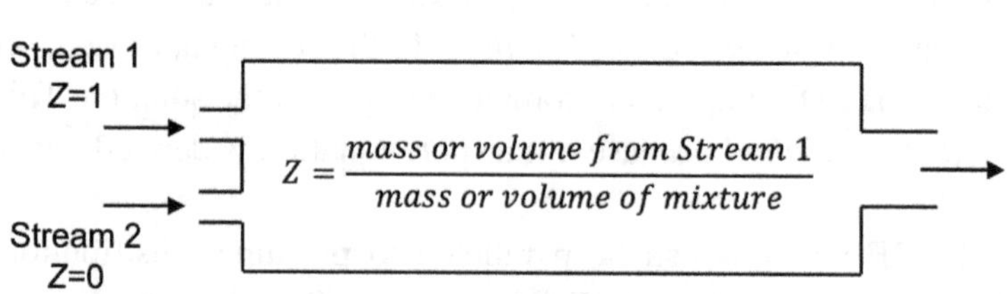

Figure 5.1 Definition of mixture fraction for a two-stream mixing problem.

jet flame, the fuel enters as a jet ($Z = 1$), while in the ambient air $Z = 0$. Similarly, in a spray, $Z = 1$ in the fuel droplets; in a solid fuel furnace, $Z = 1$ in the fuel particles. The same concept can be employed in a two-stream mixing-precipitation process, such as in a double-feed stirred reactor or a T-mixer.

If there is no reaction, then the mixture fraction will be simply the mass fraction of the component associated with a given inlet. In a reacting flow, however, the mixture downstream will also comprise reaction products. The mixture fraction can then be interpreted as follows: if we identify a certain fluid element and trace its history back to the original inlet streams, then Z would be the mass fraction originating in a given stream.

Once the mixture fraction has been defined, it is usually possible to derive a differential equation for its transport and to relate the concentrations of the reactants in a unique way to it. The mixture fraction can be related to the amount of elements, under the assumption of equal diffusivities, and since that does not change during chemical reactions, the transport equation for Z is that of a passive scalar:

$$\frac{\partial(\rho Z)}{\partial t} + \frac{\partial(\rho u_i Z)}{\partial x_i} = \frac{\partial}{\partial x_i}\left(\rho D \frac{\partial Z}{\partial x_i}\right). \tag{5.59}$$

Definitions for three-stream mixing are also available; more information on the mixture fraction can be found in Peters (2000) or Fox (2003).

With the aid of the mixture fraction, the concentrations and consequently the nucleation and growth rates can be expressed as functions of Z. It suffices, then to prescribe the PDF for the mixture fraction in order to obtain closure of the unclosed terms in the PBE. A common assumption is that the mixture fraction assumes a β-PDF. The presumed PDF involves two parameters, one of which is the mean value of the mixture fraction. The other one is its variance (Eq. 5.14). As discussed in Section 5.2.2, this involves the scalar dissipation, for which a model must be employed, and a discussion of such models can be found in Peters (2000), Veynante and Vervisch (2002) and Poinsot and Veynante (2005).

A summary and classification of various types of presumed PDF methods applied to the PBE can be found in Table I of Sewerin and Rigopoulos (2017b). Below, we look at some examples of their application in fields such as soot formation and crystallisation.

An example of a study of the effect of turbulent fluctuations on soot formation with a presumed PDF method is the work of Brookes and Moss (1999), based on the flamelet model of turbulent combustion.[6] In that work, three approaches were compared. The first one is to assume that soot is uncorrelated with mixture fraction but correlated with species and temperature, thus accounting for fluctuations

[6] For more information on this model, the reader may consult Peters (2000).

in terms such as the nucleation rate, at least within the confines of the flamelet approach (which assumes that species and temperature are correlated with mixture fraction and, in that work, on radiative heat loss). The second approach is to consider the correlation of soot with mixture fraction, which enables accounting for the unclosed nonlinear terms with the presumed β-PDF method. The third approach is to neglect all fluctuations and gave the least satisfactory results. The correlated closure gave considerably better results than the uncorrelated one and, furthermore, the latter required ad hoc scaling of the oxidation rate.

Soot formation has been approached with several other presumed-PDF models, which have been developed for turbulent reacting flows mainly with the purpose of providing closure to the nonlinear reaction source terms. Further details on these models can be found in Peters (2000), Poinsot and Veynante (2005) and Veynante and Vervisch (2002). A review of models for soot formation in turbulent flows can be found in Rigopoulos (2019).

Another problem involving population balance in turbulent flow to be approached with presumed PDF methods is that of reaction crystallisation, or precipitation. In this problem, nucleation and growth are nonlinear functions of supersaturation, itself a function of the chemical species' concentrations. A DNS-PBE study of the nonlinear interactions between turbulence, nucleation and growth, as well as an evaluation of the magnitude of the unclosed terms, has been recently conducted by Tang et al. (2022, 2023) and the main findings will be shown in Section 6.3 in the form of a case study. Here, we concentrate on the use of presumed PDF methods for the closure of these terms.

Bałdyga and Orciuch (1997, 2001) proposed a closure for precipitation processes based on the concept of a mixture fraction and the presumed PDF approach. In these processes, the reactants enter via two separate streams into a mixing equipment such as a stirred tank, a pipe or a T-mixer. It was thus assumed that the mixture fraction assumes a β-PDF, a widely used choice also in combustion. With this method, they were able to obtained a closed form of the population balance moment equations. This approach was used in several subsequent works on precipitation (Bałdyga et al., 2003, 2007; Vicum and Mazzotti, 2007). A related approach is the finite-mode PDF employed by Piton et al. (2000) and Marchisio et al. (2001c,b), where the PDF of mixture fraction is given as a sum of delta functions. The relationship of this approach to presumed PDF methods is discussed by Fox (1998). Gavi et al. (2007a,b) employed the DQMOM-IEM approach, which is related to the DQMOM method for the closure of moment equations for solving the PBE (4.2.5), indicating the links between presumed PDF methods for turbulence and PBE closure. All of the aforementioned works employ the method of moments for the solution of the population balance. More works in this field are reviewed by Tang et al. (2023).

5.5 Transported PDF Methods

Transported PDF methods proceed by deriving and solving a transport equation for the PDF itself. The motivation for doing so originates not in an inherent interest in the PDF itself, but rather in the aim to obtain its low-order moments (particularly the mean values) while overcoming the closure problem associated with nonlinear source terms.

To ensure that every nonlinear source term appears in closed form in the PDF transport equation, all the variables present in that term must be included as dependent variables in the PDF. For example, if a nucleation rate depends on the concentrations of chemical species and temperature, the joint PDF of species and temperature must be considered. This brings up the main drawback of the transported PDF approach, which is the increase in the dimensionality of the equation.

In the case of several scalar variables, the increase in dimensionality is such that it becomes too expensive to solve the PDF transport equation via discretisation methods. Stochastic (i.e. Monte Carlo) methods, which do not scale exponentially with dimensionality, must be used instead, bringing in the issue of stochastic convergence. However, since the main motivation of employing a transported PDF method is to obtain the mean values, convergence for the means is sufficient.

In the study of turbulence, transported PDF methods were first considered as an avenue towards a closure model for turbulence (Lundgren, 1967), and further research along this line is described by Pope (2000). Here, however, we are interested in the PDF approach as a method of closure for scalar transport equations with source terms, such as the discretised PBE or its moment-transformed counterpart. PDF methods for such problems have been pioneered in the study of turbulent reacting flows (Dopazo and O'Brien, 1974; Pope, 1976; O'Brien, 1980), where it is required to obtain closure for the chemical reaction source term. A related approach is the multi-fluid model of Spalding (1996). Reviews of transported PDF methods can be found in O'Brien (1980), Pope (1985, 2000), Dopazo (1994), Jones (2002), Fox (2003) and Haworth (2010).

In the field of population balance modelling, transported PDF methods were first employed in conjunction with a monodisperse soot model by Kollmann et al. (1994) and subsequently in several works based on the method of moments such as the works of Marchisio et al. (2001a) and Falk and Schaer (2001) on turbulent precipitation, Zamuner and Dupoirieux (2000) and Lindstedt and Louloudi (2005) on soot formation, and Garmory and Mastorakos (2008) on aerosol formation. Rigopoulos (2007) proposed a general framework for the extension of the PDF approach to the discretised PBE, the PBE-PDF approach, which was first applied in a RANS context to precipitation and aerosols (Di Veroli and Rigopoulos, 2009,

2010, 2011) and in an LES context to soot (Sewerin and Rigopoulos, 2018, 2019; Sun and Rigopoulos, 2022). This approach will be described in Section 5.5.2 and demonstrated in Section 6.2 (in its LES form) as part of the case study on soot formation. A related approach is the combination of PBE with Multiple Mapping Conditioning (MMC), as described in Neuber et al. (2017). For an example of an application of the transported PDF method to LES in conjunction with a moment method, the reader may consult Donde et al. (2013), while more related works are reviewed in Rigopoulos (2019).

Before considering the application of transported PDF methods to the PBE, it is necessary to explain the basic concepts of PDF methods and their solution methods in some detail. In order to make the fundamentals clear, the derivation of the PDF transport equation for a single scalar in the context of RANS will first be shown in the next section, with some of the details included in Appendix C. The PBE-PDF approach will then be described in Section 5.5.2, initially for RANS and subsequently for LES. Stochastic numerical methods are required for the solution of transported PDF equations, and these are described in Section 5.5.3. Appendix D includes more information on the derivation of the stochastic field method, which is particularly suited to LES-PBE-PDF and is used in simulation of the Sandia turbulent sooting flame (Section 6.2.8). The computational implementation of transported PDF methods is also discussed (Section 5.5.4).

5.5.1 PDF Equation for Scalar Transport with a Nonlinear Source Term

To illustrate the basic ideas involved in the derivation of PDF methods, we consider the transport equation for a single scalar, ϕ, with a nonlinear source term, $S(\phi)$, in a constant density flow:

$$\frac{\partial \phi}{\partial t} + \frac{\partial (u_i \phi)}{\partial x_i} = -\frac{\partial J_{\phi,i}}{\partial x_i} + S(\phi). \tag{5.60}$$

The following derivation will be in the context of RANS, while the extension to LES will be considered afterwards. By decomposing the velocity into mean and fluctuating parts, *but not the scalar*, we obtain the following:

$$\frac{\partial \phi}{\partial t} + \frac{\partial (\langle u_i \rangle \phi)}{\partial x_i} + \frac{\partial (u_i' \phi)}{\partial x_i} = -\frac{\partial J_{\phi,i}}{\partial x_i} + S(\phi). \tag{5.61}$$

The mean velocity will be regarded as known (i.e. computed by solving the RANS equations supplied with a turbulence model). If the decomposition was applied to the scalar as well, we would obtain Eq. 5.28, which would involve the unclosed reaction source term. Rather than this, we take Eq. 5.61 as the starting point and introduce the PDF of ϕ, $f(\psi;\mathbf{x},t)$. The objective now is to derive a transport equation for this function.

The derivation of the PDF transport equation is somewhat lengthy and shown in Appendix C; only the main steps will be outlined here. The PDF is first expressed with the aid of the *fine-grained density*, which is a function defined as follows:

$$\mathcal{F}(\psi;\mathbf{x},t) = \delta(\psi - \phi(\mathbf{x},t)). \tag{5.62}$$

The fine-grained density can be physically interpreted as the PDF of the scalar field in a single realisation.[7] Since, in that realisation, the scalar can have only one value at each spatial location and time instant, its PDF (i.e. $\mathcal{F}$) approaches infinity for that value and is zero elsewhere, and this is exactly what is expressed by the delta function. The PDF can then be expressed as an ensemble average over the individual realisations via the following identity, which is proven in Appendix C:

$$f(\psi;\mathbf{x},t) = \langle \mathcal{F}(\psi;\mathbf{x},t)\rangle. \tag{5.63}$$

The transport equation for the PDF can now be obtained in the following way:

$$\frac{\partial f(\psi;\mathbf{x},t)}{\partial t} = \frac{\partial \langle \mathcal{F}(\psi;\mathbf{x},t)\rangle}{\partial t}. \tag{5.64}$$

Using another identity (C.9) proven in Appendix C, Eq. 5.64 becomes:

$$\frac{\partial f(\psi;\mathbf{x},t)}{\partial t} = -\left\langle \frac{\partial}{\partial \psi}\left[\frac{\partial \phi(\mathbf{x},t)}{\partial t}\delta(\psi - \phi(\mathbf{x},t))\right]\right\rangle. \tag{5.65}$$

Now $\partial\phi/\partial t$ can be substituted into Eq. 5.65 using Eq. 5.61, which gives the evolution of the scalar in a single realisation. The remaining steps are shown in Appendix C. The outcome is the following equation (where the dependence of f and ϕ is omitted for brevity):

$$\underbrace{\frac{\partial f}{\partial t}}_{\text{rate of change}} + \underbrace{\langle u_i\rangle \frac{\partial f}{\partial x_i}}_{\text{transport by mean velocity}} + \underbrace{\frac{\partial}{\partial x_i}\left(\left\langle u_i' \middle| \phi = \psi\right\rangle f\right)}_{\text{turbulent transport (unclosed)}} = \underbrace{\frac{\partial}{\partial \psi}\left[\left\langle \frac{\partial J_{\phi,i}}{\partial x_i}\middle| \phi = \psi\right\rangle f\right]}_{\text{molecular mixing (unclosed)}} - \underbrace{\frac{\partial}{\partial \psi}\left(S(\psi) f\right)}_{\text{source term}}. \tag{5.66}$$

The source term appears now in closed form. As such, the PDF approach can be used to attain closure of any nonlinear source term, including the unclosed terms that arise from the interaction between turbulence and PBE kinetic processes, whether in the moment or discretised PBE. However, two other unclosed terms

[7] As a delta function, the fine-grained density has the properties of a PDF, such as normalisation to 1.

appear, introducing additional unknowns written in the form of *conditional expectations*.[8] The first of them is the turbulent transport term, while the second one represents molecular mixing. We will now consider in more detail the physical meaning and closure models available for these terms.

Closure of the Turbulent Transport Term

The turbulent transport term arises from the transport of the PDF due to the fluctuating velocity. It is first worth noting that this term would appear in closed form if the joint velocity-scalar PDF was considered. This approach has indeed been applied in turbulent reacting flows, but brings in further complications in the solution procedure associated with the computation of velocity and pressure. More information on it can be found in Pope (1985). In the following, we consider the scalar PDF only since this is the one that will form the basis for the PBE-PDF approach.

The standard approach is to close this term with a gradient transport model:

$$\frac{\partial}{\partial x_i}\left(\langle u_i'|\phi=\psi\rangle f\right) = -\frac{\partial}{\partial x_i}\left(\Gamma_t \frac{\partial f}{\partial x_i}\right), \tag{5.67}$$

where Γ_t, the eddy diffusivity, is spatially varying and must be provided by a turbulence model. Some earlier and more sophisticated approaches for modelling the turbulent transport term are discussed by Jones (2002).

Closure of the Molecular Mixing Term

The molecular mixing term in Eq. 5.66, often referred to as *micromixing*, arises from the diffusion term in the original transport equation, as shown in Appendix C. It is also shown there (Section C.4) that, assuming Fick's law and constant diffusivity, this term can be decomposed as follows:

$$\frac{\partial}{\partial \psi}\left[\left\langle \frac{\partial J_{\phi,i}}{\partial x_i}\middle| \phi=\psi\right\rangle f\right] = D\frac{\partial^2 f}{\partial x_i^2} - \frac{\partial^2}{\partial \psi^2}\left[\left\langle \left(D\frac{\partial \phi}{\partial x_i}\frac{\partial \phi}{\partial x_i}\right)\middle| \phi=\psi\right\rangle f\right]. \tag{5.68}$$

The first of these terms accounts for diffusion of the PDF due to its spatial gradient; it is often negligible in high Reynolds number flows. The second term represents changes in the PDF at a single spatial location due to molecular mixing. This becomes more clear when we observe that the term conditioned on ψ can be identified with half the scalar dissipation.

To better understand the physical meaning of this term, we may consider a two-stream mixing process between two species. Initially, the PDF will consist of two

[8] The conditional expectation of random variable A on a value β of random variable B is denoted by $\langle A|B=\beta\rangle$ and is the average of A under the condition that B has the value β.

delta functions, as either of the two species will be present at each point. Eventually, it will become a single delta function, indicating that mixing is complete. This gradual change in the shape of the PDF due to molecular mixing is what is represented by the micromixing term. Clearly, this term is very important as, without it, the PDF transport equation will represent only turbulent dispersion without molecular mixing.

The reason why the micromixing term is unclosed is that diffusion is driven by a scalar gradient. The PDF being considered refers to a single point in space and knowledge of such a PDF does not imply information about the concentration gradient at that point because a gradient involves a rate of change. Proposals have been made to include the concentration gradient (Dopazo, 1994), but this increases substantially the complexity of the formulation. The standard approach, therefore, is to employ a closure model.

A considerable amount of literature exists on the subject of mixing models for PDF methods. The earliest works are *Curl's model* (Curl, 1963), also known as *coalescence-dispersion model*, and the *Interaction by Exchange with the Mean* (IEM) (Villermaux and Devillon, 1972), also known as *Linear Mean Square Estimation* (LMSE) (Dopazo and O'Brien, 1974). Other well known models include the *Euclidean Minimum Spanning Trees* (EMST) (Subramaniam and Pope, 1998) and the *mapping closure* (Pope, 1991). Reviews can be found in Kollmann (1995) and Jones (2002), among others.

The formulation of the mixing model is closely associated with the numerical solution method, which is usually accomplished with stochastic methods. There are two types of such methods, based on stochastic particles or stochastic fields, and these will be discussed in Section 5.5.3. What is important now is that the majority of mixing models were developed in the context of the stochastic particle method. However, the presentation in this chapter will focus on the stochastic field method due to the advantages that it offers for implementation in LES. Therefore, only the IEM model will be described here, as it is the main model that has been employed in conjunction with LES and stochastic fields, particularly in the context of PBE-PDF. This model will be used in the case study on soot formation (Section 6.2). It also offers the advantage of being the simplest one.

The IEM model provides the following closure of the micromixing term:

$$-\frac{\partial^2}{\partial\psi^2}\left[\left\langle\left(D\frac{\partial\phi}{\partial x_i}\frac{\partial\phi}{\partial x_i}\right)\middle|\,\phi=\psi\right\rangle f\right]=\frac{1}{2\tau_\phi}\frac{\partial}{\partial\psi}\left[(\psi-\langle\phi\rangle)f\right], \tag{5.69}$$

which includes one model parameter, the *mixing timescale*, τ_ϕ, estimated as follows (in RANS):

$$\frac{1}{\tau_\phi} = C_d \frac{\epsilon}{k}, \tag{5.70}$$

where C_d is a parameter usually taken to be equal to 2. This relationship is based on the assumption that the ratio of the scalar mixing timescale to the turbulence timescale (k/ϵ) is constant. The value of the constant is not universal but reasonable for flows such as those in thin shear layers where the generation and dissipation rates of turbulence kinetic energy and scalar variance are approximately in equilibrium. For more information on the general modelling requirements that must be fulfilled by mixing models or for a description of the more advanced models, the reader is referred to Kollmann (1995), Jones (2002) or the other reviews on PDF methods mentioned at the end of Section 5.5. The micromixing model is particularly important for RANS simulations, but its importance diminishes in LES, along with other models, as will be discussed in Section 5.5.2.

Finally, it should be mentioned that while micromixing is particularly vital for chemical reactions, as they do not occur at all if the reactants do not mix at the molecular scale, it is less important for some of the kinetic processes in the PBE, as the diffusivity of particles is very small or even negligible. Nevertheless, many population balance problems in flows involve reactions by precursors in the nucleation and growth processes, hence the joint PDF of species and number density will need to be considered, so micromixing is still important for the overall formulation.

Modelled PDF Transport Equation

Once the gradient transport and IEM models have been substituted into Eq. 5.66, the following equation is obtained:

$$\frac{\partial f}{\partial t} + \langle u_i \rangle \frac{\partial f}{\partial x_i} = \frac{\partial}{\partial x_i}\left(\Gamma \frac{\partial f}{\partial x_i}\right) + \frac{1}{2\tau_\phi}\frac{\partial}{\partial \psi}\left[(\psi - \langle \phi \rangle) f\right] - \frac{\partial}{\partial \psi}\left(S(\psi) f\right), \tag{5.71}$$

where $\Gamma = \Gamma_t + D$. In the above, $\langle \phi \rangle$ is related to the PDF as $\langle \phi \rangle = \int_{-\infty}^{\infty} \psi f d\psi$, while $\langle u_i \rangle$ and Γ are regarded as known (having been calculated from the RANS equations and turbulence model), hence the equation is now closed.

It should be noted, however, that the PDF transport equation is not usually solved in this form, as the number of scalars involved is usually greater than one and the resulting increase in dimensionality renders conventional discretisation methods unfeasible. Solution is thus obtained with stochastic methods that simulate a stochastic system with the same statistics and typically reproduce only the low-order moments of the PDF with accuracy (which are sufficient for most purposes). These methods will be described in Section 5.5.3.

Partially Stirred Reactor

For the transport of an inert scalar in a homogeneous turbulent flow, the pdf transport equation (Eq. 5.66) is simplified to:

$$\frac{\partial f}{\partial t} = -\frac{\partial^2}{\partial \psi^2}\left[\left\langle \left(D\frac{\partial \phi}{\partial x_i}\frac{\partial \phi}{\partial x_i}\right)\middle| \phi = \psi \right\rangle f\right] - \frac{\partial}{\partial \psi}\left(S(\psi)f\right). \tag{5.72}$$

This is the *partially stirred reactor (PaSR)* equation. It was first mentioned in Section 2.6, in the context of ideal reactor models. It can be complemented with inlet/outlet terms and used to describe an imperfectly mixed system without employing a full CFD formulation. However, the PaSR relies entirely on the mixing model in order to be effective. If the IEM is employed, the resulting equation is:

$$\frac{\partial f}{\partial t} = \frac{1}{2\tau_\phi}\frac{\partial}{\partial \psi}\left[(\psi - \langle \phi \rangle) f\right] - \frac{\partial}{\partial \psi}\left(S(\psi)f\right). \tag{5.73}$$

For a PaSR with flow in/out, the above equation becomes:

$$\frac{\partial f}{\partial t} = \frac{f_{in} - f}{\tau_{res}} + \frac{1}{2\tau_\phi}\frac{\partial}{\partial \psi}\left[(\psi - \langle \phi \rangle) f\right] - \frac{\partial}{\partial \psi}\left(S(\psi)f\right), \tag{5.74}$$

where f_{in} is the distribution at the inlet and τ_{res} is the residence time, a second parameter of the model.

PaSR models are often used with stochastic particle-based mixing models. Networks of PaSRs can also be employed for more complex configurations. A comprehensive study of the PaSR model has been presented by Chen (1997). An example of a population balance model (for carbon black) in a PaSR can be found in Balthasar et al. (2002). The PaSR model will be discussed again in the context of the PBE-PDF approach in Section 5.5.3.

Variable Density Flow

For variable density flows of low Mach number, density weighting can be applied to the PDF transport equation. The Favre decomposition (Section 5.2.7) is applied to the velocity field, while a density-weighted PDF, $\widetilde{f}$, is defined in the following way:

$$\widetilde{f} = \frac{\rho f}{\langle \rho \rangle}. \tag{5.75}$$

In reacting flows, the scalar is usually a mass fraction, Y, with PDF $f(\psi; \mathbf{x}, t)$. The density-weighted PDF transport equation, then, is:

$$\langle\rho\rangle\frac{\partial\widetilde{f}}{\partial t}+\langle\rho\rangle\widetilde{u}_i\frac{\partial\widetilde{f}}{\partial x_i}+\frac{\partial}{\partial x_i}\left(\langle\rho\rangle\left\langle u_i''|Y=\psi\right\rangle\widetilde{f}\right)$$
$$=\frac{\partial}{\partial\psi}\left[\left\langle\frac{\partial j_{Y,i}}{\partial x_i}\middle|Y=\psi\right\rangle\widetilde{f}\right]-\langle\rho\rangle\frac{\partial}{\partial\psi}\left(S(\psi)\widetilde{f}\right),\tag{5.76}$$

where the diffusion flux in terms of mass fraction includes the density, as given by (cf. Eq. A.15):

$$j_{Y,i}=-\rho D\frac{\partial Y}{\partial x_i}.\tag{5.77}$$

Extension to LES

The transported PDF framework has been extended to LES (Gao and O'Brien, 1993). Since an LES is an individual realisation of a turbulent flow, we cannot use the concept of a PDF as before. Instead, we define a *subgrid probability density function* (also called *filtered density function*) of a scalar ϕ, $f_{sgs}(\psi;\mathbf{x},t)$, such that $f_{sgs}d\psi$ is the probability of finding a value $\psi<\phi<\psi+\delta\psi$ within the subgrid scales in the filter volume, at a given spatial location and time instant.

To derive the LES-PDF transport equation, the subgrid PDF is written in terms of the fine-grained density defined as in Eq. 5.62, to which a filtering operation is applied (Eq. 5.18):

$$f_{sgs}(\psi;\mathbf{x},t)=\int_\Omega\mathcal{F}(\psi;\mathbf{x}',t)G(\mathbf{x}-\mathbf{x}')d\mathbf{x}'.\tag{5.78}$$

This equation is the equivalent of Eq. 5.63 for the PDF in RANS, where averaging rather than filtering was applied. The derivation proceeds with a procedure similar to the one shown in Appendix C in the context of RANS, and the reader may consult Gao and O'Brien (1993) for the details. The following equation is obtained for a single scalar in constant density flow:

$$\frac{\partial f_{sgs}}{\partial t}+\overline{u}_i\frac{\partial f_{sgs}}{\partial x_i}+\frac{\partial}{\partial x_i}\left(\overline{u_i\mathcal{F}}-\overline{u}_if_{sgs}\right)=\frac{\partial}{\partial\psi}\left(\overline{\frac{\partial J_{\phi,i}}{\partial x_i}\mathcal{F}}\right)-\frac{\partial}{\partial\psi}\left(S(\psi)f_{sgs}\right).\tag{5.79}$$

This equation, like Eq. 5.66, contains two unclosed terms, one of them associated with subgrid transport and one with molecular mixing. The modelling of these terms can be attained with methods analogous to those used in the PDF transport equation in RANS and will be discussed in the next section, in the context of the LES-PBE-PDF approach.

5.5.2 The PBE-PDF Approach

The Joint Species-Number Density PDF

In the PDF transport equation, Eq. 5.66, any nonlinear source term that was initially unclosed appears now in closed form. This means that the transported PDF approach can be employed to obtain closure of the unclosed terms arising from the kinetic processes in the PBE. In Section 5.3.3, we saw that the averaging of the PBE results in four types of unclosed terms. Of these, the first one (correlation between fluctuating velocity and number density) will still be unclosed in a PDF transport equation and can be addressed with a gradient transport model. The other three types of correlations, which are unique to the PBE and involve nonlinear functions of number density and/or other scalars such as concentrations and temperature, can all be brought to closed form with a PDF transport equation for the PBE. In the present section, we will outline the main steps of the formulation of this equation, following Rigopoulos (2007).

In formulating a transported PDF method, the first question is which variables should be included in the PDF. If it is desired to bring a certain term in closed form, then all variables present in that term should be included in the PDF, so that the average of that term may be computed as per Eq. 5.58. The unclosed terms in the PBE include the number density and possibly certain other variables such as concentrations of precursor species and temperature.

The number density, however, is not a single variable but rather a function of particle volume (with reference to the example PBE of Eq. 5.39, though it could also be another variable). Strictly speaking, the PDF of number density is a functional (a function of a function). Before proceeding, the number density must therefore be discretised into a finite set of number densities. This is also an essential step for applying one of the discretisation methods discussed in Chapter 4. At present, however, it is not needed to choose a method; it is sufficient to know that the number density, $n(v)$, has been discretised into a set of values $(N_1, N_2, \ldots, N_n)$. The joint PDF of these values will be denoted as $f(n_1, n_2, \ldots, n_n)$ (here capitals and lower case letters are used to distinguish between the variables and their ranges of values in the PDF), or $f(\mathbf{n})$ where n is a vector containing the values n_i. This joint PDF can be interpreted as the probability of finding an entire (discretised) size distribution at a certain spatial location and time instant.

The joint PDF of discretised number densities permits the closure of the correlations between number density fluctuations, that is, the type 4 unknown terms in the classification discussed at the end of Section 5.3.1. It is also important to obtain closure for type 2 and 3 unclosed terms associated with nucleation and growth kinetics that are functions of precursor species and temperature. To attain this, we need to include into the PDF any variables participating in these terms. We thus consider the joint species-number density PDF, $f(\mathbf{y}, \mathbf{n}; \mathbf{x}, t)$ (where $\mathbf{y}$ is a vector containing the species mass fractions), such that $fdy_1 \ldots dy_m dn_1 \ldots dn_n$ is the probability that the mass fraction of each species, Y_α, has a value in the range $y_\alpha < Y_\alpha < y_\alpha + \delta y_\alpha$ and each discretised number density, N_i, has a value in the range $n_i < N_i < n_i + \delta n_i$, at location $\mathbf{x}$ and time t. Other variables, such as temperature, may need to be included if they are required by the kinetics.

The PBE-PDF Transport Equation

To derive the PBE-PDF equation, a fine-grained density for the joint PDF of species mass fractions and number densities is first defined, in a manner akin to Eq. 5.62:

$$\mathcal{F}(\mathbf{y}, \mathbf{n}; \mathbf{x}, t) = \prod_{\alpha=1}^{m} \delta(y_\alpha - Y_\alpha(\mathbf{x}, t)) \prod_{i=1}^{n} \delta(n_i - N_i(\mathbf{x}, t)). \tag{5.80}$$

From this, the PBE-PDF transport equation can be obtained as follows:

$$\frac{\partial f(\mathbf{y}, \mathbf{n}; \mathbf{x}, t)}{\partial t} = \frac{\partial \langle \mathcal{F}(\mathbf{y}, \mathbf{n}; \mathbf{x}, t) \rangle}{\partial t}. \tag{5.81}$$

The remaining steps have been shown by Rigopoulos (2007), and the procedure is similar to the derivation of the PDF equation for a single scalar with a nonlinear source term shown in Section 5.5.1 and Appendix C. The final equation is:

$$\begin{aligned} \frac{\partial f}{\partial t} + \langle u_j \rangle \frac{\partial f}{\partial x_j} + \frac{\partial}{\partial x_j} \left(\left\langle u_j'' | \mathbf{Y} = \mathbf{y}, \mathbf{N} = \mathbf{n} \right\rangle f \right) \\ = \sum_{\alpha=1}^{m} \frac{\partial}{\partial y_\alpha} \left(\left\langle \frac{\partial J^y_{\alpha,j}}{\partial x_j} \middle| \mathbf{Y} = \mathbf{y} \right\rangle f \right) + \sum_{i=1}^{n} \frac{\partial}{\partial n_i} \left(\left\langle \frac{\partial J^n_{i,j}}{\partial x_j} \middle| \mathbf{N} = \mathbf{n} \right\rangle f \right) \\ - \sum_{\alpha=1}^{m} \frac{\partial}{\partial y_\alpha} (S_{r,\alpha} f) - \sum_{i=1}^{n} \frac{\partial}{\partial n_i} (S_{n,i} f). \end{aligned} \tag{5.82}$$

In this equation, summation is explicitly indicated for the concentrations of chemical species ($\alpha = 1, \ldots, m$) and the discretised number densities ($i = 1, \ldots, n$). The vectors $\mathbf{J^y}$ and $\mathbf{J^n}$ indicate the diffusive fluxes of species mass fraction and particle number density, respectively. The source term $\mathbf{S_r}$ denotes the chemical reaction kinetics, while $\mathbf{S_n}$ includes all of the unclosed source terms in the PBE, that is, the second, third and fourth types of unclosed terms mentioned in Section 5.3.3.

As with the previous PDF transport equations, Eq. 5.82 includes unclosed terms for turbulent transport and micromixing. The former can be modelled as before with the gradient transport model, Eq. 5.67. The latter includes two terms that can be approximated with the aid of the IEM model as follows:

$$\sum_{\alpha=1}^{m} \frac{\partial}{\partial y_\alpha}\left(\left\langle \frac{\partial J^y_{\alpha,j}}{\partial x_j}\middle| \mathbf{Y}=\mathbf{y}\right\rangle f\right) = \frac{1}{2\tau_y}\sum_{\alpha=1}^{m} \frac{\partial}{\partial y_\alpha}\left[(y_\alpha - \langle Y_\alpha\rangle) f\right], \tag{5.83}$$

$$\sum_{i=1}^{n} \frac{\partial}{\partial n_i}\left(\left\langle \frac{\partial J^n_{i,j}}{\partial x_j}\middle| \mathbf{N}=\mathbf{n}\right\rangle f\right) = \frac{1}{2\tau_n}\sum_{i=1}^{n} \frac{\partial}{\partial n_i}\left[(n_i - \langle N_i\rangle) f\right], \tag{5.84}$$

where τ_y and τ_n are the micromixing timescales for the chemical species and particles, respectively. To accommodate the lower diffusivity of particles, a lower timescale may be employed for them, although there is no absolute guideline on this; more discussion on this matter can be found in the next section in the context of the implementation of the method in LES.

The PBE-PDF method was first applied in the context of RANS by Di Veroli and Rigopoulos (2009, 2010) to precipitation in a turbulent pipe flow and by Di Veroli and Rigopoulos (2011) to aerosol formation in a turbulent jet. Later work is based on LES and will be discussed below.

The LES-PBE-PDF Equation

The PBE-PDF method has been extended to LES by Sewerin and Rigopoulos (2017b) for constant density flow, with an application to aerosol condensation, and by Sewerin and Rigopoulos (2018) to low Mach number variable density flow, with an application to soot formation. In Section 6.2, the results of Sun and Rigopoulos (2022) with this method will be shown as part of a case study. The basic equations are shown below.

The extension of the PBE-PDF approach to LES involves the definition of a subgrid joint scalar-number density PDF, $f_{sgs}(\mathbf{y}, \mathbf{n}; \mathbf{x}, t)$. This should be interpreted as the probability of finding simultaneously values in the range $y_\alpha < Y_\alpha < y_\alpha + \delta y_\alpha$ for the concentrations and $n_i < N_i < n_i + \delta n_i$ for the discretised number densities within the subgrid scales in the filter volume at a specific spatial location and time instant, in a certain realisation of the flow. To derive the equation, a fine-grained density is first defined as in Eq. 5.80. Subsequently, f_{sgs} is expressed as follows:

$$f_{sgs}(\mathbf{y}, \mathbf{n}; \mathbf{x}, t) = \int_\Omega \mathcal{F}(\mathbf{y}, \mathbf{n}; \mathbf{x}, t) G(\mathbf{x} - \mathbf{x}') d\mathbf{x}'. \tag{5.85}$$

The remaining steps of the derivation have been shown in detail by Sewerin and Rigopoulos (2017b, 2018). Here, we show only the final equation in the variable density form, as it is the one that will be employed in the case study in Section 6.2. Before that, we first define the density-weighted subgrid PDF in a way similar to Eq. 5.75:

$$\overset{\approx}{f}_{sgs}(\mathbf{y}, \mathbf{n}; \mathbf{x}, t) = \frac{\overline{\rho(\mathbf{x}, t) f_{sgs}(\mathbf{y}, \mathbf{n}; \mathbf{x}, t)}}{\overline{\rho}(\mathbf{x}, t)}. \tag{5.86}$$

The final LES-PBE-PDF equation is:

$$\overline{\rho}\frac{\partial \overset{\approx}{f}_{sgs}}{\partial t} + \overline{\rho}\overset{\approx}{u}_j \frac{\partial \overset{\approx}{f}_{sgs}}{\partial x_j} + \frac{\partial}{\partial x_j}\left[\overline{\rho}\left(\overline{u_j \mathcal{F}} - \overline{u}_j f_{sgs}\right)\right] = \sum_{\alpha=1}^{m} \frac{\partial}{\partial y_\alpha}\left(\overline{\frac{\partial J^{y}_{\alpha,j}}{\partial x_j}\mathcal{F}}\right)$$
$$+ \sum_{i=1}^{n} \frac{\partial}{\partial n_i}\left(\overline{\frac{\partial J^{n}_{i,j}}{\partial x_j}\mathcal{F}}\right) - \sum_{\alpha=1}^{m} \frac{\partial}{\partial y_\alpha}(\overline{\rho} S_{r,\alpha} f_{sgs}) - \sum_{i=1}^{n} \frac{\partial}{\partial n_i}(\overline{\rho} S_{n,i} f_{sgs}). \tag{5.87}$$

The LES-PBE-PDF equation requires closures for the subgrid turbulent transport and micromixing. The former can be carried out with a gradient transport model:

$$\frac{\partial}{\partial x_j}\left[\overline{\rho}\left(\overline{u_j \mathcal{F}} - \overline{u}_j f_{sgs}\right)\right] = -\frac{\mu_{sgs}}{Sc}\frac{\partial f_{sgs}}{\partial x_i}, \tag{5.88}$$

where the subgrid scale turbulent viscosity, μ_{sgs}, can be provided by the Smagorinsky or another model in the variable density formulation; more details can be found in Jones and Prasad (2010).

In RANS-PDF, the micromixing model determines the decay rate of quantities such as mixture fraction variance, which is particularly important for reacting flows, and it is often beneficial to employ more advanced mixing models, such as EMST. In LES, however, the energetic motions are resolved, while the effect of the mixing model is confined to the subgrid-scale scalar variance. Therefore, the IEM model (which is the simplest one) may be adequate in many practical situations. It has been used extensively in LES-PDF simulations of turbulent reacting flows, such as Jones and Prasad (2010).

In the work of Sun and Rigopoulos (2022) that will be presented in Section 6.2, an augmented IEM micromixing model proposed by McDermott and Pope (2007) and applied to LES-PBE-PDF by Sewerin and Rigopoulos (2017b, 2018) is employed, providing the following closure to the unclosed terms:

$$\frac{\partial}{\partial y_\alpha}\left(\overline{\frac{\partial J^{y}_{\alpha,j}}{\partial x_j}\mathcal{F}}\right) = \frac{1}{2\tau_{sgs,y}}(\overset{\approx}{Y}_\alpha - y_\alpha) + \frac{1}{\overline{\rho}}\frac{\partial}{\partial x_j}\left(\overline{\rho}\overset{\approx}{D}\frac{\partial \overset{\approx}{Y}_\alpha}{\partial x_j}\right), \tag{5.89}$$

$$\frac{\partial}{\partial n_i}\left(\overline{\frac{\partial J^{n}_{i,j}}{\partial x_j}\mathcal{F}}\right) = \frac{1}{2\tau_{sgs,n}}(\overset{\approx}{N}_i - n_i) + \frac{1}{\overline{\rho}}\frac{\partial}{\partial x_j}\left(\overset{\approx}{D}_{p,i}\frac{\partial \overline{\rho}\overset{\approx}{N}_i}{\partial x_j}\right). \tag{5.90}$$

Note that the number density, N_i, is now a mass-based concentration, to accommodate the variable density flow, such that ρN is the number density per unit volume, hence the appearance of $\overline{\rho}\overset{\approx}{N}_i$ in the gradient of Eq. 5.90. In the above, $\tau_{sgs,y}$ and

$\tau_{sgs,n}$ denote subgrid mixing timescales and can be given by an expression of the following form:

$$\frac{1}{\tau_{sgs}} = C_d \frac{\mu_{sgs}}{\overline{\rho}\Delta^2}, \tag{5.91}$$

where the subgrid scale mixing constant is $C_d = 2$ for gas-phase species. The diffusivity of particles is much lower than that of species, and this differential diffusion can be accommodated in LES in a straightforward manner at the grid level. For the subgrid closure, it is possible to introduce different micromixing frequencies for species and discretised number densities, hence the use of two timescales in Eqs. 5.89 and 5.90.

5.5.3 Numerical Methods for Solving the PBE-PDF Equation

We now consider the numerical methods that can be used for solving PDF transport equations and the PBE-PDF equation in particular. As will be shown below, while discretisation methods can in principle be applied, their use has been limited to cases where only a few scalars are considered. The majority of applications have been based on stochastic methods, which have already been encountered in Section 4.2.9 in the context of the solution of the PBE. These methods simulate a system of notional entities whose statistics are the same as the distribution being sought. The main advantage of stochastic methods for the PDF transport equation is that they overcome the dimensionality problem, since the computational effort scales linearly with the number of dimensions, rather than exponentially, which is the case with discretisation methods.

The stochastic methods involve *stochastic processes* and *stochastic differential equations* (SDEs). Some information on these is provided in Appendix D and the references therein. Briefly, a stochastic process is a set of random variables indexed by time, which can also be space dependent, that is, $X(\mathbf{x},t)$. A *sample path* of a stochastic process is the temporal evolution of the random variable in an individual realisation of the process. An SDE is an equation that yields an increment of the sample path of a stochastic process. By simulating a number of sample paths, the statistics of the stochastic process can be obtained. The objective of stochastic methods is to devise a stochastic process with statistics equivalent to those of the PDF transport equation.

In the following, we first describe briefly discretisation methods and then focus on the two main classes of stochastic approaches, the stochastic particle and field methods. The latter is described in more detail as it is used in the context of LES-PBE-PDF and will be demonstrated in the case study on soot formation (Section 6.2).

Discretisation Methods

Eq. 5.71, describing the evolution of the PDF of a single scalar, can be solved by discretisation methods, as was indeed done in the early work of Dopazo and O'Brien (1974). However, if several scalars are considered and the PDF is already spatially dependent, the dimensionality increase is too high to be handled by conventional numerical methods. An attempt to address this issue is the adaptive mesh refinement scheme proposed by Olivieri et al. (2009), but the number of scalars considered in that work is still very small (two or three). In the PBE-PDF method, a considerable number of variables are needed (typically at least thirty) to represent the PSD, hence discretisation methods cannot be considered.

Stochastic Particle Methods

In *stochastic particle* methods, the system being constructed is an ensemble of notional particles, each of which carries a value of the independent variables of the PDF. The essence of the method is to let the particle properties evolve in such a way that their distribution is an approximation of the PDF being sought. The particles should not be thought of as actual fluid elements but rather as notional entities.

The simplest example of a stochastic particle method can be described in the context of the partially stirred reactor. In Section 5.5.1, we showed Eq. 5.72 which is a non-spatially dependent PDF equation describing the evolution of the PDF of a single scalar in a homogeneous turbulence. This equation can be extended to the PBE-PDF and also to include flow in and out. To solve it with a stochastic particle method, the contents of the reactor are represented with a number of stochastic particles, each of them carrying a chemical composition and a discretised particle size distribution. The inlet and outlet are emulated by exchanging particles selected in random with new particles featuring the inlet composition. The general procedure is detailed in Chen (1997), while the PBE-PDF equation for the PaSR can be found in Rigopoulos (2007). A mixing model must be employed, and the PaSR allows the use of any of the models mentioned in Section 5.5.1 (this is the case with stochastic particle methods in general).

An application of the PaSR with the purpose of demonstrating the impact of fluctuations in the PBE-PDF equation was shown in Rigopoulos (2007). Two cases were investigated, one for a nucleation growth problem (precipitation of $CaCO_3$) to investigate the effect of fluctuations of the second and third type according to the classification in Section 5.3.3, and one on coagulation, to expose fluctuations of the fourth type. In the former case, the two types of fluctuations countered each other, hence a model neglecting them appeared to give the correct result (for the wrong reason). A similar finding was observed by Di Veroli and Rigopoulos (2009) in a RANS-PDF study of precipitation in a pipe flow.

It is much more complex to develop a stochastic particle method for solving the spatially dependent PDF transport equation. The first method of this type was proposed by Pope (1981) and is an Eulerian method where a number of stochastic particles are located within each CFD cell. In order to simulate the effect of flow, the particles are being exchanged between cells. This method has found applications in turbulent reacting flows and in a RANS context, but it has several disadvantages (see e.g. the discussion in Jones, 2002) and has largely been superseded by the Lagrangian approach that will be described below. It has not been applied to PBE–PDF (to the author's knowledge).

In the Lagrangian particle method (Pope, 1985), the stochastic particles are moving throughout the physical domain in a continuous manner. This approach has found numerous applications in turbulent reacting flows in both RANS and LES, a review of which can be found in Haworth (2010). It has also been adapted to the PBE–PDF approach in the context of RANS by Di Veroli and Rigopoulos (2010), and the algorithm developed in that work has been employed to simulate precipitation (Di Veroli and Rigopoulos, 2009, 2010) and aerosol formation (Di Veroli and Rigopoulos, 2011).

The transport of particles in the Lagrangian approach is carried out according to an SDE containing a deterministic and a random part. The former is computed in accordance with the fluid velocity and turbulence properties as computed by the finite volume CFD code. The stochastic particle method poses several issues from the implementation point of view, mainly associated with tracking the particles (in the case of unstructured grids) and ensuring that enough of them are present in each cell. Apart from RANS, it has also been adapted to LES and employed in simulations of reacting flows. More details on stochastic particle methods can be found in Pope (1985, 2000), Jones (2002), Fox (2003) or Haworth (2010).

The Stochastic Field Method

In the *stochastic field* method, the entities being simulated are fields rather than particles. They should not be thought of as realisations of the turbulent flow, but rather as notional fields whose statistics are the same as the PDF being sought. The method is also referred to as *Eulerian stochastic field*, to juxtapose it with the Lagrangian stochastic particle method (although particle methods can also be Eulerian).

The stochastic field method offers several advantages from the implementation point of view that make it a good choice for LES–PDF formulations. One of the main ones is its simplicity: the fields obey equations similar to the original transport equations and can thus be treated in a very similar manner within an existing CFD code (with some amendments, as will be shown in the following and in Section 5.5.4). This is contrasted with particle methods, where the stochastic particles need to be moved and tracked and thus require an additional code to run concurrently with the CFD code. Furthermore, the issue of ensuring a sufficient number of particles everywhere in the domain does not exist in the field method. It must be stressed that the above constitute merely implementation advantages; both stochastic particle and field methods are subject to similar convergence requirements. However, it is usually not required to reproduce the PDF itself, but rather to approximate its low-order moments. For this purpose, a relatively small number of fields (about eight) has been found to be enough in most cases (Mustata et al., 2006; Jones and Navarro-Martinez, 2007). By contrast, with stochastic particles, one runs the risk of having an uneven number of particles per cell, unless action is taken to prevent it.

The stochastic field method was derived by Valiño (1998) and Sabel'nikov and Soulard (2005) with different methods. A derivation based on the Valiño approach is shown in Appendix D, for a single scalar in constant density flow and in the context of RANS, corresponding to the modelled PDF with the gradient transport and IEM models (Eq. 5.71). The IEM model provides continuous and differentiable solutions (unlike models such as Curl's and its variants) and is thus commonly employed in stochastic field implementations. The main steps of the derivation procedure are briefly outlined here.

The stochastic fields are realisations of a spatially dependent stochastic process, $\theta(\mathbf{x},t)$, with sample space z, which is continuous and differentiable in space and continuous but not differentiable in time. The PDF of θ can be expressed as an average of a fine-grained density, as in Eq. 5.63:

$$f(z;\mathbf{x},t) = \langle \delta(z-\theta(\mathbf{x},t)) \rangle \,. \tag{5.92}$$

An SDE is then derived for $\theta(\mathbf{x},t)$. The remaining steps are shown in Appendix D. The final equation (with the arguments omitted) is[9]:

$$\begin{aligned} d\theta = &-\frac{\partial(\langle u_i\rangle\theta)}{\partial x_i}dt + \frac{\partial}{\partial x_i}\left(\Gamma\frac{\partial\theta}{\partial x_i}\right)dt - \left[\frac{1}{2\tau_\phi}(z-\langle\theta\rangle)\right]dt \\ &+ Sdt + (2\Gamma)^{1/2}\frac{\partial\theta}{\partial x_i}dW, \end{aligned} \tag{5.93}$$

where dW is the *Wiener process* which, as explained in Appendix D, is an archetypical stochastic process representing a random walk.

To obtain an individual sample path of θ, a number of increments $d\theta$ must be calculated. To compute each of these increments, one must first simulate an

[9] It should be noted that the derivation shown in the present book is based on the concepts of Itô calculus, following Valiño (1998); an alternative is the use of Stratonovich calculus. The description of these two approaches of stochastic calculus is out of the scope of this text and can be found in Gardiner (1985) or Kloeden and Platen (1992). For a derivation based on Stratonovich calculus, the reader may consult Sabel'nikov and Soulard (2005).

increment dW of the Wiener process, which is a standard procedure and involves a random number, and then substitute into Eq. 5.93 to compute the increment $d\theta$. The involvement of random numbers means that each path thus calculated differs from the others. The statistics of the sample paths can then be collected to obtain an approximation of those of the original variable, whose evolution in this case is described by Eq. 5.71. The equivalence of the one-point moments of the stochastic fields and modelled filtered subgrid PDF is shown in the appendix of Readshaw et al. (2021).

The derivation of the stochastic field equations for solving the LES-PBE-PDF equation has been shown in Sewerin and Rigopoulos (2017b) for constant density flow and in Sewerin and Rigopoulos (2018) for variable density, and is based on extensions of the same principles as those used for deriving Eq. 5.93 in Appendix D. The variable density form is the one that will be used in Section 6.2 and is shown below:

$$\begin{aligned}\overline{\rho}d\theta^l_\alpha &= -\overline{\rho}\widetilde{\widetilde{u}}_i\frac{\partial\theta^l_\alpha}{\partial x_i}dt + \frac{\partial}{\partial x_i}\left(\overline{\rho}\Gamma\frac{\partial\theta^n_\alpha}{\partial x_i}\right)dt \\ &+ \overline{\rho}(2\Gamma)^{1/2}\frac{\partial\theta^l_\alpha}{\partial x_i}dW^l_i + \overline{\rho}\left(M^l_\alpha + S^l_\alpha\right)dt, \qquad (5.94)\end{aligned}$$

where the species mass fractions and discretised number densities have been concatenated into a single vector, $\boldsymbol{\theta}^l$, while l is an index indicating the stochastic field ($l = 1, N_f$, where N_f is the number of fields). Furthermore M^l_α and S^l_α are the micromixing and source term (including reactions and all PBE sources), respectively. The filtered values of any scalar, $\widetilde{\widetilde{\phi}}_\alpha$, corresponding to θ_α, can then be computed by averaging across the fields:

$$\widetilde{\widetilde{\phi}}_\alpha = \frac{1}{N_f}\sum_{l=1}^{N_f}\theta^l_\alpha. \qquad (5.95)$$

5.5.4 Computational Implementation

While the implementation of discretisation methods for solution of the PDF is straightforward but impractical, the implementation of stochastic methods needs some further discussion.

Since the PBE-PDF transport equation does not include the velocity as an independent variable, the stochastic method should be combined with a CFD code that computes the velocity and pressure.[10] Stochastic methods operate using the

[10] As mentioned in Section 5.5.1, it is possible to include velocity statistics in the PDF. Computational implementations for that formulation have been developed for turbulent reacting flows but, to the author's knowledge, have not been applied to PBE-PDF.

principle of fractional steps, which was explained in Section 4.4.4. By splitting the PDF transport equation into fractional steps for convection/diffusion, micromixing, reaction and PBE, it is possible to couple the stochastic part with a conventional CFD code based on a finite volume or other similar method.

In the Lagrangian stochastic particle method, a code for transporting the particles and solving the chemistry and PBE within each of them should be developed. If there is no feedback from the particles to the flow field, as is the case for non-inertial particles and constant density flow, the solution of the PBE-PDF via the stochastic particle code can be accomplished as a post-processor, once a steady-state flow field has been computed. This is the approach that was adopted by Di Veroli and Rigopoulos (2009, 2010, 2011). It is more complex to implement stochastic particle methods in the context of a variable density problem, such as soot formation or flame synthesis of aerosols, because the reactions change the fluid density. The stochastic particle code should then run concurrently with the CFD code.

To apply the micromixing model at each CFD cell, the stochastic particles present there must first be located. This can be quite complex in unstructured meshes, and a review of algorithms for that purpose can be found in Haworth (2010). Another complex issue is ensuring that each cell has a sufficiently high number of particles. Several solutions have been developed for that purpose; the algorithm employed in the PBE-PDF implementation of Di Veroli and Rigopoulos is detailed in Di Veroli (2009).

From the implementation point of view, the stochastic field method is more straightforward. Most of its applications have been in an LES context. The treatment of the scalar fields is divided into fractional steps for convection/diffusion, micromixing, reaction and PBE, as shown in Fig. 5.2. The base CFD code is used to compute the flow field and the convection/diffusion fractional step for several instances of the scalar fields, with the addition of a stochastic term. The micromixing step is carried out at each cell involving the values of all fields for that cell. The reaction and PBE fractional steps are then carried out with their respective solvers. More details on the coupling of the method with a finite volume LES CFD solver can be found in Prasad (2011).

As mentioned in the previous section, it is sufficient to reproduce the low-order moments of the PDF, and for this purpose eight fields have been found to be sufficient in most cases (Mustata et al., 2006; Jones and Navarro-Martinez, 2007). In the limit of one field, the stochastic field method reduces naturally to a solution of the problem without a model for the turbulence–chemistry and turbulence–PBE interactions. While this is bound to induce some error, it can still provide a reasonably good initial approximation.

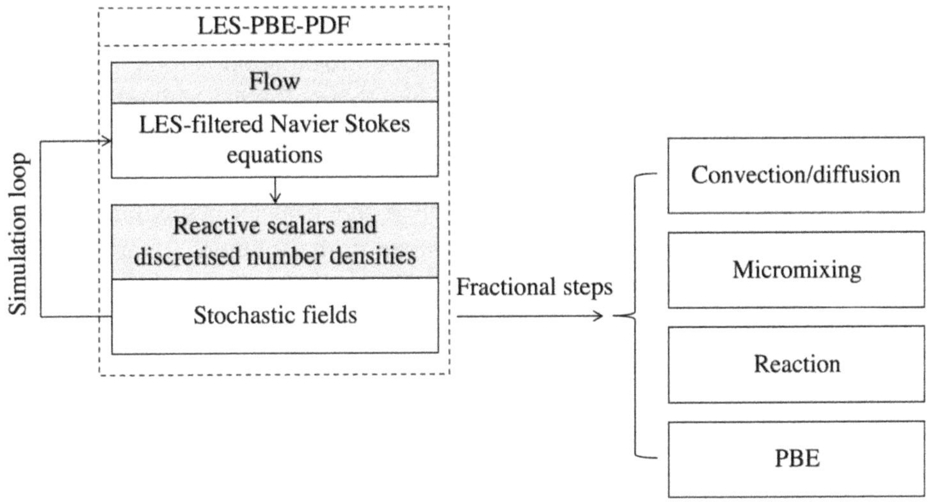

Figure 5.2 Overview of an LES-PBE-PDF code implementation. Reproduced from Sun and Rigopoulos (2022) under CC-BY 4.0 license.

Whether in the stochastic particle or field method, the solution of the PBE enters as a plug-in within its own fractional step. Thus, any of the methods shown in Chapter 4 can be employed in the same way as they would be in an ideal reactor problem.

5.6 Concluding Remarks

In the present chapter, we have considered the issues arising from the interaction of turbulent flow with the population balance. Approaches have been described for integrating the PBE within RANS, LES and DNS simulations.

The chapter dealt with kinetic processes in non-inertial particles, where the closure of turbulent transport processes is straightforward. Inertial particle dispersion requires consideration of the velocity distribution and is treated with discretisation of velocity space, multi-fluid or Lagrangian models that are discussed in multiphase flow treatises.

The unclosed terms were classified into four types: fluctuating velocity-number density correlations, unclosed nucleation and growth source terms, fluctuating growth rate-number density correlations and fluctuating number density correlations. Of these, the first one can be treated with a gradient transport model. The remaining ones are unclosed in both moment and discretisation methods, unless the PDF of the particle properties and other variables involved in the PBE (such as precursor concentrations and temperature) is taken into account.

In presumed PDF methods, the effect of turbulence on the PBE is described via a few variables, such as the mixture fraction. An assumption is then made about the shape of the PDF of these variables, on the basis of which a closed model is obtained. These methods can be parallelled with presumed distribution moment methods for the solution of the PBE.

In transported PDF methods, the joint PDF of chemical species and moments or discretised number densities is computed as part of the solution. An approach for extending the transported PDF concept to PBE discretisation methods, the PBE-PDF approach, and its extension to LES, have been presented. Since the number of independent variables in the joint PDF is large, stochastic methods are needed for solution of this approach.

In Chapter 6, the effect of turbulence on the PBE will be investigated in two case studies. In Section 6.2, an LES-PBE-PDF simulation of a turbulent sooting flame is presented. Subsequently, in Section 6.3, a DNS-PBE approach is adopted to study the effect of turbulent fluctuations on a nucleation growth problem.

6

Case Studies of CFD-PBE Application

In the present chapter, we will see how the theory and methodology presented so far can be put together and applied to problems that can be solved with the population balance. The cases chosen are from the research of the author's group and involve coupling of PBE and fluid dynamics, thus bringing together the book's main themes. A brief overview of the case studies and issues tackled in each of them is given below.

The first case study involves silica nanoparticle synthesis in a laminar flame. The dominant kinetic processes are nucleation, aggregation and sintering. The fluid dynamics formulation is that of a variable density gas-phase reacting flow. A laminar flame is investigated, in order to put aside the issues of turbulence–chemistry and turbulence–PBE interaction. Fractal aggregates are formed and emphasis is placed on particle morphology, as it is an important property of the product that must be tailored. The focus of this case study is on comparing alternative PBE models for prediction of particle morphology and on validation with experimental results from the literature.

The second case study is about soot formation. Like silica, soot is also an aerosol and the two cases have several elements in common with silica synthesis. However, unlike silica, the motivation for the study of soot formation is mitigation rather than production, and therefore the prediction of the particle properties is of interest insofar as they are important process variables rather than being an objective in itself. It must be mentioned, though, that there are also valuable carbonaceous nanoparticles whose production must be tailored to applications. The kinetic processes in this case study include nucleation, condensation, growth, oxidation, aggregation and coalescence of primary particles. Currently, the modelling of turbulent sooting flames presents a major challenge due to uncertainties in both kinetics and turbulence–chemistry-soot interaction. The case study presented here involves the simulation of both a laminar and a turbulent sooting flame with the same kinetics, in order to shed light into the nature of these uncertainties. This is the most complex of the case studies and brings together all of the themes in the book.

The third case study is the simulation of precipitation in a turbulent flow. Unlike the previous two cases, this one takes place in a constant-density liquid flow. The approach chosen here is DNS-PBE and the objective is to provide a fundamental study of flow and PBE interaction, rather than a computationally efficient modelling tool. The kinetic processes involved are nucleation and growth, and one of the objectives is to study their relative importance and interaction with flow. Another aim is to examine the fluctuations resulting from the interaction of turbulence and population balance, which are the motivation for the methods discussed in Chapter 5. An attempt is also made to draw insight into the effect of sub-Kolmogorov scales that appear in high Schmidt number liquid flows, although this remains an open problem.

It should be mentioned that these problems could be approached with several of the models and solution methods presented in the previous chapters, and the approaches and implementations presented here are not intended to be regarded as optimal but rather as case studies. The main objective of this chapter is educational, that is, to demonstrate the application of the methodology of putting together a population balance model, coupling it with fluid dynamics and applying a solution method. In each case, the population balance methodology will be deployed in the form of the four steps given in Section 1.1.3.

6.1 Silica Nanoparticle Synthesis in a Laminar Flame

6.1.1 Introduction

Overview

Nanoparticle synthesis is a good example of a problem where the coupling of fluid dynamics and population balance is the key for predicting the process outcome. Flame synthesis is one of the main methods employed for the manufacturing of nanoparticles that have a range of applications in fields such as gas sensors, energy storage materials, nanotoxicity studies, catalysts, biomaterials and electroceramics (Meierhofer and Fritsching, 2021). The size and morphology distribution of the nanoparticles is of crucial importance for the desired application. Therefore, population balance modelling opens the way for tailoring the product to specific applications.

The case to be presented here involves silica synthesis in a laminar diffusion flame and is based on the work of Tsagkaridis et al. (2023), in which simulations of the experiments of Camenzind et al. (2008) were conducted. It is instructive to juxtapose the case study in this section with the Santoro sooting flame that will be shown in the next section, as both are based on laminar flames. Nanoparticle synthesis has several elements in common with soot formation; in fact, several industrially important nanoparticles are carbonaceous (e.g. carbon black, carbon quantum dots). In the case of inorganic nanoparticles, such as silica (SiO_2) and titania (TiO_2), the flame provides the energy for the reaction to take place. However, the precursor is injected rather than being formed during the combustion process. As such, the synthesis of such particles does not involve the complex PAH chemistry that is associated with soot nucleation. Furthermore, in the silica synthesis case to be presented here, the dominant mechanisms are aggregation and sintering, as opposed to soot formation where surface growth and oxidation are of major importance.

Reviews of nanoparticle synthesis can be found in Pratsinis (1998), Kruis et al. (1998), Kodas and Hampden-Smith (1999), Pratsinis (2010), Buesser and Pratsinis (2012), Buesser and Gröhn (2012), Raman and Fox (2016) and Meierhofer and Fritsching (2021). A review of works on silica synthesis, particularly those that involve a coupling of PBE and fluid dynamics, can be found in Tsagkaridis et al. (2023).

Objectives of the Case Study

Since the present case study involves a laminar flame, the issues of turbulence, turbulence–chemistry and turbulence–PBE interaction do not come into play and the emphasis is on the prediction of particle morphology. This problem is, therefore, a good example of coupling PBE with CFD for laminar flow, with the kinetics being dominated by aggregation and sintering. There are two main objectives in this case study, which are described below.

The first objective is to apply and compare three PBE approaches for predicting aggregate morphology. The first of these is the monodisperse model, which was described in Section 2.5. This model has been used extensively in previous works (cf. Section 6.1.1) due to its ease of use and speed of computation, particularly if coupling with CFD is required. The second one is the two-PBE model discussed in Section 2.4.2 and also employed in the case study on soot formation (Section 6.2). This model is the most comprehensive of the three models compared here. The third one is the extended one-PBE model that was introduced in Tsagkaridis et al. (2023) with the objective of reducing the computational time needed by the two-PBE model in light of future application to turbulent flames. The three models will be compared in terms of both accuracy and computational efficiency.

The second objective is to validate the results with the *in-situ* small-angle X-ray scattering (SAXS) measurements of Camenzind et al. (2008). By excluding the uncertainties introduced by the presence of turbulence, the modelling uncertainties are limited to the selection of aerosol kinetic models and the results allow shedding light on the nature of discrepancies between predictions and experimental results.

The presentation initially follows the four steps involved in deploying the population balance methodology, as outlined in Section 1.1.3. Afterwards, the results are presented and discussed.

6.1.2 Formulation of the Population Balance Model

Preliminary Considerations

Following the steps in Section 1.1.3, the formulation of the population balance model involves the choice of the appropriate form of the PBE and the identification of the physical and chemical processes to be included.

The prediction of aggregate morphology is one major objective of this case study. As with soot, it is an essential part of a model with predictive capability; furthermore, in nanoparticle synthesis, it is desired to control both the size and morphology of the product. At the same time, the PBE approach must be computationally efficient in order to be coupled with CFD. For this reason, the three approaches for predicting morphology mentioned in the previous section will be applied and compared.

Regarding the description of aggregate morphology, Eq. 3.38 is employed for the aggregate collision diameter with $k_f = 1.4$ and $D_f = 1.91$, corresponding to ballistic cluster–cluster aggregation in the free molecule regime (Ball and Jullien, 1984; Friedlander, 2000).

Kinetic and Transport Processes

The kinetic processes that will be included in the problem of silica synthesis are:

- Nucleation, which occurs via the dimerisation of two SiO_2 molecules.
- Condensation, which involves deposition of molecules on the silica particle surface.
- Aggregation, taking place in the free molecule, transition and continuum regimes, as dictated by the range of particle sizes observed in the experiment.
- Sintering of aggregates.

The transport processes are the same as in the case of the laminar sooting flame (Section 6.2.2). Also, as in that case, the particles under consideration are nanoparticles with a Stokes number close to zero, hence there is no need to consider particle acceleration. Therefore, the transport processes to be included are:

- Convection of particles by the laminar flow field.
- Particle diffusion.
- Thermophoresis, incorporated as an additional convective velocity.

We can now proceed to the formulation of the population balance models. In the following, the models chosen so far will be incorporated in their general forms, as described in Chapters 2 and 3. The specific kinetic models to be used in each process will be shown in Section 6.1.3.

Monodisperse Model

The monodisperse model of nanoparticle synthesis via aggregation and sintering, as originally derived by Kruis et al. (1993), was presented in Section 2.5. It is probably the most widely used model for describing flame synthesis of inorganic nanoparticles due to its computational efficiency and the fact that this process is dominated by aggregation and sintering. The model employed here is a variation of that concept, the difference being that an equation is solved for the concentration of primary particles as opposed to the surface area, in addition to the equations for the number concentration and volume of aggregates. This choice was made in order to make the model similar to the two-PBE and extended one-PBE models, which also operate in terms of the number of primary particles, and the equivalence with the surface area-based model has been shown in the supplementary material of Tsagkaridis et al. (2023). The first equation to be solved is based on Eq. 2.36, with the introduction of spatial transport and nucleation:

$$\frac{\partial N}{\partial t} + \frac{\partial[(u_j + u_j^t)N]}{\partial x_j} = \frac{\partial}{\partial x_j}\left(D_p(v,T)\frac{\partial N}{\partial x_j}\right) + J(\mathbf{Y},T) - \frac{1}{2}\beta(v_m, v_m)N^2, \qquad (6.1)$$

where J is the nucleation rate (note that this is notated different from B in Eq. 6.4 because it represents number of particles, rather than number density, per unit time). The convection term includes the thermophoretic velocity, u_j^t (cf. Section 3.6.3).

The second equation is for the number of primary particles, N_p, and replaces Eq. 2.37 for the surface area:

$$\begin{aligned}\frac{\partial N_p}{\partial t} + \frac{\partial[(u_j + u_j^t)N_p]}{\partial x_j} &= \frac{\partial}{\partial x_j}\left(D_p(v,T)\frac{\partial N_p}{\partial x_j}\right)\\ &\quad + J(\mathbf{Y},T) - \frac{3}{\tau_f}\left(\frac{V}{v_{p,av}} - \frac{m_{2/3}}{v_{p,av}{}^{2/3}}\right),\end{aligned} \qquad (6.2)$$

where V is the particle volume fraction and τ_f is the sintering timescale.

Finally, an equation for the former closes the system:

$$\frac{\partial V}{\partial t} + \frac{\partial[(u_j + u_j^t)V]}{\partial x_j} = \frac{\partial}{\partial x_j}\left(D_p(v,T)\frac{\partial V}{\partial x_j}\right) + J(\mathbf{Y},T)v_0 + C_v(\mathbf{Y},T), \qquad (6.3)$$

where C_v stands for the source due to condensation in terms of volume fraction rate (to be distinguished from the sources expressing number density rates appearing in Eqs. 6.4 and 6.5) and v_0 is the volume of a nucleus particle.

Note that the average primary particle size, $v_{p,av}$, and the average number of primary particles per aggregate, $n_{p,av}$, can be calculated from the variables of the monodisperse model as V/N_p and N_p/N, respectively.

Two-PBE Model

The two-PBE model was discussed in Section 2.4.2 as a model for describing aggregate morphology. It comprises two equations, one for the number density of particles – whether they are spherical or fractal aggregates – and one for the number of primary particles within an aggregate. Given the choice of kinetic and transport processes made, the first equation can be written as follows[1]:

$$\frac{\partial n}{\partial t} + \frac{\partial[(u_j + u_j^t)n]}{\partial x_j} = \frac{\partial}{\partial x_j}\left(D_p(v,T)\frac{\partial n}{\partial x_j}\right) + B(\mathbf{Y},T)\delta(v - v_0) + C(\mathbf{Y},v)$$
$$+ \frac{1}{2}\int_0^v \beta(w, v-w)n(w)n(v-w)dw - n(v)\int_0^\infty \beta(v,w)n(w)dw, \quad (6.4)$$

where B and C are the nucleation and condensation rates, respectively, both expressed in terms of number density per unit time. This equation is similar to Eq. 6.20 that will be employed in the case study on soot, but without the surface growth (and oxidation) term.

The second equation is based on Eq. 2.34, to which we introduce the kinetic and transport models chosen and is shown below:

$$\frac{\partial n_p}{\partial t} + \frac{\partial[(u_j + u_j^t)n_p]}{\partial x_j} = \frac{\partial}{\partial x_j}\left(\frac{n_p}{n}D_p(v,T)\frac{\partial n}{\partial x_j}\right) + C_p(\mathbf{Y},v)$$
$$+ \frac{1}{2}\int_0^v \beta(w, v-w)n(w)n(v-w)\left(\frac{n_p(w)}{n(w)} + \frac{n_p(v-w)}{n(v-w)}\right)dw$$
$$- n_p(v)\int_0^\infty \beta(v,w)n(w)dw - \frac{3}{\tau_f}\left[\frac{n_p}{n} - \left(\frac{n_p}{n}\right)^{2/3}\right]n, \quad (6.5)$$

where we are reminded that n_p is the number density function for primary particles and C_p is the condensation rate for primary particles. The term corresponding to sintering was discussed in Section 3.2.3 (Eq. 3.46). The nucleation term is omitted here because this equation holds for particles with volume greater than the critical volume below which instantaneous coalescence occurs. This equation is similar to Eq. 6.21 that will be used in the case study on soot, the difference again being the absence of surface growth and oxidation.

[1] In this and the following equations, the arguments will be omitted except wherever necessary, such as the dependence of n in the aggregation terms.

The variables that will be compared with the SAXS measurements can be calculated from n and n_p in the following way:

$$N_p(t) = \int_{v_d}^{\infty} n_p(v,t)dv, \tag{6.6}$$

$$d_{p,av}(t) = \frac{\int_{v_d}^{\infty} d_p(v)n_p(v,t)dv}{N_p(t)}, \tag{6.7}$$

$$n_{p,av}(t) = \frac{N_p(t)}{N(t)}, \tag{6.8}$$

$$R_{g,av}(t) = \frac{1}{2}\frac{\int_{v_d}^{\infty} d_g(v)n(v,t)dv}{N(t)}, \tag{6.9}$$

where $d_{p,av}$ is the average primary particle diameter, $R_{g,av}$ is the average aggregate radius of gyration and $n_{p,av}$ is the average number of primary particles per aggregate. Furthermore, d_p is particle diameter and d_g is the aggregate collision diameter (cf. Section 3.2.3). The volume of the smallest detectable particle by SAXS measurements, v_d, is taken to be 1 nm (Camenzind et al., 2008). Finally, a critical diameter is introduced into the two-PBE model to distinguish particles that coalesce instantaneously from those to which finite-rate sintering is applied; this was selected to be 1 nm in the present case study.

Extended One-PBE Model

The extended one-PBE model, discussed in Section 2.4.2, was first proposed in the work of Tsagkaridis et al. (2023) on which the present case study is based. It aims to reduce the computational cost of the two-PBE model, while still allowing the incorporation of finite-rate sintering and avoiding the need for a predefined cut-off size separating primary particles and aggregates, as with the one-PBE model that will be used in the soot formation case study (Section 6.2; see also Section 2.4.2). Like the monodisperse model, it takes advantage of the fact that aerosol synthesis of inorganic nanoparticles is dominated by aggregation and sintering. Its use would be more questionable in a problem featuring surface growth, such as soot formation, due to the assumption of monodisperse primary particle size.

The population balance equations for the extended one-PBE model are now formulated as follows. The first one is the PBE for the particles (whether spherical or aggregates), and this is the same as for the two-PBE model (Eq. 6.4). The second equation is for the total number of primary particles, and the principles on which it is based were explained in Section 2.4.2. With the introduction of the kinetic models, this equation is the same as Eq. 6.2. It can also be obtained by integrating Eq. 6.5 over particle volume, where the condensation and aggregation terms vanish since they do not alter the total number of primary particles.

6.1.3 Kinetic Models

Nucleation

An approach similar to that of Shekar et al. (2012) is adopted for the model of the nucleation rate. It is assumed that two SiO_2 molecules collide to form a dimer, which is regarded as the first primary particle. The rate of collision between molecules is described as coagulation in the free molecule regime. The nucleation rate, therefore, is:

$$J = \frac{1}{2}\beta_{SiO_2} N_A^2 C_{SiO_2}^2, \tag{6.10}$$

where N_A is the Avogadro number, C_{SiO_2} is the concentration of SiO_2 molecules and β_{SiO_2} is the coagulation kernel for collision between two molecules given by:

$$\beta_{SiO_2} = 4E_F \sqrt{\frac{\pi kT}{m_{SiO_2}}} d_{SiO_2}^2, \tag{6.11}$$

where E_F is the Van der Waals enhancement factor for which the value of 1.88 is adopted (Kennedy, 1990), k is the Boltzmann constant, and m_{SiO_2} and d_{SiO_2} are the mass and diameter of a SiO_2 molecule, respectively. The diameter of SiO_2 molecules is assumed to be $d_{SiO_2} = 0.44$ nm, while the density of silica particles is $\rho_p = 2196$ kg/m^3.

Condensation

Condensation is treated as a collision process between silica molecules and particles. The condensation rate in the monodisperse model is:

$$C_v(\mathbf{Y}, T) = \beta(v_m, v_{SiO_2}) N \frac{p_{SiO_2}}{kT} v_{SiO_2}, \tag{6.12}$$

where v_{SiO_2} is the volume of a SiO_2 molecule, v_m is the particle mean volume (given by V/N) and p_{SiO_2} is the partial pressure of SiO_2. As mentioned, this rate is given in terms of volume fraction per unit time.

The approach adopted here for implementing condensation in the context of the PBE discretisation methods is similar to the one that will be used for the soot formation case (Section 6.2.3), based on the work of Rodrigues et al. (2018). The rates that constitute the total condensation rate (in terms of number density per unit time) are:

(a) Condensation of molecules on particles of section i, $C_{0\to i}$:

$$C_{0\to i} = C_{SiO_2} N_A \int_{v_{i-1}}^{v_i - v_s} \beta_{0,i} n(w) dw. \tag{6.13}$$

(b) Condensation of molecules on particles of section $i-1$ that will enter into section i, $C_{i-1\to i}$:

$$C_{i-1\to i} = C_{\mathrm{SiO_2}} N_A \int_{v_{i-1}-v_s}^{v_{i-1}} \beta_{0,i-1} n(w) dw. \tag{6.14}$$

(c) Condensation of molecules on particles of section i that will go to section $i+1$, $C_{i\to i+1}$:

$$C_{i\to i+1} = C_{\mathrm{SiO_2}} N_A \int_{v_i - v_s}^{v_i} \beta_{0,i} n(w) dw, \tag{6.15}$$

where the collision frequency of a molecule and a particle in section i is given by:

$$\beta_{0,i} = 1.28 \sqrt{\frac{\pi k T}{2\rho_p}} (d_{\mathrm{SiO_2}} + d_{g,i})^2 \sqrt{\frac{1}{v_{\mathrm{SiO_2}}} + \frac{1}{v_{m,i}}}, \tag{6.16}$$

where $v_{m,i}$ is the volume at the middle of section i, $d_{g,i}$ is the collision diameter of particles at section i (cf. Section 6.1.2) and the amplification factor of 1.28 is due to Van der Waals interactions for silica (Kennedy, 1990). The condensation source terms in Eqs. 6.4 and 6.5 for a section i, therefore, are:

$$C_i(\mathbf{Y}, v) = (C_{0\to i} + C_{i-1\to i} - C_{i\to i+1}) \frac{v_{\mathrm{SiO_2}}}{v_{m,i} dv_i}, \tag{6.17}$$

$$C_{p,i}(\mathbf{Y}, v) = \left(C_{0\to i} \frac{n_{p,i}}{n_i} + C_{i-1\to i} \frac{n_{p,i-1}}{n_{i-1}} - C_{i\to i+1} \frac{n_{p,i}}{n_i} \right) \frac{v_{\mathrm{SiO_2}}}{v_{m,i} dv_i}. \tag{6.18}$$

Aggregation and Sintering

The aggregation regime ranges from free molecular to continuum. Therefore, both the free molecule regime and the Brownian kernel are used, the latter with the Cunningham slip factor, and Eq. 3.37 is used for the transition regime (cf. Sections 3.2.2 and 3.2.3).

Viscous flow is considered to be the dominant mechanism for sintering of silica nanoparticles (Frenkel, 1945; Friedlander, 2000). Several expressions for the sintering characteristic time of silica nanoparticles have appeared in the literature, and a discussion and tests of them in a perfectly stirred reactor can be found in Goudeli et al. (2015). The expression used by Tsagkaridis et al. (2023) is due to Tsantilis et al. (2001) and is the following:

$$\tau_f = 6.5 \cdot 10^{-13} d_p \exp\left[\frac{8.3 \cdot 10^4}{T} \left(1 - \frac{d_c}{d_p} \right) \right], \tag{6.19}$$

where d_p and d_c are in m, T is in K and τ_f is in s. Regarding d_c, it is a critical particle diameter below which sintering can be considered instantaneous, and a value of 10^{-9} m was proposed by Tsantilis et al. (2001) as a first approximation. Tsagkaridis et al. (2023) showed results with other expressions too, but the above model provided the best results, at least for the flame conditions in that study.

Table 6.1 *Global two-step reaction mechanism for HMDSO combustion. The reaction rate constant is given in the form* $k = AT^n \exp[-E_a/(RT)]$, *where* E_a *stands for the activation energy. The units of the parameter A are expressed in terms of cm, s and mol, while* E_a *is given in cal/mol.*

Step	Reaction	A	n	E_a
G1	$HMDSO + OH \longrightarrow 2\,SiO + 6\,CH_3 + H$	6×10^{12}	0.46	15,000
G2	$SiO + H_2O \longrightarrow SiO_2(g) + H_2$	8.5×10^{10}	0	5,650

Adapted from Feroughi et al. (2017) with permission from © 2016 The Combustion Institute. Published by Elsevier Inc. All rights reserved.

6.1.4 Fluid Dynamics and Transport Phenomena

The general forms of the equations of reacting flow are provided in Appendix A. The formulation of the equations of reacting flow for this case study is the same as for the laminar sooting flame (Section 6.2.4).

A variable density formulation is needed since, in a flame, the chemical reaction changes significantly the density of the gas mixture. Furthermore, the flow is at a low Mach number, which means that it can be treated as an incompressible, variable density flow. The appropriate forms of the continuity and Navier–Stokes equations, therefore, are Eqs. A.1 and A.22, respectively.

The species transport equation that is employed here is Eq. A.17, where multicomponent diffusion is modelled approximately with Fick's law. The reaction and population balance source terms are given by Eqs. A.20 and A.21, respectively. The Gas Research Institute (GRI) 1.2 mechanism (Frenklach et al., 1995) for methane combustion, which contains 175 steps and 31 species, is used for the combustion kinetics. Regarding the HMDSO decomposition and oxidation kinetics, a reduced two-step mechanism due to Feroughi et al. (2017) is employed and shown in Table 6.1. This global mechanism has also been used by Rittler et al. (2017) to study silica nanoparticle synthesis via flame spray pyrolysis.

The energy transport equation is written in terms of the sum of chemical and thermal energy, as shown in Section A.3.5. The radiative heat flux is given by the optically thin model (cf. Section A.3.4) Regarding species, radiation from CH_4, H_2O, CO_2, and CO is considered (Eq. A.38). Based on the study of Lee (2011) on the emission properties of silica particles, the model for emissions from particles (Eq. A.39) is employed, with the value $C_s = 1{,}100\ \mathrm{m^{-1}K^{-1}}$ which is lower than the one used for soot ($1{,}307\ \mathrm{m^{-1}K^{-1}}$). More information on the radiation model can be found in Tsagkaridis et al. (2023), where comparison with a simulation without radiation was also carried out.

6.1.5 Numerical Methods

The numerical methods employed in this case study are similar to those used for the laminar Santoro flame in Section 6.2.6. More details are given below.

Discretisation of the PBE

The PBE is solved with the conservative finite volume discretisation method (Liu and Rigopoulos, 2019; O'Sullivan and Rigopoulos, 2022), which has been detailed in Section 4.2.8. The extension of this method to the two-PBE formulation has been carried out by Sun and Rigopoulos (2022) and detailed in Section 4.2.11. This method is conservative for the first moment of the PBE, which is proportional to the mass of particles and is thus an important property for ensuring mass balance. An exponential grid comprising 60 intervals is employed, covering a particle size range of 0.44 nm–1 μm in diameter. Convergence studies were carried out with a perfectly stirred reactor to ensure that the PBE grid was sufficient.

Coupling with CFD

For solving the coupled fluid dynamics-PBE formulation, the conservative finite volume discretisation method is coupled with the CFD code BOFFIN (Jones et al., 2002), which is based on a variable-density low Mach number formulation of the three-dimensional Navier–Stokes equations and employs a finite volume scheme. A fractional step method is employed, hence the PBE solver can be implemented via its own fractional step, in the manner described in Section 4.4.4. A second-order central difference scheme and a TVD scheme with van Leer's limiter are employed for the diffusion and convection terms, respectively. The second-order accurate Crank–Nicolson method is applied for the temporal integration of the convection and diffusion processes. The chemical reaction fractional step involves solving a stiff system of ODEs, hence its integration is carried out with a backward Euler method, while a modified Newton–Raphson scheme is employed for solving the resulting system of nonlinear equations. Finally, the temporal integration of the discretised PBE is carried out with the explicit Euler method, as these equations are not stiff.

6.1.6 Simulation of Silica Synthesis in a Laminar Flame: The Camenzind et al. Case

Description of Experiment and Computational Setup

The flame that will be simulated corresponds to the case described as 'S-2'[2] in Camenzind et al. (2008), the experimental setup for which is shown in Fig. 6.1.[3]

[2] The notation 'S-y' denotes a series of flames with different O_2 flow rates, where y is the flow rate in l/min.

[3] The help of Sotiris Pratsinis on the comparison with the experimental results from his group in the Camenzind et al. (2008) paper and the insightful discussions with him on silica synthesis are gratefully acknowledged.

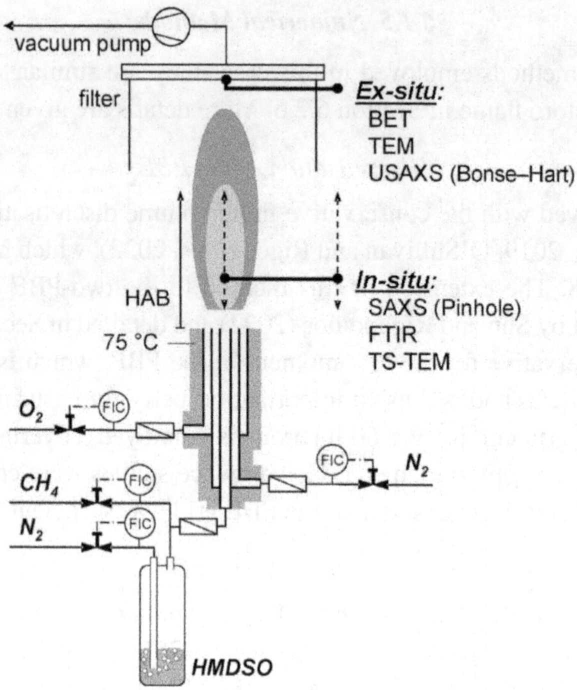

Figure 6.1 Experimental setup for the synthesis of the SiO_2 nanoparticles with a co-flow diffusion flame burner. Reprinted from 'Nanostructure evolution: from aggregated to spherical SiO_2 particles made in diffusion flames', by Camenzind et al. (2008). Copyright Wiley–VCH GmbH. Reproduced with permission.

It is a co-flow diffusion flame where silica synthesis takes place via oxidation of hexamethyldisiloxane (HMDSO), a vapour precursor. The Reynolds number, based on the bulk velocity and the outer diameter of the reactor, was $Re \approx 1414$. One major advantage of this experiment is that an extensive experimental dataset is provided for particle properties at several heights above the burner (HAB) obtained via *in-situ* SAXS measurements.

The computational setup comprises a two-dimensional axisymmetric domain with an axisymmetric boundary condition being applied along the centreline, a symmetry boundary condition being applied at the opposite side and a zero-gradient boundary condition at the outlet boundary in the streamwise direction. Laminar flow velocity profiles were applied to the inflow of the inner jet and to the annuli. The size of the computational domain is $60D \times 15D$, where D is the inner tube diameter. The grid comprises 264×162 cells with spacing increasing exponentially by factors 1.0132 and 1.03 in the axial and radial directions, respectively. Grid refinement was employed at the height of the second annulus to ensure that the inlet velocity profile is resolved. A time step of $\Delta t = 10^{-7}$ s was employed, resulting in a maximum Courant-Friedrichs-Levy (CFL) number of 0.03.

For more details on the experimental setup, operating conditions and simulation parameters, the reader may refer to Camenzind et al. (2008) and Tsagkaridis et al. (2023).

Temperature Prediction

The temperature is the driving force for most of the kinetic processes involved in particle formation, hence its accurate prediction is the cornerstone of a flame synthesis model. Camenzind et al. (2008) provided experimental data for temperature obtained using Fourier transform infrared (FTIR) spectroscopy (Morrison Jr. et al., 1997; Griffiths and De Haseth, 2007), which is a line-of-sight technique over the total flame width. The FTIR data are representative of a radial-average temperature in the region close to the burner tip where reactants are not mixed (Camenzind et al., 2008), while further downstream the radial average is close to the centreline value. Therefore, a comparison is made in Fig. 6.2 between the simulation results for the radial maximum temperature value and the experimental results at each HAB.

Since the emitted radiation (relevant to FTIR measurements) scales with the fourth power of temperature, it can be expected that regions with the highest temperature would dominate in the FTIR data analysis. Results are also shown for a radial average of the temperature in the cells within the flame zone (estimated

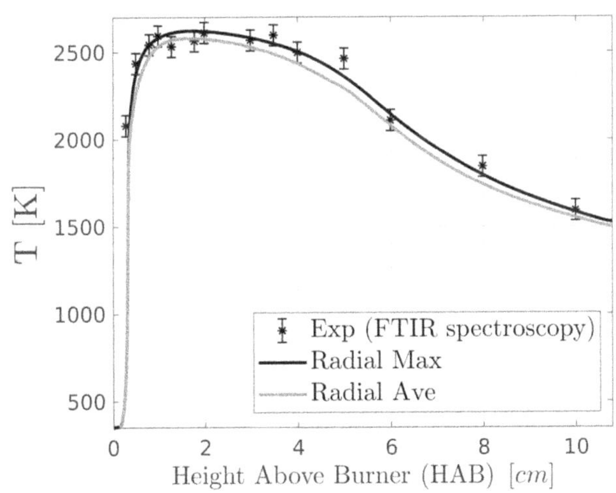

Figure 6.2 Comparison of numerical predictions (two-PBE model) with the experimental FTIR data of Camenzind et al. (2008) for flame temperature along the centreline and effect of radial averaging. An average experimental error of ±60 K given by Kammler et al. (2002) was used in the figure for the FTIR data. Reproduced from Tsagkaridis et al. (2023) under CC-BY 4.0 license.

via the cells whose value of CO_2 concentration is greater than 80% of the radial maximum), which is very close to the radial maximum. Both simulation results are very close to the experimental data, in most cases lying within the average experimental error of ±60 K (Kammler et al., 2002). This indicates that the flame temperature was predicted well, providing a good foundation for the remaining results. The good prediction can be attributed to the direct coupling of detailed chemistry with flow, the absence of modelling uncertainties involved in turbulent flame simulations (turbulence modelling and turbulence–chemistry interaction) and the inclusion of the model for the thermal radiation from particles. Regarding the last point, Tsagkaridis et al. (2023) conducted also two alternative simulations, one neglecting all thermal losses due to radiation and one considering radiation from gas species only. In both cases, the results were in good agreement close to the burner but subsequently exhibited a significant overestimation of the experimental data, thus indicating that radiative losses, particularly from particles, have an important effect on the evolution of the temperature field.

Particle Properties and Comparison with SAXS Measurements

The predictions of particle properties will now be compared with the experimental SAXS data of Camenzind et al. (2008). The SAXS measurements involve a monochromatic X-ray beam penetrating the flame and being collected on a SAXS detector located a few meters away from the flame. The experimental data refer to the thin flame zone where the scattered intensity should be maximum – see also Sztucki et al. (2007), where it was found that the maximum scattered intensity was achieved when the X-ray beam penetrated the flame in the soot growth zone located slightly off the centreline of the flame. Therefore, the comparison is made with data along the pathline of maximum silica volume faction. Furthermore, since the smallest detectable particle size by the SAXS measurements was 1 nm, particles smaller than that were not taken into account when post-processing the simulation results.

Figs. 6.3a–d show the average primary particle diameter, the average aggregate radius of gyration, the number concentration of primary particles and the average number of primary particles per aggregate, respectively, as obtained from the three PBE models and the measurements. The most comprehensive of the three models, the two-PBE model, exhibits reasonably good agreement with the measurements for the primary particle diameter. The trend of the experimental data is correctly captured and the predictions are quantitatively accurate at low HABs, while further downstream (HAB > 5 cm) there is an overprediction of about 15%. This is an indication of a good representation of aggregation and sintering by the model. Regarding N_p, the predictions are in qualitative agreement but the value is overestimated by more than one order of magnitude. This suggests an overprediction of the particle formation rate that can be attributed to uncertainties in the SAXS measurements, precursor decomposition kinetics and nucleation model. It should

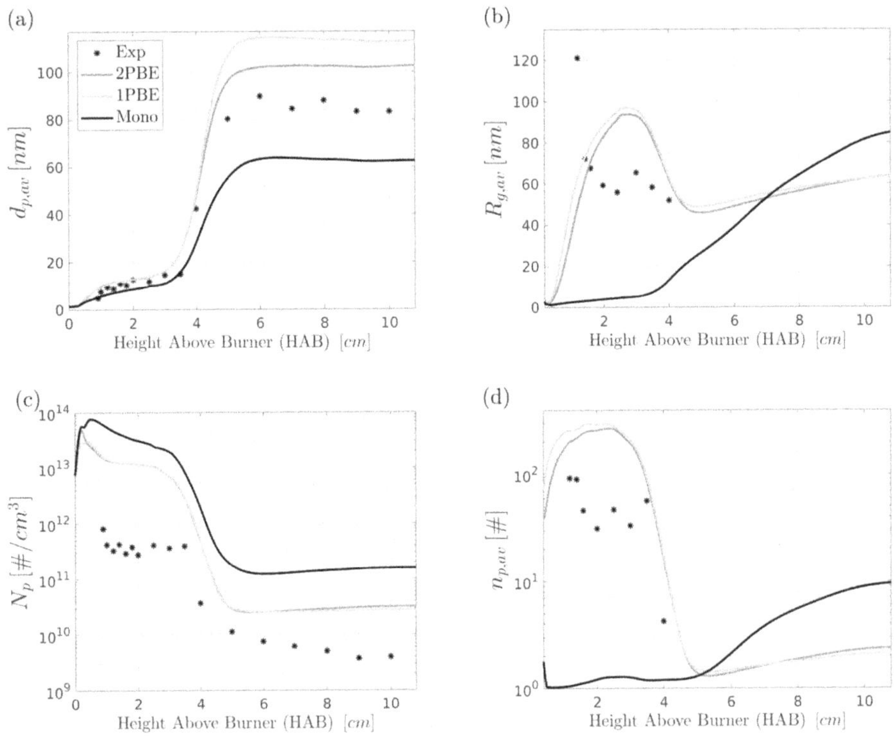

Figure 6.3 Comparison of numerical predictions with the experimental SAXS data of Camenzind et al. (2008) for (a) primary particle diameter, (b) radius of gyration of aggregates, (c) primary particle number concentration and (d) average number of primary particles per aggregate along the pathline of maximum silica volume fraction. Results were obtained with the two-PBE ('2PBE'), the extended one-PBE ('1PBE'), and the monodisperse model ('Mono'), while 'Exp' refers to experimental results. Reproduced from Tsagkaridis et al. (2023) under CC-BY 4.0 license.

be mentioned that the two-step reaction mechanism of Feroughi et al. (2017) was developed with data obtained from low-pressure flames and low HMDSO concentrations, while the flame under consideration is at atmospheric pressure and with relatively high precursor concentration. However, the good agreement with experiments for $d_{p,av}$ is more significant, due to the practical importance of this quantity. Regarding R_g, the overestimation is again likely to be associated with uncertainties involved in the kinetic models, as well as the description of fractal aggregates via Eq. 3.38 and fixed parameters at the early stages of particle formation. The results for $n_{p,av}$ suggest that aggregates are formed at low HABs (where values over 200 are observed), while further downstream the primary particles fuse and $n_{p,av}$ is reduced to just a few particles, a trend that is evident also in the SAXS data at HAB = 4 cm. The overestimation of $n_{p,av}$ is consistent with the overprediction of N_p.

Table 6.2 *Comparison of CPU time required by each of the three PBE models used in the present case study. The total CPU time per time step includes all processes (flow field, scalar convection/diffusion, gas phase reaction, and PBE solution). Tabulation of the aggregation kernel was employed for the one-PBE model.*

Model	Two-PBE	Extended one-PBE	Monodisperse
No. of equations	116	61	3
PSD of aggregates	polydisperse	polydisperse	monodisperse
PSD of primary particles	polydisperse	monodisperse	monodisperse
CPU time (s) / time step	1.14	0.32	0.187

Reproduced from Tsagkaridis et al. (2023) under CC-BY 4.0 license.

We now examine the results obtained with the extended one-PBE and the monodisperse model (also presented in Fig. 6.3). In most cases, the extended one-PBE model provides results very close to the two-PBE model. In particular, the results for $d_{p,av}$ coincide up to HAB = 4 cm while afterwards the former exhibits a higher deviation from the experimental results. For all other results, the predictions of the two models are very close throughout the burner. On the other hand, the monodisperse model shows considerably different behaviour. The prediction of $d_{p,av}$ has the correct qualitative trend but exhibits considerable underprediction, while N_p also shares the same trend but is overpredicted to a greater degree. R_g and $n_{p,av}$, however, show very different trends from the PBE model predictions and are severely underpredicted compared with the experimental data. More results, as well as an analysis of the spatial distribution of rates and timescales, are shown in Tsagkaridis et al. (2022).

Computational Performance of PBE Models

It is instructive to compare the computational performance of the three PBE models employed in this case study by considering the total CPU time per time step for each of the three models (Table 6.2). The extended one-PBE model requires less than a third of the CPU time needed by the two-PBE model, despite the fact that it involves about half the number of equations. This extra speed gain is due to the tabulation of the aggregation kernel, which was not carried out for the two-PBE model because it was more memory-intensive.[4] Considering also the results shown in the previous section, the extended one-PBE model comes across as an efficient way for simulating aerosol dynamics.

[4] Tabulation of the kernel in the one-PBE requires storing a three-dimensional matrix (for two volumes and d_p), while the two-PBE requires a four-dimensional matrix since every section has a different d_p.

Remarkably, the time required for computation with the extended one-PBE model is comparable with that of the monodisperse model (less than twice). This indicates that the PBE does not result in a computational overload compared with the other processes that must be carried out in each time step of a coupled CFD-PBE simulation.

6.1.7 Concluding Remarks

The case study presented in the present section involved a laminar flame for synthesis of silica nanoparticles. As such, it has several elements in common with the Santoro flame that will be investigated in Section 6.2.7. However, the kinetics here are dominated by aggregation and sintering, unlike the Santoro flame where surface growth and oxidation play an important role. Furthermore, in nanoparticle synthesis there is more emphasis on the prediction of particle morphology because it is an important property of the material to be produced.

Three PBE approaches were compared in this case study: a monodisperse, an extended one-PBE and a two-PBE model. The two-PBE model is the same as the one used in the Santoro flame, while the monodisperse model has been used extensively in the literature on inorganic nanoparticle synthesis. The extended one-PBE model was first presented in the work of Tsagkaridis et al. (2023) on which this case study is based; it aims at providing a computationally efficient alternative to the two-PBE model while retaining the capacity to account for finite-rate sintering, which is not possible with the one-PBE model used in the sooting flames case study. The results showed that the extended one-PBE model provides a description of the evolution of particle morphology than the monodisperse model, with results close to those of the two-PBE model, at least in a problem that does not feature surface processes, such as inorganic nanoparticle synthesis. Furthermore, the computational cost of the CFD-PBE simulation with the extended one-PBE model is less than twice the time needed for the CFD-monodisperse model; this efficiency is largely due the tabulation of the aggregation kernel.

Unlike the case of soot, where surface chemistry is a major uncertainty and no reliable model for the fusing of soot particles is available, the aerosol kinetics employed for simulation of silica synthesis are more robust. However, uncertainties exist regarding radiation properties and precursor decomposition kinetics, as well as nucleation (which is almost always a major uncertainty). Comparison with experimental results showed that the flame temperature profile could be well reproduced, as long as radiation is taken into account. Good agreement with SAXS data was obtained for the primary particle size as predicted by the two-PBE and extended one-PBE models, while aggregate sizes and number of primary particles per aggregate were overestimated. The monodisperse model did not capture the evolution of particle morphology at the early stages of the synthesis. The particle formation rate was overpredicted by all models, and, consequently, the concentration of primary particles was overestimated. The precursor kinetics and nucleation model are more likely the reasons for this discrepancy, along with uncertainties in the experimental data.

The case study in this section was an example of CFD-PBE coupling without the complications of turbulent flow, with an emphasis on the prediction of aggregate properties. It should be noted that industrial processes for nanoparticle production are usually turbulent, hence the methods employed in the case study of soot formation that follows can also be employed here. It should also be mentioned that the emphasis on morphology is relevant in sooting flames too, because there are carbonaceous nanoparticles being produced for use as advanced materials, in which case morphology is also an important variable to be predicted.

6.2 Soot Formation in Laminar and Turbulent Flames

6.2.1 Introduction

Overview

Soot is particulate material that is formed during incomplete combustion of hydrocarbons. Soot emissions are harmful for human health, as they have carcinogenic potential. The adverse effects of soot particulates depend on particle size (Kreyling et al., 2006). On the other hand, there are several manufactured carbonaceous nanoparticles such as carbon black, which is used in tyres, inks, batteries and solar cells, and new carbonaceous nanomaterials of exceptionally high value such as carbon quantum dots and carbon-coated nanoparticles for use as magnetic biofluids (Kelesidis et al., 2017). The value of such nanoparticles depends on their size and morphology. As such, the formation of soot and carbonaceous nanoparticles is one of the best examples of application of population balance modelling.

Soot is formed within combustion devices such as gas turbines and diesel engines, while flame synthesis is one of the main technologies for manufacturing nanoparticles. In both cases, therefore, the chemical reactions taking place within a flame are of key importance. These reactions are very fast, with the reaction zone having the form of a thin layer (flame). As a result, the composition of the mixture is highly heterogeneous and the process cannot be adequately approximated by ideal reactor models. A comprehensive account for fluid dynamics, transport phenomena and thermochemistry is required in order to predict the distribution of the chemical species that act as soot precursors. Hence, CFD-based models of combustion must be coupled with population balance models to predict the concentration and properties of the resulting soot particles. Two-way coupling is effected by species consumption due to soot formation (for the species transport equations) and radiation (for the energy transport equation, as well as indirectly for the flow field via the density. Soot formation is, therefore, an excellent example of the coupling between fluid dynamics and population balance modelling and brings together many of the concepts discussed so far in the present book.

A considerable body of literature is devoted to soot formation. A general overview of the field is provided by Haynes and Wagner (1981) and Glassman (1988). For aspects of soot chemistry and physics, reviews include Harris and Weiner (1985), Richter and Howard (2000), Stanmore et al. (2001), Frenklach (2002b), Mansurov (2005), D'Anna (2009) and Wang (2011). Particular applications in fields such as engines and coal combustion are covered in Fletcher et al. (1997), Kittelson (1998) and Tree and Svensson (2007). For reviews focussed on models of soot formation, the reader may refer to Kennedy (1997), Raman and Fox (2016) or Rigopoulos (2019, 2020); the three last ones cover the later works involving population balance and fluid dynamics and are thus the most relevant to this case study.

It should also be mentioned that soot formation models involve the theory of reacting flow. An elementary presentation of the basic equations is provided in Appendix A, while the reader may refer to Peters (2000), Veynante and Vervisch (2002), Poinsot and Veynante (2005) or Cant and Mastorakos (2008) for more details.

Objectives of the Case Study

In this case study, it will be demonstrated how the general approach of population balance modelling together with a selection of the models expounded in the earlier chapters can be applied to formulate a model for a problem of particle formation within a fluid flow and how the numerical methods for solution of the PBE can be combined with computational fluid dynamics (CFD) to provide a scheme for the coupled fluid dynamics-population balance problem. Soot formation in flames is a problem that combines a multitude of kinetic processes (nucleation, condensation, surface growth and oxidation, aggregation and coalescence) and features strong coupling of population balance and flow, thus providing an excellent example for the integration of the methods presented so far. The presentation here is based on two works by the author's group (Sun et al., 2021; Sun and Rigopoulos, 2022), where a laminar and a turbulent sooting flame were simulated with the same models, and the objectives of the investigation are summarised below.

The first objective is to explore the application of one-PBE and two-PBE models to account for aggregate morphology. One-PBE models have been employed in soot modelling due to their computational simplicity, as the overall problem is very computationally expensive due to the need to incorporate detailed chemistry. However, they are subject to inherent limitations when it comes to problems involving fractal aggregates (cf. Section 2.4.2). On the other hand, the modelling of the fusing of primary particles in soot is impeded by the lack of kinetic information, unlike the case of inorganic nanoparticles which will be discussed in Section 6.1. In this study, the two-PBE approach is implemented via the conservative finite volume method discussed in Section 4.2.8 and applied to the laminar Santoro sooting flame, where it is compared with the one-PBE approach. It is shown that the two-PBE approach provides a better description of morphology, but the one-PBE approach can still predict reasonably well several process variables, hence it is employed in the much more computationally expensive turbulent flame.

The second objective is to investigate the effect of kinetic parameters in the laminar flame, where the issues of turbulence, turbulence–chemistry and turbulence–PBE interaction do not come into play. As discussed in the previous sections, there is considerable uncertainty about many aspects of soot kinetics, such as nucleation and surface growth. A comprehensive model for all kinetic processes is employed here and its sensitivity to some of the most important parameters is explored in the context of the two aforementioned PBE approaches. Furthermore, the co-flow diffusion flame requires the full coupling of PBE with an unsteady CFD code, unlike one-dimensional formulations, thus preparing the ground for the subsequent turbulent flame simulation.

Third, a turbulent sooting flame is simulated with the same kinetic models and parameters employed in the laminar flame. The objective here is twofold: first, to shed light on the nature and origin of discrepancies in the prediction of turbulent sooting flames by leveraging the knowledge acquired from the simulation of the laminar flame, and second, to investigate the mechanisms of soot formation in flames. The LES-PBE-PDF method, detailed in Chapter 5, is employed for the treatment of turbulence–chemistry and turbulence–soot interaction.

In deploying the population balance methodology, we will follow the four steps given in Section 1.1.3: formulation of the population balance model, selection of kinetic models, coupling with models for flow and species/energy transport and application of a solution method. Subsequently, selected results will be shown and discussed.

6.2.2 Formulation of the Population Balance Model

Preliminary Considerations

In model selection for a study on soot formation, an important factor to consider is the CPU time demands, which are typically dominated by the chemistry. The main classes of soot kinetic models were mentioned in Section 3.4.3. Among them, non-precursor models do not require any chemistry, while semi-empirical precursor-based models require a gas-phase mechanism sufficiently detailed to provide good predictions of the precursors (usually C_2H_2). In detailed models, the chemistry required for prediction of PAHs is very complex and often the most expensive part of the model, which is a concern when coupling the model with CFD, and even more so in turbulent flows. Many of the studies with detailed models have been conducted on ideal reactors or one-dimensional laminar flames. However, there is an increasing number of studies that incorporate detailed chemistry in CFD, such as the case study to be presented here. The population balance model in the present study is a comprehensive one, in line with the objectives of the study, which involve exploring the capacity of such models for predicting soot in both laminar and turbulent flames.

Among the classes of solution methods described in Chapter 4, the ones that have been employed for soot formation are moment, discretisation and Monte Carlo methods. Furthermore, the prediction of morphology is often included in detailed models because it is an important process variable that determines the available surface area for growth and oxidation processes. In the case of carbonaceous nanoparticle synthesis, morphology is important per se because it affects the properties of the product formed. The model to be employed in this case study is based on a discretisation method. In the context of such a method, the presence of both spherical particles and fractal aggregates can be accounted for via a modified one-PBE model with an imposed cut-off size between the two types of particles or via a coupled PBE approach (cf. Section 2.4.2). One of the main objectives of the case study is to compare these two approaches. It should be mentioned that the extended one-PBE approach that was employed successfully in the silica synthesis case study is not recommended here, because of the considerable role of surface growth in soot formation.

Eq. 3.38 is employed for the description of aggregate morphology. A value of $k_f = 1.94$ is adopted for the fractal prefactor (Link et al., 2011). Regarding the fractal dimension, D_f, a constant value is usually assumed (see e.g. Zhang et al., 2009b; Dworkin et al., 2011; Saggese et al., 2015), and a value of 1.8 is employed here, based on experiments (e.g. Köylü et al., 1997; Brasil et al., 1999; Link et al., 2011; Schenk et al., 2015).

Kinetic and Transport Processes

The kinetic processes involved in soot formation include the following:

- Nucleation, which occurs via the formation of larger and larger PAHs.
- Condensation, which involves deposition of PAHs on the soot surface.

- Surface growth, occurring via reaction of C_2H_2 on the particle surface.
- Oxidation, which occurs mainly via O_2 and OH, while some models include also O.
- Aggregation, taking place in the free molecule, transition and continuum regimes, as dictated by the range of particle sizes observed in the experiments.
- Particle coalescence and fusing. Young soot particles that are liquid-like and coalesce to spherical ones, while mature soot particles fuse gradually.

In addition, the following transport processes should be considered:

- Convection of particles by the flow, which is laminar in the Santoro flame and turbulent in the Sandia flame. Since the particles in this case study are nanoparticles, the Stokes number is almost zero and the particles follow the flow.
- Particle diffusion, which diminishes with increasing particle size. In turbulent flows, its effect is negligible.
- Thermophoresis should be considered in laminar flames, and therefore will be included in the case of the Santoro flame. It can be incorporated as an additional convective velocity.

One-PBE Model

As we have seen in Chapter 2, the prediction of morphology can be accomplished via two-dimensional or coupled PBEs. The former have been employed mainly in the context of moment methods because a discretisation would result in too many variables to be coupled with CFD. The simplest way to employ a discretisation method is to use a single PBE and assume that all primary particles are identical, so that the collision kernel for aggregates is employed above a cut-off size. This approach has several limitations that will be discussed below.

We first formulate the one-PBE model of soot formation. Soot and other nanoparticles have negligible inertia owing to their small size, hence the appropriate form of a space dependent PBE is Eq. 2.21. The nucleation, growth and condensation are all functions of particle size, selected chemical species and temperature. In this and the other case studies, the vector **Y** will be used to denote all species present in the system. For brevity, the space and time dependence of the variables will be omitted in the equations below, but the dependence on all other variables will be explicitly shown. The thermophoretic velocity will be denoted as u_i^t. We can now write the PBE including all the processes listed in the previous section as follows:

$$\frac{\partial n}{\partial t} + \frac{\partial \left[(u_i + u_i^t) n \right]}{\partial x_i} + \frac{\partial (G(v, \mathbf{Y}, T) n)}{\partial v}$$
$$= \frac{\partial}{\partial x_i} \left(D_p(v, T) \frac{\partial n}{\partial x_i} \right) + B(\mathbf{Y}, T) \delta(v - v_0) + C(v, \mathbf{Y}, T)$$
$$+ \frac{1}{2} \int_0^v \beta(w, v - w) n(w, t) n(v - w, t) dw - n(v, t) \int_0^\infty \beta(v, w) n(w, t) dw. \quad (6.20)$$

Note that the growth function, $G(v, \mathbf{Y}, T)$, denotes the net effect of surface growth and oxidation. These two processes are combined in the same term because both are described by a surface chemistry mechanism. By contrast, condensation, $C(v, \mathbf{Y}, T)$, is included as a separate process in the PBE, because its mechanism is different from that of surface growth and oxidation.

As discussed in Section 2.4.2, when the one-PBE approach is to be applied to a problem that involves fractal aggregates such as soot formation, a cut-off size is imposed between primary (spherical) particles and fractal aggregates, while the latter are assumed to be composed of a number of equally sized primary particles. Primary particles within aggregates are assumed to have a diameter d_p equal to the cut-off size. This approach enables the one-PBE to predict the main soot properties, but not the morphological ones. It should be mentioned that the morphological characteristics are often not of primary interest in the case of soot, unlike nanoparticle synthesis (such as the case study in Section 6.1) where they are determining the characteristics of the product.

Even in the case of soot, however, the simplified one-PBE approach has two main limitations (cf. Section 2.4.2). First, it does not allow accounting for finite-rate fusing and instead employs the cut-off size, which must be prescribed. Second, it involves an inconsistent calculation of the growth and oxidation rates. This is because the size of the primary particles is assumed to remain fixed once they have reached the cut-off size. However, growth and oxidation continues to take place and change the primary particle size, and the rate of these processes depends on the surface area, which is calculated on the basis of a fixed primary particle size. One of the objectives of this study is, therefore, to assess the impact of these limitations.

The value employed for the cut-off size (which is the only parameter in the one-PBE model) in the study of Sun et al. (2021) is $d_c = 30.8$ nm, a value also adopted by Liu et al. (2020) based on experimental observations and is also in agreement with the average size of 29 nm reported by Tian et al. (2006), with both studies having been also on laminar co-flow ethylene diffusion flames. The effect of this parameter was also investigated and results will be shown in Section 6.2.7 (Fig. 6.9).

Two-PBE Model

The coupled PBE approach allows for a more realistic description of aggregation and its discretisation requires only twice the number of variables than a single PBE, unlike the two-dimensional PBE, thus rendering it suitable for coupling with CFD. The formulations used for soot usually employ a second PBE for the number of primary particles per aggregate. Park et al. (2005) employed such a model for soot formation in a jet stirred reactor – plug flow reactor (JSR-PFR), based on earlier work by Park and Rogak (2003) on aerosol synthesis, while further works employed this approach for the JSR-PFR (Wen et al., 2005, 2006) and for laminar flames (Zhang et al., 2008, 2009a,b). Sun et al. (2021) employed a two-PBE formulation and investigated the effect of the model employed for the fusing of primary particles by conducting a parametric analysis on models used in inorganic nanoparticles, as at present there is scant knowledge on this process in soot. This is the approach that is reported in the present case study.

The two-PBE model was presented in Section 2.4.2 as a model for describing aggregate morphology and was also used in the case study on silica (cf. Section 6.1.2). It comprises two equations, one for the number density of particles – whether they are spherical or fractal aggregates – and one for the number of primary particles within an aggregate. The first equation is the same as Eq. 6.20. The second equation is based on Eq. 2.34 and is shown below:

$$\begin{aligned}\frac{\partial n_p}{\partial t}+\frac{\partial (u_j n_p)}{\partial x_j}+\frac{\partial \left(G(v,\mathbf{Y})n_p\right)}{\partial v}&=\frac{\partial}{\partial x_j}\left(\frac{n_p}{n}D_p(v,T)\frac{\partial n}{\partial x_j}\right)+C_p(v,\mathbf{Y})\\&+\frac{1}{2}\int_0^v \beta(w,v-w)n(w,t)n(v-w,t)\left(\frac{n_p(v-w,t)}{n(v-w,t)}+\frac{n_p(w,t)}{n(w,t)}\right)dw\\&-n_p(v,t)\int_0^\infty \beta(v,w)n(w,t)dw-\frac{3}{\tau_f}\left[\frac{n_p}{n}-\left(\frac{n_p}{n}\right)^{2/3}\right]n, \quad (6.21)\end{aligned}$$

where C_p is the rate of condensation on primary particles. Compared with Eq. 2.34, we can see that Eq. 6.21 includes also terms for surface growth, condensation and particle fusing, as well as spatial transport. As with Eq. 6.5, no nucleation term is included here because this equation holds particles with volume greater than the critical volume below which instantaneous coalescence occurs.

The two-PBE approach overcomes the limitations of the one-PBE formulation when the latter is applied to problems including fractal aggregates and accounts for finite-rate particle fusing. However, it also introduces the problem of obtaining kinetics for the fusing process. Its first applications were in the field of inorganic nanoparticles (cf. Section 2.4.2) where sintering laws are more established. In soot, there is no such experimental information and applications have employed expressions for silica (Chen et al., 2013) or titania (Veshkini et al., 2016) nanoparticles. In the study of Sun et al. (2021), these two expressions were both tested and the sensitivity to the modelling parameters was investigated; these results will be reported in Section 6.2.7.

6.2.3 Kinetic Models

In the present section, we describe the models used in the PBE, namely nucleation, condensation, growth, oxidation, aggregation and fusing. The kinetics of the gas-phase chemistry, including PAH formation, will be discussed in Section 6.2.4.

Nucleation and Condensation

The model of Blanquart and Pitsch (2009) for nucleation and condensation is employed, in line also with the choice of a model gas-phase and PAH kinetics by the same group (cf. Section 6.2.4). The description of the kinetics is here intertwined with concepts relating to the PBE solution method. The original work of Blanquart and Pitsch (2009) was formulated in the context of a moment method. The implementation shown here, in the context of a discretisation method, follows the work of Rodrigues et al. (2018).

Nucleation is considered to occur via collision of two PAHs to form a dimer, which can be regarded as an intermediate state between PAHs and solid nuclei. Subsequently, the dimers may coalesce to form nuclei or condense on the surface of soot particles. Eight PAHs, from naphthalene ($C_{10}H_8$) up to cyclopenta[cd]pyrene ($C_{18}H_{10}$), are considered for this dimerisation process and the collision rate is given by:

$$\dot{\omega}_{PAH,i} = \gamma_i \sqrt{\frac{4\pi k_B T}{m_i}} d_{PAH,i}^2 [PAH]_i^2 N_A{}^2, \tag{6.22}$$

where γ_i is the sticking coefficient factor for PAH_i, N_A is the Avogadro number, k is the Boltzmann constant, m_i is the mass of PAH_i, and $d_{PAH,i}$ is the diameter of PAH_i, given by (Frenklach and Wang, 1994):

$$d_{PAH,i} = D_A \sqrt{\frac{2n_{c,i}}{3}}, \tag{6.23}$$

where D_A is the size of a single aromatic ring of $1.395\sqrt{3}$ Å and $n_{c,i}$ is the number of C atoms in a PAH_i molecule. The PAHs considered in the dimerisation and their corresponding parameters can be found in Table 1 of Blanquart and Pitsch (2009).

The formation of soot nuclei takes place via coalescence of two dimers. The dimer concentration can be obtained by assuming that they are in a quasi-steady state due to their high collision rate with other dimers and soot particles. Therefore, their concentration is obtained by solving the following quadratic equation:

$$a_{nuc}[\text{dimer}]^2 + b_{cond}[\text{dimer}] = \sum_{i=1}^{8} \dot{\omega}_{\text{PAH},i}, \tag{6.24}$$

of which the positive root is accepted. The coefficients a_{nuc} and b_{cond} are determined by the dimer consumption due to nucleation and condensation, respectively, and the expressions for their determination will be shown below.

The nucleation rate, $\dot{w}_{nuc}$, takes the following form:

$$\dot{w}_{nuc} = E_F \left(\frac{3}{4\pi}\right)^{1/6} \sqrt{\frac{6kT}{\rho_s}} 4\sqrt{2} v_{\text{dimer}}^{1/6} [\text{dimer}]^2 N_{\text{A}}{}^2, \tag{6.25}$$

where $E_F = 2.2$ is the Van der Waals enhancement factor (Harris and Kennedy, 1988) and $\rho_s = 1{,}800$ kg/m^3 is the density of the soot particles. The volume of dimers, v_{dimer}, is an average volume defined as follows:

$$v_{\text{dimer}} = \frac{\sum_{i=1}^{8} (\dot{\omega}_{\text{PAH},i} \cdot 2v_{\text{PAH},i})}{\sum_{i=1}^{m} \dot{\omega}_{\text{PAH},i}}. \tag{6.26}$$

Note that v_{dimer} depends on the local PAH concentrations. The nucleation source term as employed in Eq. 6.20 is, therefore, given by:

$$B = \dot{w}_{nuc} \frac{2v_{\text{dimer}}}{dv_0 v_{m,nuc}}, \tag{6.27}$$

where $v_{m,nuc}$ is the representative volume (midpoint) in the section that covers the volume of nuclei and dv_0 is the interval of that section, respectively. The coefficient a_{nuc} in Eq. 6.24 can now be written as follows:

$$a_{nuc} = 2\frac{\dot{w}_{nuc}}{[\text{dimer}]^2}, \tag{6.28}$$

where the factor 2 accounts for the fact that two dimers are consumed to form a soot nucleus.

Condensation takes place when a dimer collides with a soot particle. The same approach as the one shown in Section 6.1.3 for silica is adopted. For each section, i, there are three contributions to the rate of change of number density (Rodrigues et al., 2018):

(a) Condensation such that particles remain in section i:

$$\begin{aligned} C_{\text{dimer}\to i} &= [\text{dimer}]N_{\text{A}} \int_{v_{i-1}}^{v_i - v_{\text{dimer}}} \beta_{d,i} N(w)dw \\ &= [\text{dimer}]N_{\text{A}} \beta_{d,i}(dv_i - v_{\text{dimer}})N_i. \end{aligned} \tag{6.29}$$

(b) Condensation such that particles from section $i-1$ move to section i:

$$C_{i-1\to i} = [\text{dimer}]N_{\text{A}} \int_{v_{i-1}-v_{\text{dimer}}}^{v_{i-1}} \beta_{d,i-1} N(w)dw$$
$$= [\text{dimer}]N_{\text{A}}\beta_{d,i-1}v_{\text{dimer}}N_{i-1}. \tag{6.30}$$

(c) Condensation such that particles from section i move to section $i+1$:

$$C_{i\to i+1} = [\text{dimer}]N_{\text{A}} \int_{v_i-v_{\text{dimer}}}^{v_i} \beta_{d,i} N(w)dw$$
$$= [\text{dimer}]N_{\text{A}}\beta_{d,i}v_{\text{dimer}}N_i. \tag{6.31}$$

The collision frequency of a dimer and a soot particle in section (v_{i-1}, v_i) is given by:

$$\beta_{d,i} = 1.3\sqrt{\frac{\pi kT}{2\rho_s}}\left(d_{\text{dimer}} + d_{g,i}\right)^2\sqrt{\frac{1}{v_{\text{dimer}}} + \frac{1}{v_{m,i}}}, \tag{6.32}$$

where d_{dimer} is the diameter of the dimer, $d_{g,i}$ is the collision diameter of the particle and $v_{m,i}$ is the volume at the midpoint of the section, while the amplification factor of 1.3 accounts for Van der Waals interactions (Marchal, 2008).

We can now obtain the coefficient b_{cond} in Eq. 6.24 as follows:

$$b_{cond} = \frac{\sum_i \left(C_{\text{dimer}\to i} + C_{i\to i+1}\right)}{[\text{dimer}]}. \tag{6.33}$$

Surface Growth and Oxidation

The HACA concept, first proposed by Frenklach and Wang (1991) (cf. Section 6.2.1), is employed for the surface processes, according to the formulation by Blanquart and Pitsch (2009), and the parameters of the reaction rates are shown in Tables 2 and 3 of the latter paper. As discussed in Section 6.2.1, the prescription of the surface reactivity, α, is the main uncertainty in HACA and early works employed constant values (Frenklach and Wang, 1991) while later works expressed it as a function of temperature (Guo et al., 2006) and other parameters such as particle size (Appel et al., 2000), or proposed more complex models including the effects of both temperature and residence time (Veshkini et al., 2014; Khosousi and Dworkin, 2015). In the work of Sun et al. (2021) which is reported here, the approach of Guo et al. (2006) was adopted, in which the reactivity varies with temperature according to the Arrhenius-type expression $\alpha = \text{a}\cdot\exp(9000/T)$. The pre-exponential factor, a, was adjusted such that the peak value for integrated soot volume fraction could be matched to a satisfactory degree with the experimental data, as will be discussed in Section 6.2.7. The net rate that incorporates surface growth and oxidation is given by the following expression:

$$G(v, \mathbf{Y}) = \frac{a(v)}{\rho_s} R_g(\text{HACA}), \tag{6.34}$$

where $a(v)$ is the soot surface area and R_g(HACA) is the mass growth rate, which is a function of α, χ_{soot}, reaction rates and the local concentrations of species participating in the HACA mechanism (Frenklach and Wang, 1991; Blanquart and Pitsch, 2009). An estimate of the theoretical maximum value of the overall site density, χ_{soot}, is 2.3×10^{19} sites/m^2 (Frenklach and Wang, 1991).

Aggregation and Fusing of Primary Particles

Soot particles comprise both spherical particles and fractal aggregates, and the aggregation regime ranges from free molecular to continuum. The aggregation kernels are the same as in the case of silica nanoparticles (Section 6.1.3).

Soot aggregates are composed of primary particles that rapidly coalesce or gradually fuse, with an associated change in size and morphology. As mentioned in Section 3.2.1, the processes in soot are different than in inorganic nanoparticles and also differ between young particles and mature particles, hence we will use the term 'coalescence' for the formation of spherical particles between young soot and the term 'fusing' for the gradual merging of primary particles within aggregates.

While the sintering of inorganic nanoparticles has been extensively studied, such knowledge is, unfortunately, not available for soot. As a result, soot formation is often modelled with approaches that do not involve a timescale for fusing. In the one-PBE model, for example, particles below the cut-off size are assumed to coalesce instantaneously, while those above are regarded as aggregates with a 'frozen' primary particle size. More advanced approaches, such as the two-PBE model, allow for the incorporation of finite-rate fusing, but there is no straightforward way to prescribe a fusing model. In a study of laminar premixed ethylene sooting flames, Chen et al. (2013) employed a sintering model proposed for silica particles by Tsantilis et al. (2001), in which the characteristic sintering time, τ_f, is given by an equation of the following form:

$$\tau_f = A_s d_p \exp\left[\frac{E_a}{T}\left(1 - \frac{d_c}{d_p}\right)\right], \tag{6.35}$$

where A_s is a pre-exponential factor, d_p is the diameter of the primary particles and E_a is the activation energy. This model was used in the silica case study of Section 6.1 and the values of the parameters in that case were shown in Eq. 6.19. On the other hand, in a study of the Santoro flame, Veshkini et al. (2016) used the following expression, proposed for the sintering of titanium oxide Kobata et al. (1991):

$$\tau_f = A_s d_p^4 T \exp\frac{E_a}{T}. \tag{6.36}$$

Table 6.3 *Parameter values and units in the sintering models.*

Variable	Silica sintering model (Chen et al., 2013)	Titania sintering model (Kobata et al 1991; Xiong et al., 1993)
A_s	1.1×10^{-14} (s/m)	7.44×10^{16} $\left(\text{s}/\left(\text{K}\cdot\text{m}^4\right)\right)$
E_a	$1.8 \times 10^4 \leq E_a \leq 1.8 \times 10^5$ (K)	3.31×10^4 (K)
d_c	$1 \leq d_c \leq 5$ (nm)	–

Adapted from Sun et al. (2021) under CC-BY 4.0 license.

For liquid particles, such as silica, sintering is driven by viscous flow, while for titania, it is transport by diffusion; none of these expressions has been developed for soot and, furthermore, even in the context of silica, different models exhibit differences in the predicted primary particle diameter and collision diameter (Goudeli et al., 2015). Therefore, the employment of either of these models for soot is a major source of uncertainty. The results from the work of Sun et al. (2021) presented here compare three cases: the one-PBE model, which does not include a sintering model, the two-PBE model with the silica sintering model and the two-PBE model with the titania sintering model. The parameters of the models from the literature that were employed in the investigation of Sun et al. (2021) can be seen in Table 6.3. Note that, for the silica sintering model, the table shows an optimised range of parameters due to Chen et al. (2013); the values used in the results shown in Section 6.2.7 are $d_c = 5$ nm and $E_a = 10.0 \times 10^4$ K, while a parametric study of these variables was also conducted by Sun et al. (2021).

6.2.4 Fluid Dynamics and Transport Phenomena

The overall formulation of the equations of reacting flow is the same as for the case of the laminar flame for silica synthesis (Section 6.1.4). For the laminar Santoro flame, the equations of fluid dynamics are solved without further models, while for the turbulent flame, the LES approach is used, as will be described in the next section.

The detailed chemical mechanism of Blanquart et al. (2009) is employed for the gas-phase and PAH kinetics. It is a mechanism developed with focus on species important for soot formation such as acetylene, the C_3H_4 isomers, propene and butadiene and has been validated for a large set of fuels ranging from methane to isooctane, in laminar premixed and counterflow diffusion flames. It contains 149 species and 1,651 reactions and includes kinetics for formation of larger PAH species up to cyclopenta[cd]pyrene ($C_{18}H_{10}$).

6.2.5 Turbulence, Turbulence–Chemistry and Turbulence–PBE Interactions

Most of the soot formation processes of practical interest take place in turbulent flows, and accounting for the turbulence–chemistry interaction is vital for the accurate prediction of a turbulent flame (cf. Section 5.2.6). Turbulence also interacts with the kinetic processes in the PBE, as discussed in Section 5.3. Since soot kinetics are heavily dependent on concentration of precursors and temperature, having the right flame structure and comprehensive modelling of chemistry, turbulence–chemistry and turbulence–PBE interactions are prerequisites for the successful prediction of soot formation in a turbulent flow.

The earlier soot modelling studies were based on RANS. Such models have a low computational cost but rely heavily on modelling choices in order to account for the effect of turbulence on the mean flow and for the turbulence–chemistry PBE interaction. At the same time, the soot models that have been employed in RANS-based studies are typically simpler, not only because comprehensive models may have not been available at that time but also because the soot model should be at a level of detail commensurate with that of the rest of the simulation. More recent research has explored LES for sooting flames and combustors, while DNS studies have also appeared on simplified configurations and moderate Reynolds numbers. A review of these works can be found in Rigopoulos (2019).

The LES approach with the standard Smagorinsky model is employed as the basis of the approach shown here, and the basic equations have been shown in Section 5.2.5. Turbulence-chemistry and turbulence–PBE interactions are accounted for via the transported PBE-PDF approach, which has been detailed in Section 5.5.2. The LES-PBE-PDF equation employed in the present study is Eq. 5.87. The micromixing model employed is an augmented IEM model proposed initially by McDermott and Pope (2007) and incorporated into the LES-PBE-PDF approach by Sewerin and Rigopoulos (2017b, 2018), and the equations are given by Eqs. 5.89 and 5.90.

As discussed in Section 5.5.2, the impact of the micromixing model is important in RANS-PDF but diminishes in the case of LES. Its use for nanoparticles, such as soot, is more questionable due to their low diffusivity. Differential diffusion of both species and soot is considered at the grid level. At the subgrid level, one way of accounting for it would be to employ a reduced micromixing frequency for particles. In the work of Sun and Rigopoulos (2022) reported here, the objective was to employ all models without adjustments, hence the micromixing frequency was the same as in the majority of turbulent flame simulations. It must be acknowledged that the modelling of micromixing for soot is an open problem at the time of writing and improved models may be developed in the future.

6.2.6 Numerical Methods

Discretisation of the PBE

The PBE is solved with the conservative finite volume discretisation method (Liu and Rigopoulos, 2019; O'Sullivan and Rigopoulos, 2022), described in Section 4.2.8, which was extended for the two-PBE formulation by Sun and Rigopoulos (2022) as discussed in Section 4.2.11. The same approach was employed in the silica synthesis case study, although the present case also includes surface growth and oxidation. These processes are included in a combined growth term, which is treated numerically via the TVD scheme due to Koren (1993), first employed in the context of the PBE by Qamar et al. (2006) and discussed in Section 4.3.4 (Eqs. 4.114–4.116). In both laminar and turbulent flames, 60 intervals are employed in the PBE grid, corresponding to a particle size range of 0.72 nm to 1.24 μm; grid convergence was tested via successive refinements.

Coupling with CFD

The laminar Santoro flame in this case study resembles the one in the silica synthesis of Section 6.1, and therefore the numerical methods employed are the same as the ones described in Section 6.1.5.

The turbulent flame requires further numerics for the LES-PBE-PDF method and stochastic fields. The implementation follows the approach described in Section 5.5.4 and shown schematically in Fig. 5.2. For the temporal integration of the stochastic terms, an Euler-Maruyama scheme is adopted (for more details, see Kloeden and Platen, 1992 or Pavliotis, 2014). A dynamic load-balancing scheme (Sewerin and Rigopoulos, 2019) is applied to the reaction fractional step to accelerate performance in parallel computers. Regarding the number of stochastic fields to be employed, convergence studies have shown that eight fields are sufficient to obtain convergence for the means and RMS of the scalars; Mustata et al. (2006) used eight and sixteen stochastic fields for a turbulent diffusion flame but found that the changes in the mean and RMS of the major scalars were very small, and a similar conclusion was reached by Jones and Navarro-Martinez (2007). In accordance with these findings, eight fields were employed in the simulations of Sun and Rigopoulos (2022).

6.2.7 Simulation of Soot Formation in a Laminar Flow: The Santoro Flame

Problem Description and Computational Setup

Investigations of soot formation in laminar flows are often based on one-dimensional premixed and non-premixed flames. In these configurations, soot formation can often be decoupled and applied as a post-processor after the steady-state solution for the flame has been obtained, which makes them suitable for testing and calibrating soot kinetics. Another configuration is the co-flow non-premixed burner used in the Santoro flames (Santoro et al., 1983, 1987). In this case, a two-dimensional simulation coupling CFD, radiative heat transfer and PBE must be carried out. As such, the simulation of such flames is a good intermediate step for the application of models to turbulent flames. A considerable body of experimental data is available for these flames (Megaridis and Dobbins, 1989; Köylü et al., 1997; Iyer et al., 2007) and, furthermore, they have been the subject of several modelling studies.[5] For these reasons, this type of flame was selected for the laminar sooting flame simulation in Sun et al. (2021).

Several flame experiments have been presented in the aforementioned publications, but the target flame here is a non-smoking ethylene-air diffusion flame at atmospheric pressure and temperature. The burner and computational domain are shown schematically in Fig. 6.4, while the details can be found in Sun et al. (2021).

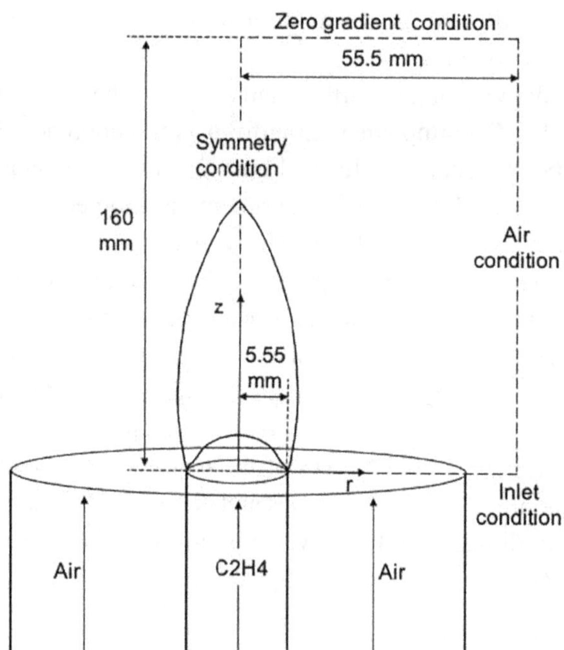

Figure 6.4 Illustration of the burner and computational domain for the Santoro laminar sooting flame. Reproduced from Sun et al. (2021) under CC-BY 4.0 license.

[5] See Zhang et al. (2009a); Dworkin et al. (2011); Veshkini et al. (2016); Akridis and Rigopoulos (2017); Liu and Rigopoulos (2019), among others.

Table 6.4 *Summary of experimental data on soot aggregates in the flame investigated in the present case study and their sources.*

Data	Centreline	Pathline with maximum f_v
f_v, ppm	Santoro et al. (1983, 1987)	Santoro et al. (1983, 1987)
d_p, nm	–	Megaridis and Dobbins (1989), Iyer et al. (2007)
N, m^{-3}	–	Santoro et al. (1987)
N_p, m^{-3}	–	Iyer et al. (2007)
n_{av}	–	Iyer et al. (2007)

Adapted from Sun et al. (2021) under CC-BY 4.0 license.

The grid comprises 200 and 100 cells with stretching by factors of 1.01 and 1.03 in the axial and radial directions, respectively. Regarding the inlet condition, it has been suggested (Guo et al., 2002) that an increase in the fuel temperature is employed to account for the heating of the flame by the burner nozzle, with the amount of preheating being chosen so as to obtain the best match with experimental measurements; this approach has been adopted in several studies on this flame (Dworkin et al., 2011; Veshkini et al., 2016). In the work of Sun et al. (2021) reported here, the objective was to maintain the number of adjustments to the minimum and focus on the PBE model rather than on the best matching with experiments, hence preheating was not employed and the inlet conditions were 300 K for both fuel and air, along with uniform velocity profiles.

Results and Discussion

We now present the simulation results and the comparison with the experiments. The temperature, axial velocity, concentration and integrated soot volume fraction at different heights above burner (HAB) are shown first, followed by an investigation of soot morphology along the flame centreline and the pathline of maximum soot volume fraction at each axial location. Table 6.4 summarises the experimental data available on soot morphology and their sources. The implementations of the one-PBE and two-PBE models, both with the conservative finite volume method, will be denoted as CFV-1PBE and CFV-2PBE, respectively.

Fig. 6.5 shows radial profiles of temperature, axial velocity, OH and C_2H_2 at different HAB; note that both PBE models yielded similar results for these variables, hence only results from the one-PBE model are shown. Good agreement is obtained for temperature and velocity, indicating that the structure of the flame

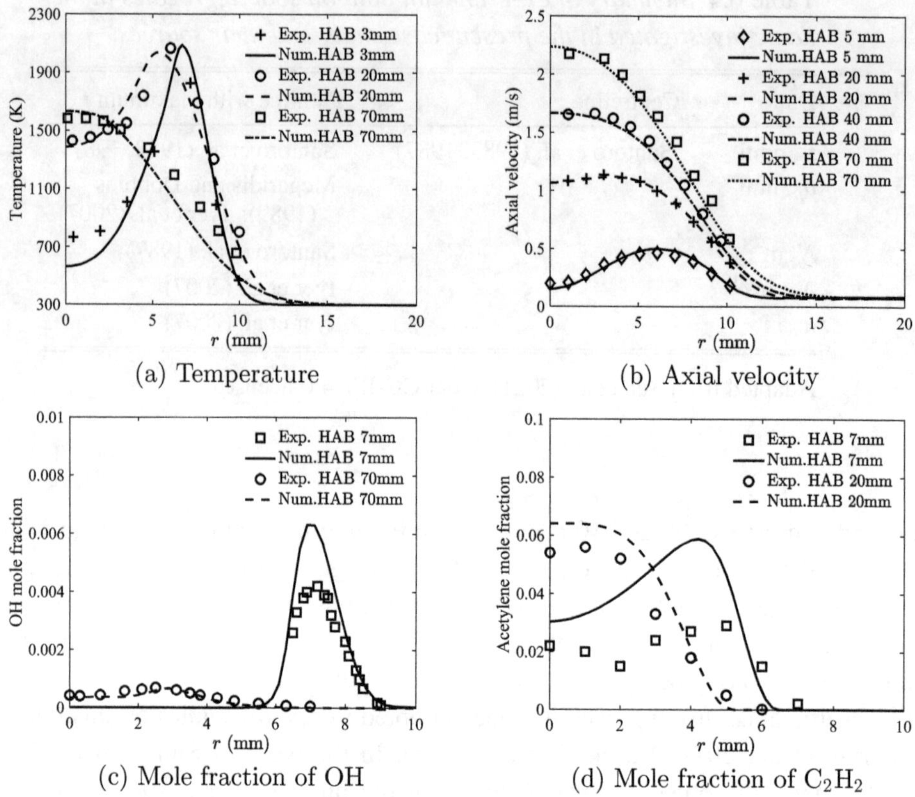

Figure 6.5 Comparison of numerical predictions and experimental data in radial direction for (a) temperature (measurements from Santoro et al. (1987)), (b) axial velocity (measurements from Santoro et al. (1987)), (c) OH mole fraction (measurements from Santoro et al. (1987)), and (d) acetylene mole fraction (measurements from Kennedy et al. (1996)). Only results from CFV-1PBE are shown, as both PBE models yielded similar results. Reproduced from Sun et al. (2021) under CC-BY 4.0 license.

is well predicted. The underprediction of temperature close to the burner (3 mm HAB and $r < 4$ mm) is probably due to heat transfer (it should be reminded that no modification of inlet temperature was applied to remedy for this effect). The OH profiles also show good agreement, apart from an underprediction of the peak value at 7 mm HAB. Given that the experimental errors for OH measurement with laser-induced fluorescence (LIF) can be as much as 50% (Kennedy et al., 1996), the level of agreement can be deemed reasonable. The C_2H_2 mole fraction profiles show good agreement with experiments at 20 mm HAB, but at 7 mm there is a considerable overprediction; a similar level of overprediction can be found in the work by Zhang et al. (2009b), Veshkini et al. (2014) and Akridis and Rigopoulos (2017).

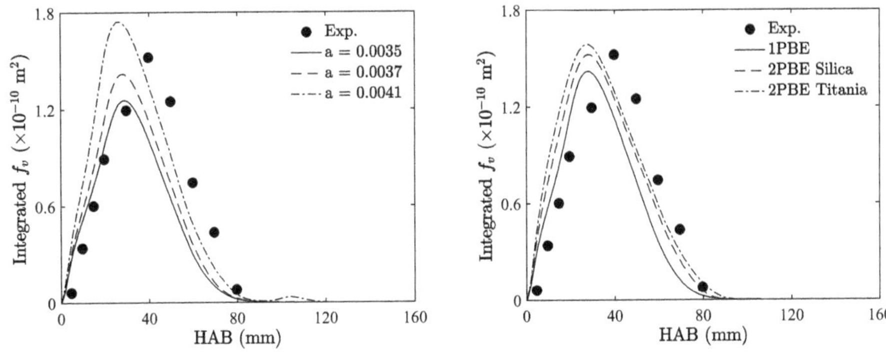

Figure 6.6 Integrated soot volume fraction along the axial direction. The left figure shows predictions with the CFV-1PBE method for different values of the prefactor a in the expression for the surface reactivity. The right figure shows a comparison of the CFV-1PBE and the CFV-2PBE method, the latter with two different models for fusing, all computed with a = 0.0037. In both plots, comparison is made with the experimental data of Santoro et al. (1987). Reproduced from Sun et al. (2021) under CC-BY 4.0 license.

It should be noted that, while C_2H_2 is the main species contributing to soot mass growth, the rate-limiting reactions in the HACA mechanism employed here are the hydrogen abstraction reactions and therefore, an accurate prediction of H and, to some extent, OH radicals (which also contribute to oxidation) is more important for surface growth (Veshkini et al., 2014).

Fig. 6.6 (left) shows the calculated and measured integrated soot volume fraction along the flame axis for different values of the prefactor a in the expression for the surface reactivity. As mentioned in Section 6.2.3, this is the main unknown parameter in HACA, with the common site density $\chi = 2.3 \times 10^{19}$ sites/m^2 reported by Frenklach and Wang (1991) being viewed as a theoretical limit and no exact value currently available (Chen and Luo, 2020). Therefore, in Sun et al. (2021), a value of a = 0.0037 was chosen for the pre-exponential factor such that the peak value of the integrated soot volume fraction from the simulation is close to the experimental value, while the values of all the other parameters of the soot model were adopted from the literature without modification. It can also be observed that the profile is slightly shifted towards the fuel nozzle, a fact also observed by Blanquart and Pitsch (2009). Furthermore, Fig. 6.6 (right) shows the integrated soot volume fraction as calculated by the CFV-1PBE and CFV-2PBE (with two sintering models), with the aforementioned value of surface reactivity; it can be seen that all three methods yield similar results.

Fig. 6.7 shows the soot volume fraction and temperature along the centreline. All three CFD-PBE models yield similar results. The temperature predictions are

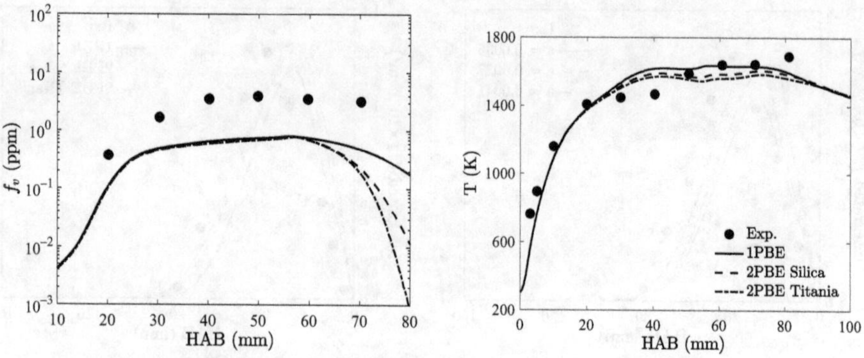

Figure 6.7 Comparison of numerical predictions by CFV-1PBE and CFV-2PBE with experimental data along the centreline for (left) soot volume fraction (measurements from Santoro et al., 1983, 1987), (right) temperature (measurements from Santoro et al., 1987; Köylü et al., 1997). Reproduced from Sun et al. (2021) under CC-BY 4.0 license.

in good agreement with the experimental results, while the soot volume fraction is underpredicted by an order of magnitude. This underprediction has also been observed by Dworkin et al. (2011) and may be associated with uncertainties in experimental measurements; the measurements by Santoro et al. (1983, 1987) employed the laser extinction/scattering approach with a constant refractive index (RI) without taking into account the soot composition and morphology, which might lead to an overestimation of soot volume fraction by as much as 100% (Kelesidis and Pratsinis, 2021). The uncertainties in soot kinetics are another possible reason; this will be explored further in the following section where the contributions of individual kinetic processes will be computed and compared.

At this point, it is worth comparing the results of the CFV-1PBE with those of a CFV-2PBE formulation with no sintering model and the same critical diameter ($d_c = 30.8$ nm) as the CFV-1PBE model. As discussed in Section 2.4.2, the one-PBE model involves an inconsistent calculation of surface growth and oxidation due to the fact that the size of the primary particles is assumed to be frozen once they reach the cut-off size. Fig. 6.8 shows the comparison of integrated soot volume fraction and temperature along the centreline, and it can be seen that the difference is very small. This indicates that the aforementioned inconsistency has a minor effect, due to the fact that surface processes have a minor effect on particle properties beyond such a high cut-off size. It must be reminded that a formulation without finite-rate fusing cannot account properly for the morphology and shaping of the PSD; however, these results indicate that the one-PBE model is a reasonable

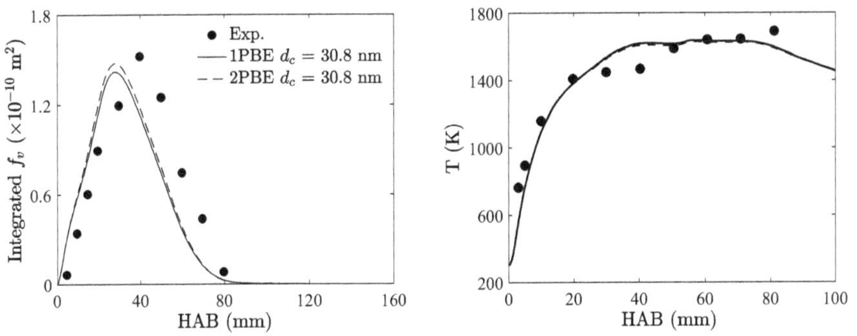

Figure 6.8 Comparison of numerical predictions by CFV-1PBE and by CFV-2PBE without finite-rate fusing of primary particles with the experimental data of Santoro et al. (1987) for integrated soot volume fraction (left) and with those of Santoro et al. (1987); Köylü et al. (1997) for temperature (right) along the centreline. Reproduced from Sun et al. (2021) under CC-BY 4.0 license.

choice when predictions of such properties are not required. Furthermore, the two-PBE is more expensive to compute, which may be important in turbulent flames, where often temperature and volume fraction are the main measured variables to compare with.

We now compare the results of all PBE models with the experimental results along the pathline of maximum soot volume fraction, focussing on aggregation-related quantities. We first look at the impact of the cut-off size, the parameter that determines soot morphology in the one-PBE. The top row of Fig. 6.9 shows results from CFV-1PBE with $d_c = 30.8$ nm and $d_c = 20$ nm. In the initial region, the total number density decreases with decreasing cut-off, due to the earlier formation of aggregates. On the other hand, the volume fraction increases due to the fact that more surface area is available and thus there is increased surface growth. In the bottom row of Fig. 6.9, the impact of the key parameters in the two sintering models on the two-PBE model is investigated. With the titania model, soot aggregates tend to form earlier in the case of slow sintering rate, the number density is smaller and the soot volume fraction is larger due to a greater number of primary particles and larger surface area. A similar pattern can be seen in the silica model.

Particle Size Distribution

While no experimental data is available for the PSD, it is instructive to look at its evolution. In Fig. 6.10, the PSD of soot is shown at three locations along the centreline, characteristic of the nucleation, growth and oxidation regions. We first note

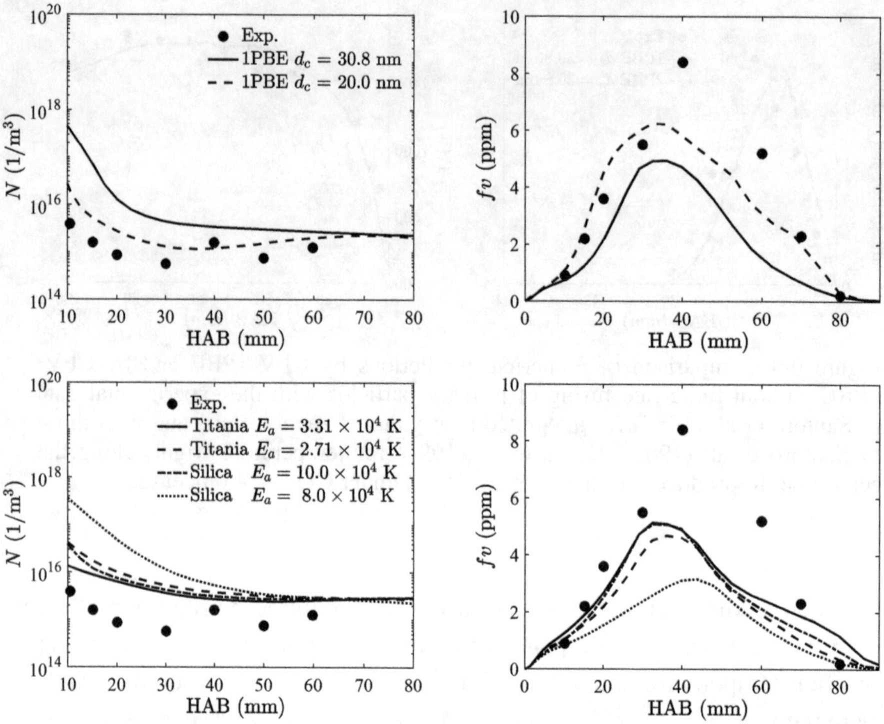

Figure 6.9 Comparison of model predictions for soot number density (left) and volume fraction (right) along the annular pathline containing maximum soot volume fraction with experimental data. Top: comparison of CFV-1PBE simulations with experiments and effect of cut-off size. Bottom: comparison of CFV-2PBE simulations with experiments and effect of activation energy in the silica and titania sintering models. Table 6.4 shows the sources of experimental data. Reproduced from Sun et al. (2021) under CC-BY 4.0 license.

that the three PBE-CFD models show more pronounced differences here than was the case with the soot volume fraction. In the nucleation region, the PSD is more or less unimodal and the differences between the models are small. In the growth region, where both growth and aggregation are present, bimodal PSDs are predicted by the CFV-2PBE model. At the oxidation zone, the PSD becomes unimodal again. These results show that, while the one-PBE model may be adequate for predicting the soot volume fraction, it can exhibit considerable difference when it comes to the size distribution.

Spatial Distribution of Rates of Kinetic Processes

Soot kinetics is one of the main sources of uncertainty in soot models and may be partly responsible for the discrepancies observed in the soot volume fraction

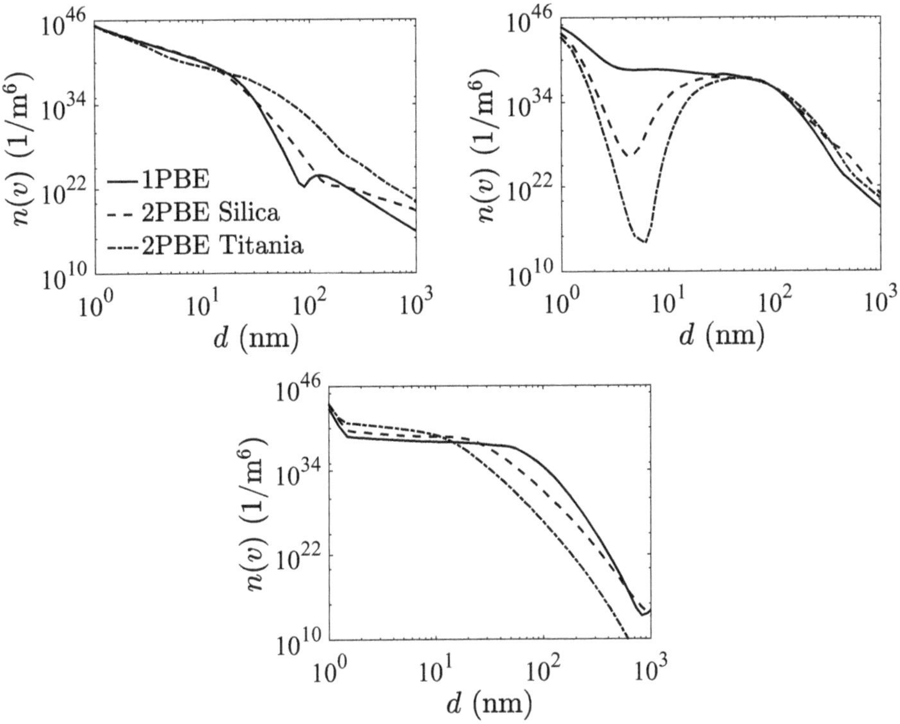

Figure 6.10 Comparison of the numerical predictions of soot aggregate number density distribution by CFV–1PBE and CFV–2PBE of two sintering models at three probe locations along the centreline: (top-left) 15 mm HAB (nucleation region), (top-right) 50 mm HAB (growth region), and (bottom) 80 mm HAB (oxidation region). Reproduced from Sun et al. (2021) under CC-BY 4.0 license.

predictions, hence it is worthwhile to investigate this matter further by examining the spatial distribution of kinetic rates throughout the flame. Fig. 6.11a shows a breakdown of the source terms contributing to soot volume fraction, i.e. nucleation, surface growth, condensation and finally oxidation by OH and O_2. Furthermore, since nucleation and condensation depend on PAHs, Fig. 6.11b shows the distribution of the dominant PAHs along HAB.

Between HAB 20 mm and 45 mm, nucleation and condensation are the dominant mechanisms. After HAB 55 mm, there is a rapid increase in surface growth that is soon countered by oxidation due to OH radicals. The formation of small PAHs occurs early in the flame (from HAB = 5 mm to HAB = 20 mm); afterwards the PAHs are gradually consumed. The formation of OH and H starts approximately at HAB = 50 mm, and subsequently these radicals trigger the H-abstraction in the HACA mechanism, thus boosting surface growth. From HAB 60 mm, growth is

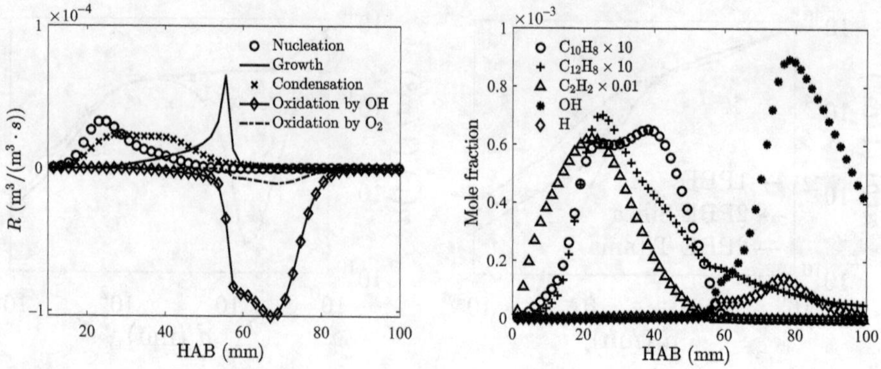

Figure 6.11 The rates of change for each soot formation process and the predicted species mole fractions along the centreline by the CFV-1PBE method. Reproduced from Sun et al. (2021) under CC-BY 4.0 license.

limited by the low C_2H_2 concentration; in the meantime, oxidation by OH counters growth. Other works (Dworkin et al., 2011; Veshkini et al., 2014) have also found soot mass addition along the centreline to occur mainly through nucleation and condensation.

6.2.8 Simulation of Soot Formation in a Turbulent Flow: The Sandia Flame

Problem Description and Preliminary Considerations

We now turn to the simulation of the Sandia turbulent flame. For this flame, the flow field will be simulated with LES and, as discussed in Section 6.2.5, the LES-PBE-PDF approach will be used to account for turbulence–chemistry and turbulence–soot interaction. Furthermore, the one-PBE model is chosen as the population balance formulation. While the two-PBE model offers a more realistic description of aggregate morphology, as we saw in the simulation of the laminar Santoro flame, the results of the two models for the temperature, chemical species and volume fraction did not differ very much, and it is the predictions of these variables that will be compared against experiments in this case. The turbulent flame is very expensive to simulate, and thus the computational advantage of the one-PBE model is important here.

As with the Santoro flame, the fuel is ethylene. The design of this burner was first described by Zhang et al. (2011). A central fuel jet is surrounded by two concentric tubes delivering a pilot flame and air. The fuel jet has inner and outer diameters of 3.2 mm and 4.6 mm, respectively, while the pilot has inner and outer diameters of 15.2 mm and 19.1 mm, respectively. Ethylene at room temperature 295 K is injected with bulk velocity $V_{fuel} = 54.7$ m/s in the central nozzle, at a Reynolds

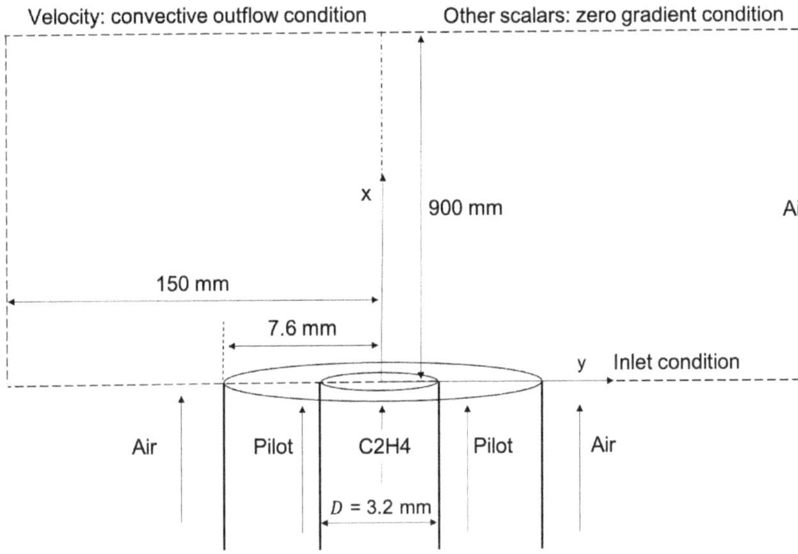

Figure 6.12 Illustration of the burner and computational domain for the Sandia turbulent sooting flame. Reproduced from Sun and Rigopoulos (2022) under CC-BY 4.0 license.

number of 20,000. The pilot comprises premixed ethylene and air at an equivalence ratio of 0.9, while the flow rate of the pilot mixture is adjusted such that the heat release corresponds to 2% of that of the fuel jet. The pilot is surrounded by a co-flow of air at 295 K at a uniform velocity $V_{air} = 0.6$ m/s, while the pressure is atmospheric.

A schematic illustration of the computational domain is shown in Fig. 6.12. In order to provide higher resolution near the burner, the grid is stretched by a factor of 1.013 axially, while Radially ten cells are allocated in the central region to resolve the fuel jet with a resolution of 0.3 mm. The grid is also stretched by a factor of 1.12 from the pilot towards the lateral boundaries. The mean axial velocities at the inflow for the fuel, jet and pilot are set to power-law profiles with exponents 1/6 and 1/2, respectively, while the air co-flow is set to a constant mean axial inflow velocity. Turbulence intensities of 6% are prescribed for all flows. The RMS inflow velocity for the fuel jet varies quadratically (at the jet rim it is four times as large as the nominal centreline value) and maintains a constant value for the pilot and co-flow. The initial composition of the pilot is set to be the adiabatic equilibrium composition of the mixture. The four lateral boundaries are set to have the same composition and velocity as the co-flow air, while entrainment/outflow is allowed. At the domain exit, convective outflow and zero gradient conditions are applied to the velocity and scalar fields, respectively.

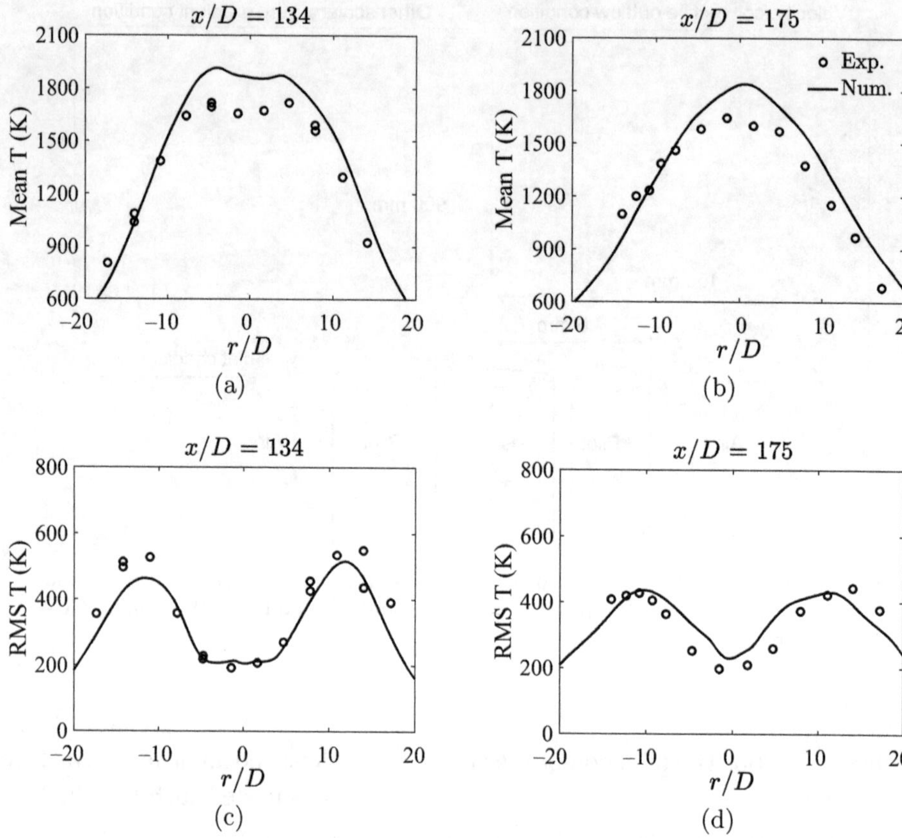

Figure 6.13 Comparison of mean and RMS temperature radial profiles between experimental data (symbols) and numerical results (lines) at $x/D = 134$ and $x/D = 175$. Reproduced from Sun and Rigopoulos (2022) under CC-BY 4.0 license.

A simplified inflow geometry is adopted, ignoring the thin wall of the tube and the jet immersion; while velocity measurements are not available to test this implementation, a similar configuration was used by Sewerin and Rigopoulos (2018) and good agreement was obtained for both mean and RMS velocity profiles in the Delft III turbulent flame. The collection of data for temporal statistics commenced after the simulation had reached a statistically steady state and data for a period of about three flow-through times were collected, thus ensuring that the statistics are time-invariant.

Results and Discussion

We will now compare the simulation results with the measurements. The experimental data available for this flame include the following quantities: temperature

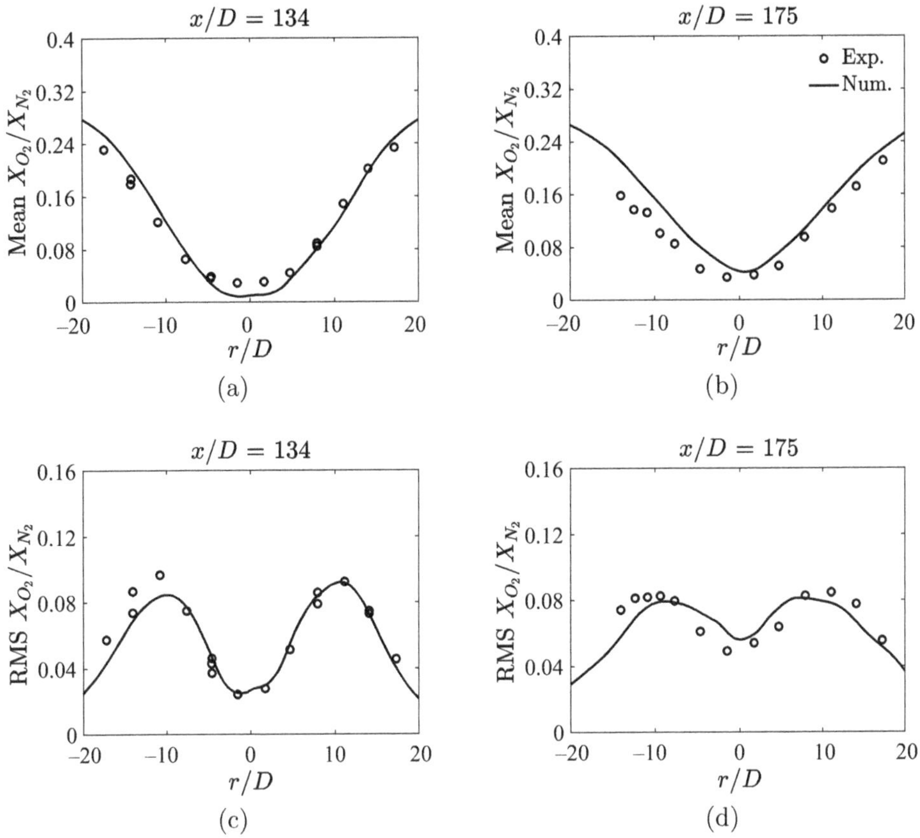

Figure 6.14 Comparison of mean and RMS X_{O_2}/X_{N_2} radial profiles between experimental data (symbols) and numerical results (lines) at $x/D = 134$ and $x/D = 175$. Reproduced from Sun and Rigopoulos (2022) under CC-BY 4.0 license.

and X_{O_2}/X_{N_2} ratio (Kearney et al., 2015), planar laser-induced fluorescence (PLIF) OH and PLIF PAH (Shaddix et al., 2010), axial profile of soot intermittency (Shaddix et al., 2010) and soot volume fraction (ISF Workshop, 2018).

Mean and RMS temperature profiles are shown in Fig. 6.13 for two heights, while mean and RMS profiles of the X_{O_2}/X_{N_2} ratio at the same heights are shown in Fig. 6.14. Overall, the predictions of the mean values are quite good, apart from an overprediction of the temperature around the centreline, while the RMS values are generally well reproduced. This overprediction of temperature is linked with an underprediction in soot volume fraction that will be discussed below, as less soot implies less radiative heat loss.

Mean OH radial profiles are shown in Fig. 6.15 (note that the measurements are PLIF signals rather than mole fractions, hence both experimental and numerical

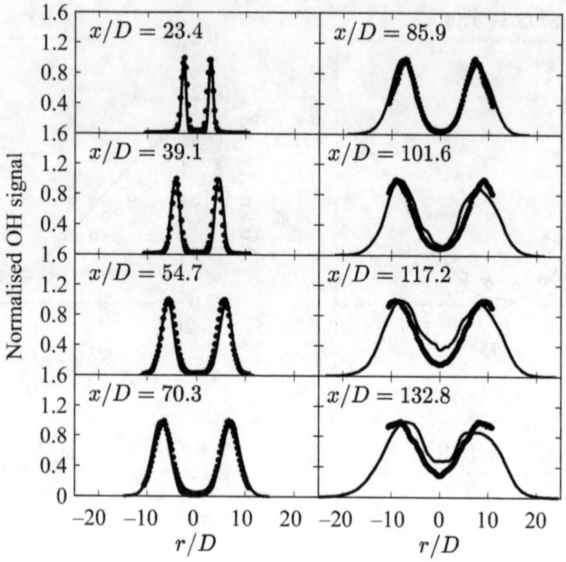

Figure 6.15 Comparison of radial profiles of normalised OH signals from experimental data (symbols) and normalised mean OH mass fraction from the simulation (lines). Reproduced from Sun and Rigopoulos (2022) under CC-BY 4.0 license.

results are normalised by their corresponding maximum values at each height). The agreement is very good, except from small discrepancies at $x/D = 117.2$ and $x/D = 132.8$, indicating a satisfactory prediction of the mean flame shape; furthermore, a good matching of OH profile is required for a proper account of oxidation and even surface growth (as OH participates in the HACA mechanism). A comparison of simulations with normalised PAH signals is carried out in Fig. 6.16; the comparison is made with the $C_{10}H_8$ (A_2) radial profile, since this was found to be the dominant (in terms of mass fraction) PAH contributing to nucleation and condensation in the simulations. The agreement is satisfactory, although the simulated profile is wider at the early stages of the flame. PAH profiles are important for nucleation and condensation, and it will be later shown that these processes are not so important in this region, hence this discrepancy is not expected to exert a significant influence on the predictions.

Fig. 6.17 shows axial mean and RMS soot volume fraction profiles along the centreline. The simulation reproduces qualitative features of the experimental measurements, such as the location of the soot formation zone and shape of profile, but the magnitude of the soot volume fraction is considerably underpredicted. Sun and Rigopoulos (2022) also showed radial profiles of mean and RMS of soot volume fraction at different heights and, likewise, the location and qualitative trends in the

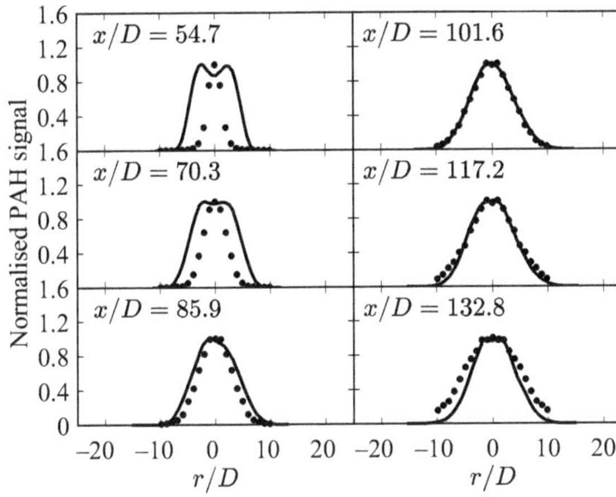

Figure 6.16 Comparison of normalised PAH signals from experimental data (symbols) and normalised mean A_2 mass fraction from the simulation (lines). Reproduced from Sun and Rigopoulos (2022) under CC-BY 4.0 license.

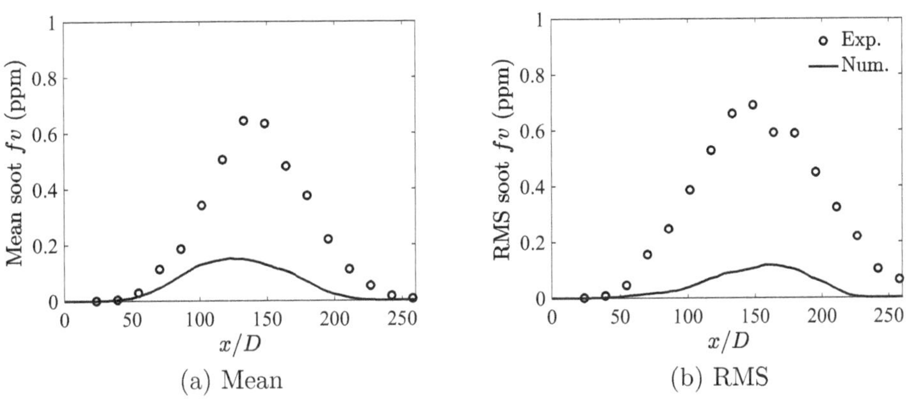

Figure 6.17 Axial profiles of the mean and RMS soot volume fraction along the flame centreline. Reproduced from Sun and Rigopoulos (2022) under CC-BY 4.0 license.

experimental data were reproduced, but the magnitude of soot volume fraction in the region away from the jet was underpredicted.

It is worthwhile to examine the results as a whole and juxtapose them with the findings from the simulation of the laminar Santoro flame. The first thing to consider is the prediction of the flame structure, which is the backbone for the population balance and soot model. While the experimental dataset for this flame does not provide flow field measurements, LES has been shown to provide an accurate description of the fluid dynamics of jet flames away from walls (Jones and

Prasad, 2010). Furthermore, we can infer the accuracy of the flow field prediction from the temperature and species profiles; these are generally well predicted apart from an overprediction of the peak temperature, which is likely to be associated with the underprediction of the soot volume fraction and thus the underestimation of radiation. The spread of the means and RMS of most quantities is also well predicted, and it can be inferred that the LES and LES-PDF approach have provided an accurate account of the turbulent flow field and turbulence–chemistry interaction.

Regarding soot, while several qualitative features are reproduced, the main discrepancy lies in the magnitude of the soot volume fraction. It should first be acknowledged that such a discrepancy is also present in other simulations of the same flame at the time of writing; for example, the works of Xuan and Blanquart (2015) and Jain and Xuan (2019) also featured underprediction, while Rodrigues et al. (2018) observed an overprediction and Yang et al. (2019) conducted several cases with different parameters associated with the conditional soot subfilter PDF model employed in that work, some of which featured underprediction and some overprediction. The prediction of soot in turbulent flames is an open problem at the moment, but some light can be shed by considering the validation of each element of the approach adopted here and the simulation of the laminar Santoro flame.

The solution of the population balance with a discretisation method contains no modelling assumptions, while the conservative finite volume method has been shown to be accurate when validated with analytical and similarity solutions (Sun et al., 2021; O'Sullivan and Rigopoulos, 2022). The convergence of the stochastic fields method for the solution of LES-PDF has also been investigated (Mustata et al., 2006; Jones and Navarro-Martinez, 2007) and the eight fields employed by Sun and Rigopoulos (2022) have been shown to be sufficient.

With respect to turbulence–PBE interaction, the PBE-PDF method with the IEM micromixing model does not have adjustable parameters other than the micromixing frequency, and the only possible modification in this respect is to employ a different micromixing constant to account for differential diffusion of soot and gas-phase species; in the present simulation, however, the standard form of the model is employed. It should be mentioned that the impact of the micromixing model is diminished in LES as compared with RANS, and simulations of turbulent jet flames of the Sandia type with LES-PDF and stochastic fields have shown that concentration profiles can be predicted quite well, except for cases with excessive local extinction (Jones and Prasad, 2010). However, the impact of micromixing on soot is more questionable due to the very low diffusivity of nanoparticles, and while the differential diffusion between species and soot can be accommodated at the grid levels, the standard IEM model does not distinguish between them. This is an issue that should be investigated in future research, either by the use of a different micromixing constant for nanoparticles, or perhaps via the development of a different micromixing model for that purpose.

We finally consider the kinetic model, for which we can leverage the findings from the laminar Santoro flame simulation with the same kinetics. While, in that flame, reasonably good predictions of the integrated soot volume fraction (with a smaller level of underprediction), this measurement is not available for the turbulent Sandia ethylene flame and comparison was made instead for the centreline profile of soot volume fraction (Fig. 6.17), which was also underpredicted to a similar extent in the Santoro flame (Fig. 6.7). Since the simulation of the laminar flame does not involve a micromixing model, it can be inferred that kinetics, and particularly surface reactivity, is likely to be the main source of uncertainty. The model employed for surface reactivity is relatively simple and accounts for the change of temperature only; more advanced models that consider the effect of soot carbonisation on surface reactivity (Veshkini et al., 2014; Kholghy et al., 2016) are available and could be investigated in future work.

Investigation of Soot Formation Mechanisms in the Turbulent Sandia Flame

We now wish to probe further into the soot formation mechanisms within the flame, by considering the spatial distribution of kinetic rates and species partaking in the soot formation processes. In Fig. 6.18, we see the distribution of temperature, soot volume fraction and soot number density. We observe that soot is located at the rich side of the mixture fraction iso-contour. The soot number density is higher in the region between $x/D = 40$ and $x/D = 100$, while the soot volume fraction peaks at a region between $x/D = 80$ and $x/D = 160$, as indicated also by Fig. 6.17; the higher number density at the earlier part of the flame is due to smaller particles that do not contribute much to volume fraction.

In order to understand how the soot spatial distribution is shaped, we consider the spatial distribution of several key species (Fig. 6.19) and of the contribution of nucleation, condensation, surface growth and oxidation rates to volume fraction (Fig. 6.20).[6] From Figs. 6.19a and 6.18c, we can see that A_2 is formed before the appearance of measurable soot in the flame and is limited to the inner jet, consistent with the experimental observations of Shaddix et al. (2010). The correlation of A_2 (which participates in the dimer formation) with nucleation and condensation is evident by juxtaposing Figs. 6.20a,b and 6.19a. Furthermore, by comparing

[6] Note that, among PAHs, only the dominant A_2 is shown, whose concentration is orders of magnitude larger than that of other PAHs contributing to dimer formation. Also, only the OH oxidation rate is shown, as it is much larger than the O_2 oxidation rate.

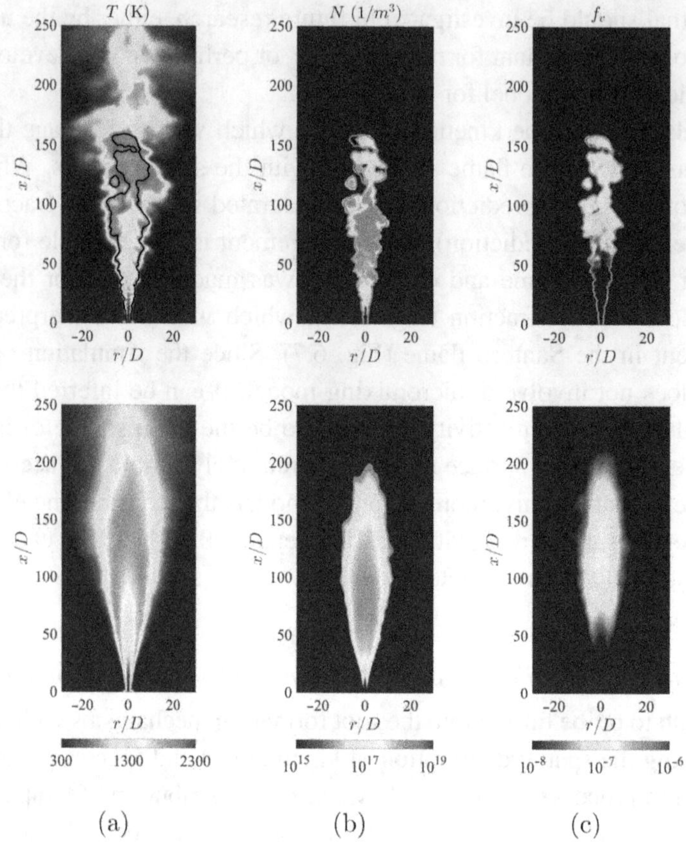

Figure 6.18 Contour plots of instantaneous (top) and mean (bottom) (a) temperature, (b) soot number density and (c) soot volume fraction computed from the LES-PBE-PDF model. The iso-contour of stoichiometric mixture fraction ($Z = 0.0634$) is also shown as a solid line in the instantaneous contour plots. Reproduced from Sun and Rigopoulos (2022) under CC-BY 4.0 license.

the rates of nucleation, condensation and growth (Figs. 6.20a–c), we can see that surface growth is responsible for most of the mass addition.

According to Shaddix et al. (2010), the structure of OH should develop from straight, thin layers at the jet exit to increasingly wrinkled, thick structures downstream from the nozzle, and this feature is evident in Fig. 6.19c. Regarding the distribution of C_2H_2, an interesting feature is that its spatial correlation with growth rate is rather weak. Early soot models (cf. Section 6.2.1) correlate soot with C_2H_2 but, in the HACA mechanism employed here, the rate-limiting reactions are the hydrogen abstraction reactions. Therefore, C_2H_2 is consumed via the acetylene addition step in regions that are rich in OH and H, as can be seen in Figs. 6.19b–d.

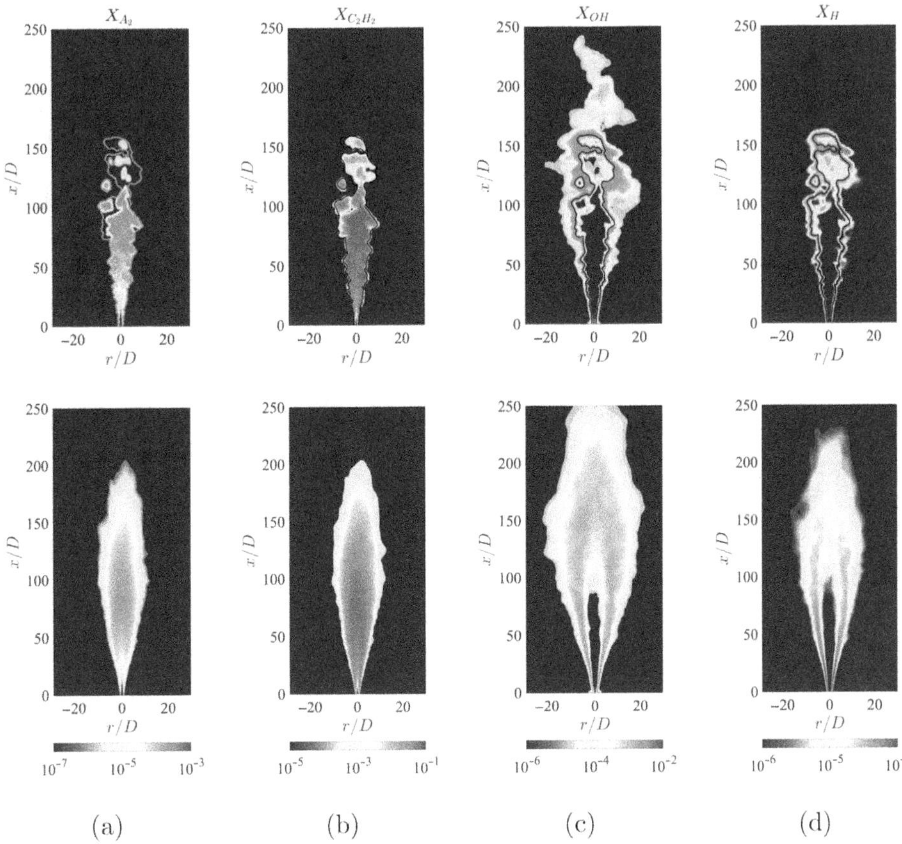

Figure 6.19 Contour plots of the instantaneous (top) and mean (bottom) (a) A_2 mole fraction, (b) C_2H_2 mole fraction, (c) OH mole fraction and (d) H mole fraction computed from the LES-PBE-PDF model. The iso-contour of stoichiometric mixture fraction is shown as a solid line in the instantaneous plots. Reproduced from Sun and Rigopoulos (2022) under CC-BY 4.0 license.

Computational Performance of the LES-PBE-PDF Method

A breakdown of the CPU time taken by various parts of the simulation, per time step and averaged over 3,000 time steps, is shown in Table 6.5. Most notably, the solution of the PBE with the conservative finite volume discretisation method requires 17.79% of the total CPU time and, therefore, does not incur a major overload on the LES-PBE-PDF simulation. Furthermore, the time taken for convection/diffusion of scalars is 12.25%, about 29% of which is associated with the PBE (since the chemical mechanism involved 149 scalars). The most expensive part is the gas-phase chemistry, which takes about half of the total CPU time. It must be mentioned that the chemistry was accelerated by approximately one order

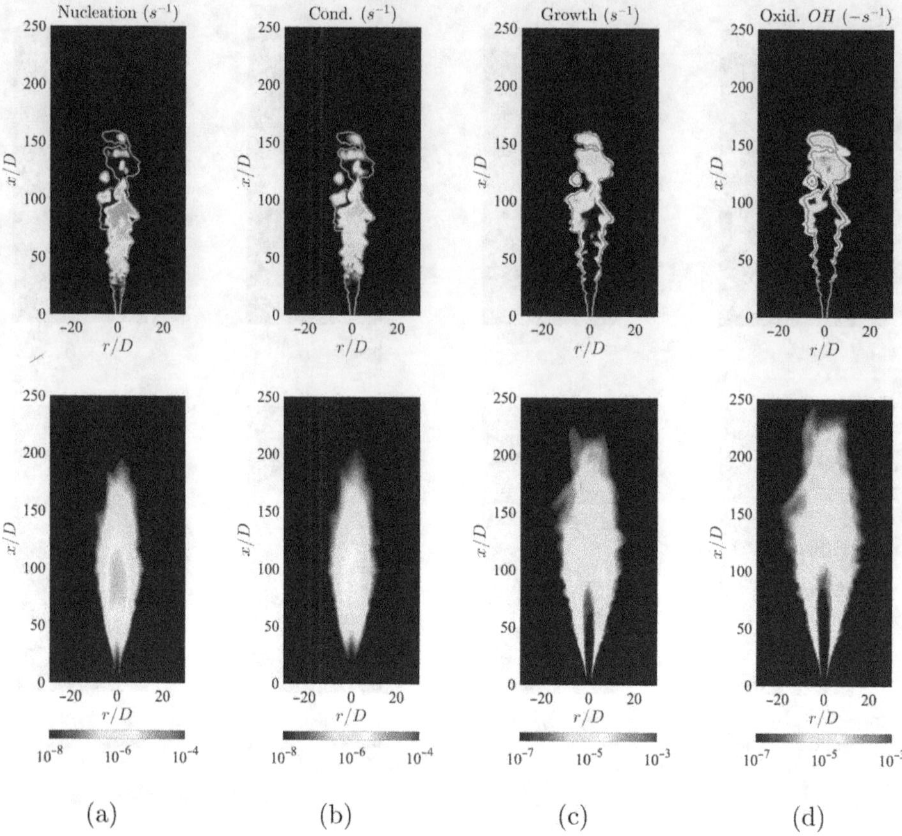

Figure 6.20 Contour plots of the instantaneous (top) and mean (bottom) (a) nucleation rate, (b) condensation rate, (c) growth rate and (d) oxidation rate due to OH computed from the LES-PBE-PDF model. The iso-contour of stoichiometric mixture fraction is shown as a solid line in the instantaneous plots. Reproduced from Sun and Rigopoulos (2022) under CC-BY 4.0 license.

of magnitude with ways that do not incur any loss of accuracy, such as the hard-coding of the mechanism. Nevertheless, it is still the main bottleneck in the present simulation, due to the use of a detailed mechanism that includes PAH formation and a PDF method that requires real-time computation of chemistry. Mechanism reduction and tabulation can be employed to reduce this cost, such as In-Situ Adaptive Tabulation (ISAT) (Pope, 1997), Hybrid Transported-Tabulated Chemistry (HTTC) (Ribert et al., 2014; Duboc et al., 2019) and machine learning methods (Chatzopoulos and Rigopoulos, 2013; Franke et al., 2017; Readshaw et al., 2021, 2023; Ding et al., 2021, 2022). Recently, machine learning has also been employed for the solution of the PBE in laminar premixed sooting flames (Seltz et al., 2021), as discussed in Section 4.4.7.

Table 6.5 *Average runtimes for the LES-PBE-PDF method to advance one time step ($\Delta t = 3 \times 10^{-6}$ s) on 20 nodes (480 MPI processes) in the ARCHER supercomputer.*

Process	Average runtime (s)	Percentage (%)
Scalar convection/diffusion	3.67	12.25
Micromixing	1.59	5.31
Gas phase reaction	14.86	49.60
PBE solution	5.33	17.79
Flow field	0.42	1.40
All other processes	4.09	13.65
Total	29.96	–

Reproduced from Sun and Rigopoulos (2022) under CC-BY 4.0 license.

6.2.9 Concluding Remarks

The first case study of population balance modelling tackled the most complex of the three applications discussed in this chapter. Soot formation in flames involves strong coupling of population balance and fluid dynamics, a wide range of kinetic processes (nucleation, condensation, surface growth, oxidation, aggregation and fusing of primary particles) and possibly turbulent flow, thus bringing together most of the themes in the book. The application of the population balance methodology followed the steps outlined in Section 1.1.3. We now summarise the main outcomes of this case study, with reference to the objectives that were set in the beginning.

Soot comprises both primary particles and fractal aggregates, and aggregate morphology must be accounted for in a comprehensive soot model. The first objective of this study was to investigate the applicability of the one-PBE and two-PBE models to soot. The one-PBE model has inherent limitations for describing problems with fractal aggregates but is widely applied to soot due to its computational simplicity. By contrast, the two-PBE model accounts for finite-rate fusing of primary particles and provides a consistent description of aggregate surface growth, but is more expensive and, furthermore, there is a lack of physical models for the fusing of soot particles. Both models were applied to the Santoro laminar sooting flame and, in the case of the two-PBE model, sintering models of inorganic nanoparticles were tested. Results showed that both models provided similar results for the main process variables such as temperature and soot volume fraction, although the one-PBE model is not able to predict morphological parameters. The inconsistent description of growth of primary particles within aggregates in the one-PBE model did not impact severely the results because surface processes had a minor effect beyond the cut-off size. However, it must be acknowledged that the two-PBE approach attains its results with a consistent description of growth and a more physical model for the fusing of primary particles. As more information becomes available from experiments or microscopic simulations about the fusing of soot particles, the two-PBE should become the preferred approach.

The second objective was to investigate soot kinetics in a laminar flame, in preparation for the simulation of a turbulent flame with the same parameters. By excluding the issues of turbulence, turbulence–chemistry and turbulence–PBE interaction, the effect of soot kinetics in a coupled CFD-PBE simulation can be brought to light more clearly. The Santoro laminar flame was chosen because it requires a full coupling of the PBE with an unsteady CFD code, unlike one-dimensional flames. Soot mass growth occurs mainly by surface growth, and hence the surface reactivity in the HACA mechanism was found to be the most critical parameter for the prediction of the soot volume fraction. At present there is no unique way for prescribing this parameter; a temperature-dependent model was used in the present work and the pre-exponential factor was adjusted on the basis of the profile for the integrated soot volume fraction. Reasonably good predictions were obtained overall, but the soot volume fraction in the centreline was still underpredicted.

Finally, a turbulent flame – the Sandia flame – was simulated with the same kinetics as the laminar one. The one-PBE model was employed due to computational efficiency and the fact that the comparisons with experimental data would not include morphological parameters for which the two-PBE model would make a difference. The LES-PBE-PDF method was employed to account for turbulence–chemistry and turbulence–soot interaction. In this approach, there is only one closure parameter, the micromixing frequency, and the standard value was employed. The results showed that the overall structure of the flame was predicted well and several features of the soot formation could be reproduced, such as the location and qualitative shape of the profiles, but the magnitude of soot volume fraction at the centreline was underpredicted. Given the sensitivity of volume fraction to surface reactivity and the fact that the underprediction at the centreline was a feature of the laminar flame simulation too, it was concluded that this is the main uncertainty in the overall CFD-PBE approach, although micromixing may also have an effect. Subsequently, the spatial distribution of kinetic rates and species within the flame was investigated and the contributions of the various processes to the shaping of soot volume fraction were analysed.

It must be acknowledged that the quantitative modelling of soot formation in turbulent combustion is, at the time of writing, still an open problem. While qualitative trends can be reproduced by the models, it is very difficult to establish a quantitatively predictive simulation that does not involve a degree of parameter adjustment, and it is also hard to conclude whether better agreement overall is due to a particular aspect of the model or to a fortuitous compensation of errors. At the moment, it is hard to pinpoint the dominant factors for this issue, although several shortcomings are known. The fact that the transition from RANS to LES has not guaranteed quantitative predictions of sooting flames indicates that fundamental uncertainties in soot chemistry, PBE kinetics and the of the above interaction with turbulence are still outstanding. Better agreement can be attained in the case of laminar sooting flames, which may suggest that turbulence and the associated interactions are the issue, but also laminar flame simulations often employ very detailed chemical mechanisms (usually involving PAH chemistry) that are too expensive to be implemented in turbulent flames. With advances in computational power, it can be expected that more detailed chemistry and PBE models can be incorporated in turbulent sooting flame simulations in the future, alongside comprehensive turbulence–chemistry and turbulence–PBE interaction models.

Soot formation is one of the most difficult problems that can be approached with population balance modelling. This is because there are considerable uncertainties regarding the modelling of chemical kinetics, particularly surface reactions, and turbulence–soot interaction. A considerable amount of future research is needed to reach the state of quantitative agreement without parameter adjustment, particularly for turbulent sooting flames. The next two case studies do not feature some of these uncertainties and focus on somewhat different aspects of the CFD-PBE approach.

6.3 Precipitation in a Turbulent T-Mixer Flow

6.3.1 Introduction

Overview

Precipitation, or reaction crystallisation, is a process for generating a particulate product via a chemical reaction in the liquid phase. It is used for the production of numerous products, such as pharmaceutical powders and pigments. The PSD of the product has significant impact on the product quality and determines its suitability for particular applications. The application of the population balance can thus aid in product engineering, that is, the manufacturing of products tailored to specific applications, as well as in process design, optimisation and scale-up.

Nucleation and growth are the kinetic processes that are present in almost every crystallisation process. Aggregation (usually called agglomeration in this context) is also present in many cases, while fragmentation (or breakage) can also occur. As was the case with soot formation and gas-phase nanoparticle synthesis, fluid dynamics also sets the stage here, as it determines the spatial distribution of supersaturation which drives nucleation and growth, as well as shear which may drive aggregation and breakage. Precipitation is, therefore, another classic example of coupling between fluid dynamics and population balance.

The case study to be presented here is based on the works of Tang et al. (2020, 2022, 2023) where an experiment on precipitation of $BaSO_4$ in a turbulent T-mixer flow by Schwarzer et al. (2006) was simulated via a coupled DNS-PBE approach. The similarities and differences of this problem with the two previous case studies in this chapter are outlined below.

The kinetic processes present in this case study are nucleation and growth, as aggregation was suppressed in the experiment that is simulated due to electrostatic repulsion. Thus, this study is complementary to the gas-phase nanoparticle synthesis case study (Section 6.1), where the process was driven by nucleation, aggregation and sintering. One common feature between precipitation of inorganic salts and combustion is that both involve fast reactions, which means that the coupling between flow, reaction and PBE is strong and may not be possible to resolve with ideal reactor models. While the flow in this case study is turbulent, a DNS approach is adopted here, which circumvents the need for modelling turbulence–chemistry and turbulence–PBE interaction (as was the case with the turbulent sooting flame in Section 6.2.8). This is a computationally expensive approach, but the objective here is to probe into the underlying mechanisms of flow and population balance interaction in turbulent precipitation, as well as to guide modelling approaches. One further issue that appears in liquid-phase flows is the high Schmidt number and the sub-Kolmogorov scales that emerge, which cannot be resolved at present by DNS; an investigation of their impact will be presented here by juxtaposing simulations at $Sc = 1$ and $Sc = 10$.

Reviews of crystallisation and precipitation can be found in Randolph and Larson (1988), Söhnel and Garside (1992), Mersmann (2001), Mullin (2001) and Myerson et al. (2019), among others. The coupling of precipitation models with turbulent flow via presumed and transported PDF methods has been discussed in Chapter 5 (Sections 5.4 and 5.5, respectively). Detailed reviews of works focussed on the modelling of precipitation in turbulent flows can be found in Tang et al. (2020, 2022, 2023).

Objectives of the Case Study

The case study in the present section, based on the works of Tang et al. (2020, 2022, 2023), has four main objectives that are summarised below.

First, an approach coupling DNS with a discretised PBE is adopted, with the following aims: (a) to account for all of the temporal and spatial scales of turbulence, (b) to solve the population balance without closure assumptions, (c) to predict the PSD and (d) to resolve the coupling of species transport and PBE via the source terms. All individual aspects, that is flow field prediction, convergence of PBE solution and source term coupling, are validated on their own before comparing the results of the overall approach with experiments, in order to prevent the compensation of errors associated with modelling and numerical solution. The outcome of the simulation is compared and validated with the experiments of Schwarzer et al. (2006). Although the approach is DNS-based, validation is important because it is only the resolution of the flow structures that can be considered as model-free; the impact of kinetic models remains to be evaluated, as well as that of mixing at sub-Kolmogorov scales.

Second, an investigation is conducted with respect to the interaction of turbulence with nucleation and growth. The presence of two timescales associated with these two processes gives some unique characteristics to precipitation. In particular, the balance between them, particularly at the mixing zone, is pivotal for shaping the particle size distribution; for example, nucleation dominance promotes the formation of a large number of small crystals that consume rapidly the supersaturation before growing to large sizes. The kinetics of nucleation are highly nonlinear, and turbulent fluctuations in supersaturation give rise to highly intermittent nucleation bursts occurring within thin zones. A related research question is about whether a separation of space and time scales between the fluid dynamic and kinetic phenomena is possible, as this has important implications for modelling. In particular, if the kinetic phenomena are much slower than the characteristic flow timescale, ideal reactor models can be employed; at the other end, if kinetic phenomena are much faster than the flow timescale, it is possible to use models that treat the process is entirely mixing-controlled. Similar questions arise in combustion, where a wealth of models have been developed (Peters, 2000; Veynante and Vervisch, 2002; Poinsot and Veynante, 2005), but such investigations have not been carried out to the same extent in precipitation.

Third, the closure problem associated with the fluctuating nucleation and growth rates is illuminated by evaluating the unclosed terms in the DNS-PBE simulation. This has important implications for RANS and LES simulations, where micromixing is not resolved and models must be employed. The importance of micromixing on precipitation has been discussed extensively in the literature (see e.g. Bałdyga and Bourne, 1999) and its modelling is a long-standing problem; for example, Di Veroli and Rigopoulos (2009) found that different unresolved terms can have competing effects, such that a model that neglected all terms in their case had seemingly good agreement with experimental data due to cross-cancellation of errors. Such studies were conducted in RANS or LES, however, and therefore no definitive conclusion could be made about the effect of unclosed terms; a DNS-based simulation can thus provide further insights to this matter.

Fourth, even though the resolution of the sub-Kolmogorov scales down to the Batchelor scale (see Section 4.4.5) at the Reynolds and Schmidt numbers of the case study is not possible, a preliminary investigation of their effect on precipitation is conducted by an additional simulation on a geometry similar to that of the case study but with $Re = 400$, at which it is possible to resolve the Batchelor scale for Sc up to 10. By comparing simulations with $Sc = 1$ and $Sc = 10$, the effect of the sub-Kolmogorov scales on the reaction zones and the precipitation outcome is elucidated.

As with the previous case studies, the presentation follows the four steps outlined in Section 1.1.3.

6.3.2 Formulation of the Population Balance Model

Preliminary Considerations

The population balance formulation chosen for this case involves a single PBE, since the particles have been observed to have an oval-like shape close to spherical (Schwarzer and Peukert, 2002), and therefore there is no need for an approach that accounts for morphology (as was the case with the previous two case studies). The choice of kinetic and transport processes and the formulation of the PBE are discussed below.

Kinetic and Transport Processes

The kinetic processes to be included in the case under consideration are:

- Nucleation, which will be considered to occur via the homogeneous nucleation route.
- Surface growth of crystals.

Some further discussion is merited at this point. While aggregation is often present in crystallisation processes, in the experiments to be simulated here it was suppressed due to electrostatic repulsion (Schwarzer and Peukert, 2005; Kucher et al., 2006). Furthermore, nucleation may occur via several routes in (homogeneous, heterogeneous, secondary), but this experiment was conducted under high supersaturation conditions at which homogeneous nucleation is expected to be the dominant mechanism (Schubert, 1998).

Regarding transport processes to be included are the following:

- Convection of particles by the flow, which is turbulent in the T-mixer flow under consideration.
- Particle diffusion.

No particle acceleration is included because the Stokes number is almost zero with the size of particles encountered in this study.

PBE Formulation

Based on the aforementioned choice of kinetic and transport processes, the PBE can now be formulated as follows:

$$\frac{\partial n}{\partial t} + \frac{\partial (u_i n)}{\partial x_i} + \frac{\partial (G(\mathbf{C})n)}{\partial v} = D_p \frac{\partial^2 n}{\partial x_i \partial x_i} + B(\mathbf{C})\delta(v - v_0), \tag{6.37}$$

where $\mathbf{C}$ is the vector containing the concentrations of the chemical species.

To be consistent with the PBEs for soot and aerosol formation, the formulation in terms of particle volume is retained here, although a formulation in terms of diameter could also have been used since aggregation and fragmentation are not present. The process is isothermal, hence the particle diffusivity does not vary with space and the temperature dependence of the kinetics is not explicitly indicated.

6.3.3 Kinetic Models

An introduction to crystallisation kinetics was provided in Section 3.4.2. The precipitation of $BaSO_4$ can be described by the following global reaction:

$$BaCl_2 + H_2SO_4 \longrightarrow BaSO_4 \downarrow + 2HCl.$$

While the initial dissolution of the reactants into Ba^{2+}, Cl^-, H^+ and HSO_4^- is complete, the dissociation of HSO_4^- proceeds according to the following equilibrium:

$$HSO_4^- \rightleftharpoons H^+ + SO_4^{2-}.$$

The supersaturation is given by the following activity-based expression (Vicum et al., 2003):

$$S = \gamma_\pm \sqrt{\frac{C_{Ba^{2+}_{(\text{free})}} C_{SO^{2-}_{4(\text{free})}}}{K_{SP}}}, \tag{6.38}$$

where the solubility product, K_{SP}, is taken from Monnin (1999), while the mean activity coefficients $\gamma_\pm$ are calculated with the modified Debye–Hückel method, as shown in Appendix A of Tang et al. (2022). Eq. 6.38 considers the contributions from the free ions, as $BaSO_4$ tends to form undissociated ion pairs (Vicum et al., 2003), and the equilibrium constant for the ion complex formation is also taken from Monnin (1999).

The choice of nucleation and growth kinetics in the present study follows Gradl and Peukert (2009), who employed the classical theories of nucleation and growth, and the expressions are from Mersmann (2001). This choice of kinetics has also been adopted by several authors under high supersaturation conditions (e.g. Marchisio et al., 2006; Gavi et al., 2007b; Metzger and Kind, 2017). At such conditions, primary homogeneous nucleation is the dominant mechanism (Schubert, 1998), and its rate is shown below:

$$B_N = 1.5 D_{AB} \left(\sqrt{K_{SP}} S N_A \right)^{7/3} \sqrt{\frac{\gamma_{CL}}{kT}} V_m \exp\left[-\frac{16}{3} \left(\frac{\gamma_{CL}}{kT} \right)^3 \frac{V_m^2}{(\nu \ln S)^2} \right], \tag{6.39}$$

where B_N is the rate of generation of particles via nucleation (note that this is related to the B in Eq. 6.37 as $B_N = B dv_0$ where dv_0 is the interval including the nuclei volume), while N_A, k, D_{AB}, V_m, γ_{CL} and ν are the Avogadro number, Boltzmann constant, apparent diffusion coefficient, volume of a $BaSO_4$ molecule, interfacial energy and dissociation number, respectively. The interfacial energy is taken from Schwarzer and Peukert (2004). The apparent diffusion coefficient, D_{AB}, is given by the following expression (Mersmann, 2001):

$$D_{AB} = \frac{(z_A + z_B) D_A D_B}{z_A D_A + z_B D_B}, \tag{6.40}$$

where D_A and D_B are the diffusivities of the two ions and z_A and z_B are the corresponding charge numbers. The nuclei size is taken to be the critical particle size, given by:

$$\ell_c = \frac{4\gamma_{CL} V_m}{\nu k T \ln S}. \tag{6.41}$$

For high supersaturation conditions as in this case, growth is transport-controlled (Angerhöfer, 1994) and given by:

$$G_\ell = \frac{k_a}{3k_v} \frac{Sh D_{AB} \sqrt{K_{SP}} MW (S-1)}{\rho_C \ell}, \tag{6.42}$$

where G_ℓ is the linear growth rate and ℓ is the particle diameter, while Sh, MW and ρ_c are the Sherwood number, molecular weight and crystal density, respectively. The Sherwood number stands for the ratio of convection to diffusion mass transfer

Table 6.6 *Kinetics parameters of $BaSO_4$ precipitation at 25°C used in the present study.*

	Parameter	Value	Ref.
D_{AB}	Apparent diffusion coefficient	$9.46 \cdot 10^{-10}$ m²/s	Mersmann (2001)
K_{SP}	Solubility product	$9.83 \cdot 10^{-11}$ kmol²/m³	Monnin (1999)
K_{IP}	Equilibrium constant of $BaSO_4$(aq) ion pair formation	$5.4 \cdot 10^{-3}$	Monnin (1999)
K_{dis}	Dissociation constant between ions HSO_4^-, H^+ and SO_4^{2-}	$1.2 \cdot 10^{-2}$	Schwarzer and Peukert (2004)
M	Molar mass	233.38 kg/kmol	–
Sh	Sherwood number	2	Schwarzer and Peukert (2004)
V_m	Molecular volume	$8.61 \cdot 10^{-29}$ m³	–
ν	Dissociation number	2	–
γ_{CL}	Interfacial energy	0.1181 J/m²	Schwarzer and Peukert (2004)
ρ_c	Crystal density	4480 kg/m³	–

Reproduced from Tang et al. (2022) under CC-BY 4.0 license.

rate and is taken to be 2 (Schwarzer and Peukert, 2004). The surface (k_a) and volume (k_v) shape factors are π and $\pi/6$, respectively, referring to spherical particles, as the experiments showed oval-like and spherical particle shapes (Schwarzer and Peukert, 2002). The volumetric growth rate employed in Eq. 6.37 is related to the one given by Eq. 6.42 as follows (cf. Eq. 3.76):

$$G = 3\left(\frac{\pi}{6}\right)^{1/3} v^{2/3} G_\ell. \tag{6.43}$$

The values of the kinetic parameters used in all of the above expressions are summarised in Table 6.6.

Since one of the main objectives of this investigation is to examine the interaction of turbulence, nucleation and growth, it is worthwhile to take a closer look at the dependence of the kinetic processes on supersaturation. Graphs of the nucleation and growth rate as functions of supersaturation are shown in Fig. 6.21. The most notable feature is the strongly nonlinear dependence of nucleation on supersaturation. This has important repercussions for the following discussion, as it means that small fluctuations in supersaturation can induce large fluctuations in the nucleation rate.

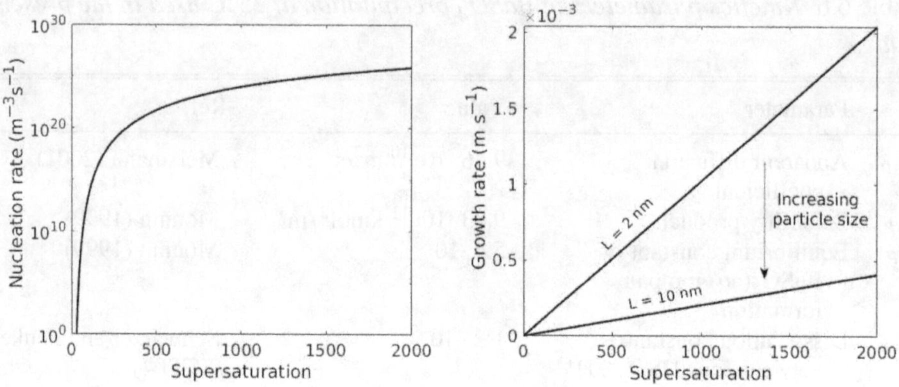

Figure 6.21 Variation of the nucleation and growth rates of $BaSO_4$ with supersaturation. Reproduced from Tang et al. (2022) under CC-BY 4.0 license.

6.3.4 Fluid Dynamics and Transport Phenomena

The equations of fluid dynamics to be employed in this case study are the continuity and Navier–Stokes equations for incompressible flow with constant density and viscosity (Eqs. A.2 and A.25), while the energy equation is not needed since the flow is isothermal. The species transport equation is also required because the nucleation and growth kinetics depend on supersaturation, which in turn depends on the species' concentrations. The appropriate form for it is Eq. A.13, with the last term denoting the species' consumption due to particle formation and providing the coupling with the PBE. The simulation is a DNS where all flow scales are resolved (cf. Section 5.2.4), hence no turbulence modelling is required.

6.3.5 Numerical Methods

Discretisation of the PBE

The PBE is discretised with a finite volume method. As there is no aggregation or fragmentation, the main choice is about the discretisation of the growth term. As discussed in Section 4.3.4, the main problem with this term is that of numerical diffusion. The TVD scheme due to Qamar et al. (2006) with the flux limiting function of Koren (1993), which has already been described in Section 6.2.6, is also employed here.

The choice of grid requires further consideration. In this nucleation-growth problem, a fine resolution is required at the low end of the particle volume spectrum to account for the nuclei and size-dependent growth. On the other hand, the number of discretised number densities must be kept to a minimum in order to reduce the

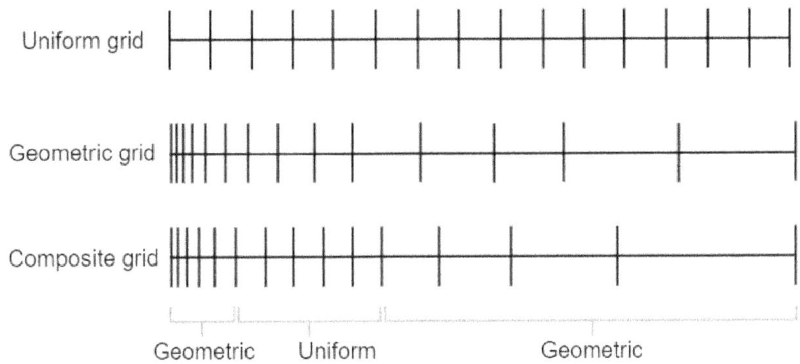

Figure 6.22 The composite grid, compared with a uniform and a geometric grid. Reprinted with permission from Tang et al. (2020).

number of scalars to be transported by the DNS. To this end, a fixed composite grid was proposed in Tang et al. (2020), combining the features of uniform and geometric grids. The particle range is divided into three parts. A high-resolution geometric grid is used to for the range of the fast-growing nuclei (0.6 nm–20 nm), followed by a uniform grid to provide reasonable resolution at the intermediate range (20 nm–150nm) where the main part of the distribution at the exit is located. After that, a coarse geometric grid is used to accommodate the largest particles and avoid spurious mass loss. The grid comprises 45 intervals in total and is shown on Fig. 6.22.

The complete discretised PBE can be written as follows:

$$\frac{\partial n_k}{\partial t} + \frac{\partial (u_i n_k)}{\partial x_i} + \frac{\partial (G(\mathbf{C}) n_k)}{\partial v} = \frac{\mu}{Sc} \frac{\partial^2 n_k}{\partial x_i \partial x_i} + \frac{B_N(\mathbf{C})}{dv_1} \delta_{k1}. \tag{6.44}$$

Coupling of PBE with DNS

The DNS is performed with code PANTARHEI, which employs the finite volume method for discretising the Navier–Stokes equations in a collocated, unstructured grid; the reader may refer to Tang et al. (2020) for the details of the numerical schemes employed in the DNS. PANTARHEI is coupled with the PBE code CPMOD, and the overall structure of the code is depicted schematically in Fig. 6.23.

The coupling between DNS and PBE is effected via the species consumption term, and it is important to ensure that the consumption term in the species transport equation is consistent with the particle formation in the PBE. The reaction source term is thus computed based on the rate of increase of particle volume in the PBE,

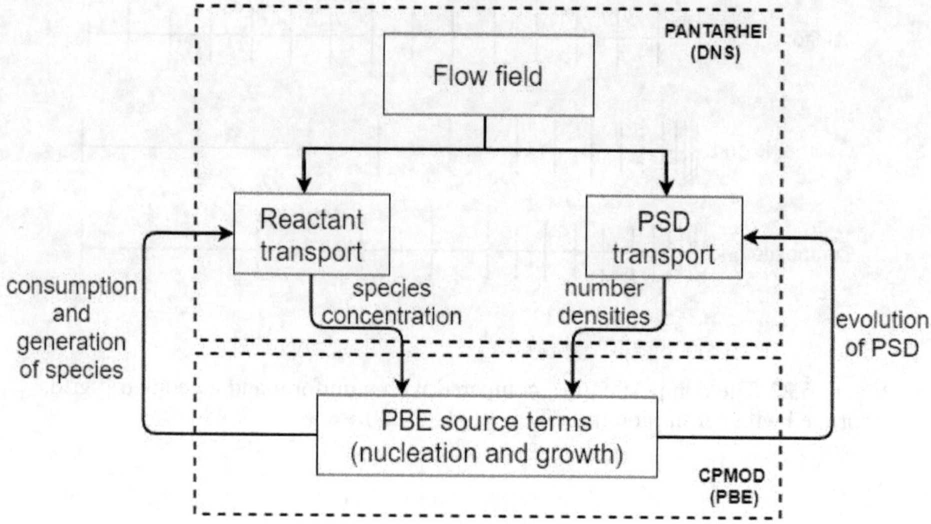

Figure 6.23 Coupling of DNS and PBE codes. Reprinted with permission from Tang et al. (2020).

which can be obtained by summing the volume of nuclei formed and the volume increase in every interval in the discretised PBE due to growth. The rate of change of the first moment, S_{m_1}, is thus given by:

$$S_{m_1} = \left[B(\mathbf{C}) v_0 + \sum_{k=1}^{m} v_{m,k} \left(\frac{\partial G(\mathbf{C}, v_k) n_k}{\partial v} \right) dv_k \right]. \tag{6.45}$$

The concentration of $BaSO_4$ is then calculated as follows:

$$C_{BaSO_4} = \frac{\rho_c m_1}{MW_{BaSO_4}}, \tag{6.46}$$

where MW_c is the molecular weight of the crystal. Hence the reaction source term for each of the reacting species in Eq. A.13 can be calculated as follows:

$$R_{\alpha,r} = \frac{\nu_\alpha \rho_c}{MW_{BaSO_4}} S_{m_1}, \tag{6.47}$$

where ν_α is the stoichiometric factor for species α.

Due to the very short timescales associated with the precipitation process, the coupled CFD-PBE simulation requires a smaller time step than a DNS for the flow field alone. Therefore, a Courant–Friedrichs–Lewy (CFL) condition for the PBE (cf. Section 4.3.4) must be defined in accordance with the one for the flow.

It should be reminded that growth acts as velocity in the PBE domain, hence a Courant number can be defined for the PBE as follows:

$$C_{PBE}(\ell_k, \mathbf{x}) = \frac{G(\ell_k, \mathbf{x})\delta t}{\delta \ell_k}, \tag{6.48}$$

where $\delta \ell_k$ is the distance between node k and the previous one in the PBE grid. The PBE CFL condition is that $\max(C_{PBE}(\ell_k, \mathbf{x})) \leq 1$. Tang et al. (2020) showed tests that demonstrate the importance of meeting this condition. A time step of $1.5 \cdot 10^{-7}$ s is chosen, which yields a C_{PBE} of 0.35.

6.3.6 Simulation of Precipitation of $BaSO_4$ in a Turbulent T-Mixer Flow: The Schwarzer et al. Case

Problem Description and Computational Setup

The first simulation to be conducted is that of the T-mixer experiment reported by Schwarzer et al. (2006). The research questions to be investigated in this case are:

- Can we predict accurately the PSD with the coupling of DNS and PBE, eliminating the uncertainties associated with turbulence–PBE interaction?
- Do the flow and precipitation timescales overlap?
- What is the thickness of the zones where precipitation takes place?
- Is the process controlled by mixing or kinetics?
- What is the importance of the fluctuation terms?

The dimensions and configuration of the T-mixer in this case is shown in Fig. 6.24. The reactants are injected from the two inlets with equal flow rates and

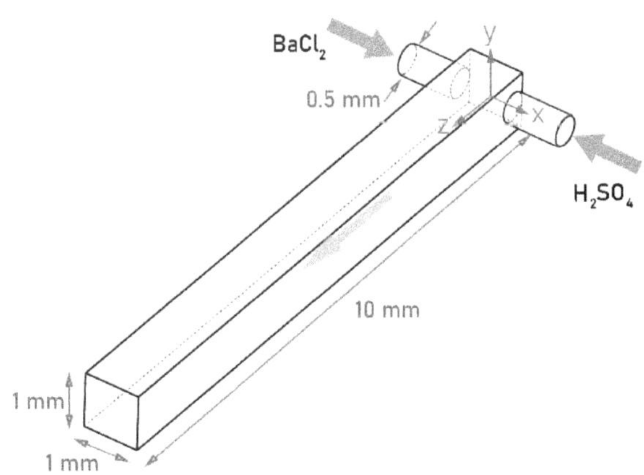

Figure 6.24 Illustration of the T-mixer setup for the Schwarzer et al. (2006) case. Reproduced from Tang et al. (2023) under CC-BY 4.0 license.

concentrations of 0.5 kmol/m^3 and 0.33 kmol/m^3 for $BaCl_2$ and H_2SO_4, respectively. The Reynolds number in the mixing channel is 1,135, which is turbulent for this T-mixer configuration (Telib et al., 2004).

The simulation is performed with a grid of 21 million cells resulting in a maximum grid-to-Kolmogorov scale ratio ($\delta x/\eta$) of about 2. This simulation is performed with $Sc = 1$; the resolution of the Batchelor scale, η_c, while maintaining the ratio $\delta x/\eta_c$ to about 2, would require a δx around 30 times finer in each direction, which corresponds to 30^3 times more cells than the aforementioned grid, i.e. 570 billion cells. Such a simulation is clearly not feasible with present-day computational resources. However, an assessment of the effect of the sub-Kolmogorov scales will be attempted in the case discussed in Section 6.3.7, where the Batchelor scale will be resolved for $Sc = 10$. A Poiseuille velocity profile is applied to the inlets, while a convective boundary condition is applied at the channel exit. The reactants are injected once the flow is statistically steady. For more details about the simulation setup, the reader may refer to Tang et al. (2020).

Flow Field in the T-Mixer

It is first worth mentioning that the DNS of the flow field was validated in Tang et al. (2020) via the simulation of a geometrically similar but 80 times scaled-up T-mixer and comparison with particle image velocimetry (PIV) measurements in that geometry in an experiment without precipitation by Schwertfirm et al. (2007). For the T-mixer on which the precipitation was carried out, the spectrum at several locations was calculated and a range of frequencies with a slope of $-5/3$ was observed; furthermore, the terms of the turbulent kinetic energy transport equation were calculated and the balance was satisfied to an acceptable accuracy (Tang et al., 2022).

The main features of the flow field in the T-mixer that are pertinent to the following discussion can be observed in Fig. 6.25. Shortly after entering the mixing channel, the flow transitions to turbulent flow. The mixing environment in the impingement zone is crucial for the induction of precipitation, as the boundaries between the two streams are where precipitation is initiated. A large helical vortex is formed at the centre of the mixing channel, making the velocity contour asymmetric. Finally, Fig. 6.26 shows the contours of the turbulent kinetic energy and its dissipation rate. The transition in the impingement zone is evidenced by the rapid increase in turbulent kinetic energy along the inlets. The contours of both quantities are asymmetric due to the helical flow pattern. The turbulent kinetic energy peaks at the impingement centre and subsequently decays within the mixing channel, while the energy dissipation is strong within the impingement area and extends to the immediate downstream region.

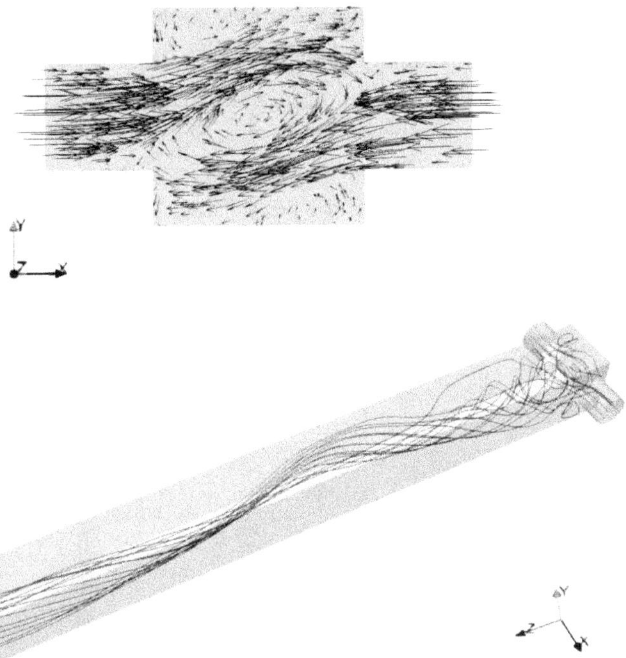

Figure 6.25 Mean velocity vectors in the impingement zone (top) and mean streamlines of the flow field in the T-mixer (bottom). Reproduced from Tang et al. (2023) under CC-BY 4.0 license.

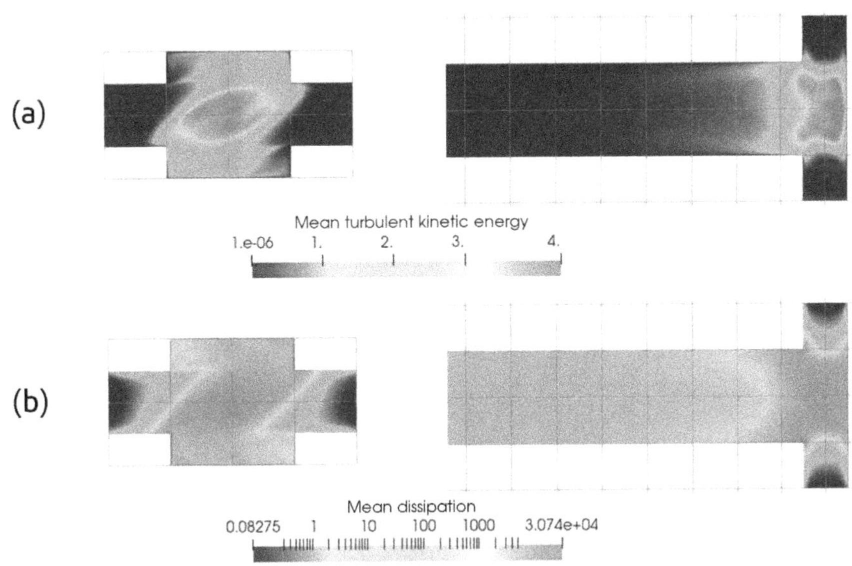

Figure 6.26 Mean turbulent kinetic energy [$m^2\ s^{-2}$] (a) and dissipation [$m^2\ s^{-3}$] (b). Left: x-y plane at z=0, right: x-z plane at y=0 (only the first 40% of the channel is shown). Reproduced from Tang et al. (2022) under CC-BY 4.0 license.

Prediction of PSD and Comparison with Experiments

The evolution of the time and cross-sectional average PSD, normalised by the zeroth moment of the distribution alongside the mixing channel is shown in Fig. 6.27. A comparison with the experiment of Schwarzer et al. (2006) is made for the exit PSD, where a measurement is available, and the agreement is very good, apart from a small overprediction of the peak value. It should be noted that the PSD resides mainly within the uniform section of the grid, where the composite grid provides good resolution. The quality of the prediction can be ascribed to the resolution of fluid dynamics and turbulence–chemistry interaction down to the Kolmogorov

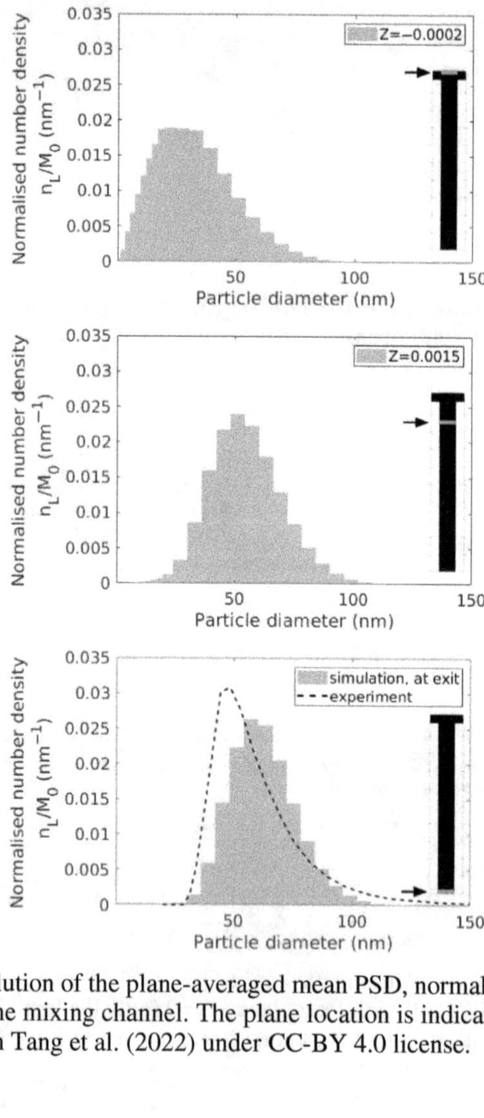

Figure 6.27 Evolution of the plane-averaged mean PSD, normalised by the zeroth moment, along the mixing channel. The plane location is indicated by the arrows. Reproduced from Tang et al. (2022) under CC-BY 4.0 license.

scale (albeit not to the Batchelor scale), the solution of the PBE without assumptions on the shape of the distribution and the nature of the experiment, which did not feature heterogeneous/secondary nucleation, aggregation or fragmentation, any of which would complicate the kinetic modelling. The lack of accounting for sub-Kolmogorov scales is the main remaining modelling uncertainty, and an attempt for its evaluation will be made in Section 6.3.7.

It can also be observed that the changes in the PSD are minor after the impingement zone, which indicates that the PSD is determined mainly by the early stage of the process. Mixing effects are critical at that stage because narrow zones of high and intermittent supersaturation are formed, as will be shown in the following discussion and in Fig. 6.28.

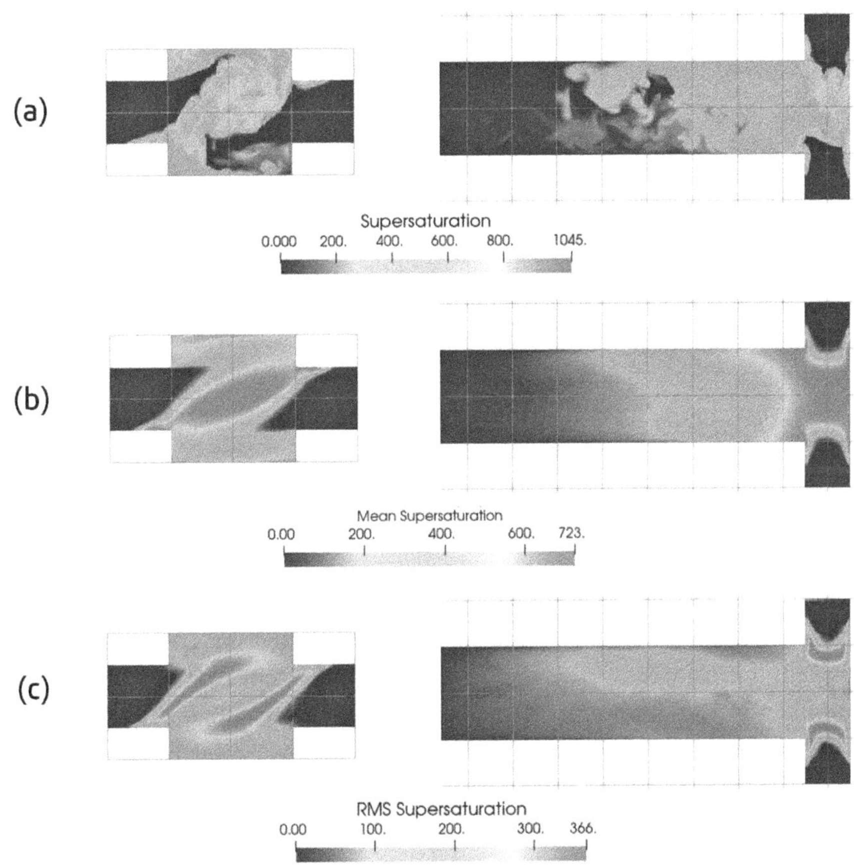

Figure 6.28 Supersaturation [−] contours (instantaneous (a), mean (b) and RMS (c)). Left: x-y plane at z=0, right: x-z plane at y=0 (only the first 40% of the channel is shown). Reproduced from Tang et al. (2022) under CC-BY 4.0 license.

Precipitation Dynamics

Fig. 6.28, shows the spatial distribution of the supersaturation, which is the cornerstone of precipitation (instantaneous, mean and RMS values are juxtaposed). The overall picture confirms the high degree of unmixedness and the presence of small-scale structures, indicative of the need for coupling PBE with fluid dynamics. The mean supersaturation pattern is slightly asymmetric due to the fact that the concentrations at the two inlets are not equal. The supersaturation decays rapidly beyond the impingement zone, the supersaturation level drops significantly and eventually reaches zero within the mixing channel, which explains why the distribution does not change significantly along it (cf. Fig. 6.27). The highest RMS values are observed at the inlet stream boundaries (up to 50% of the mean). At these locations, turbulent fluctuations are also very strong, hence the supersaturation hot spots meander and vary rapidly (as is evident in the instantaneous plot), hence the mean supersaturation is considerably lower.

Figs. 6.29 and 6.30 show the contributions of nucleation and growth, respectively, to the rate of reactant consumption. Nucleation is confined mainly within the impingement region and occurs mainly within thin zones. The spatial distribution of the consumption terms exhibits similar features with that of supersaturation, such as the appearance of RMS hot spots at the stream boundaries. In the case of nucleation, these hot spots are very localised and have magnitude comparable to the mean value. By juxtaposing the nucleation and growth rate contours with that of supersaturation, it can be seen that the nucleation consumption rate is closely matched with supersaturation at the impingement zone, while shortly afterwards nucleation is terminated and growth takes over.

While growth is a linear function of supersaturation, the spatial distribution of the consumption due to growth is not identical with that of supersaturation. This is because growth depends also on the available particle surface area. Therefore, hot spots of growth consumption rate may appear at locations with large numbers of particles with considerable surface area, despite intermediate supersaturation levels. On the other hand, regions of high supersaturation may feature low growth consumption rates due to small numbers of particles being present or particles being very small; this happens, for example, at the impingement zone.

Turbulent flow is intermittent, and it is of interest to examine the effect of this intermittency on the supersaturation and consequently the nucleation and growth rates. This is because simulations of turbulent precipitation are often based on average values, but averages can be highly non-representative of the true state of a process. Tang et al. (2022) showed that the energy dissipation and supersaturation are highly intermittent in the impingement zone, which is crucial for the shaping of the PSD. This suggests an intermittent behaviour in the kinetic processes, and

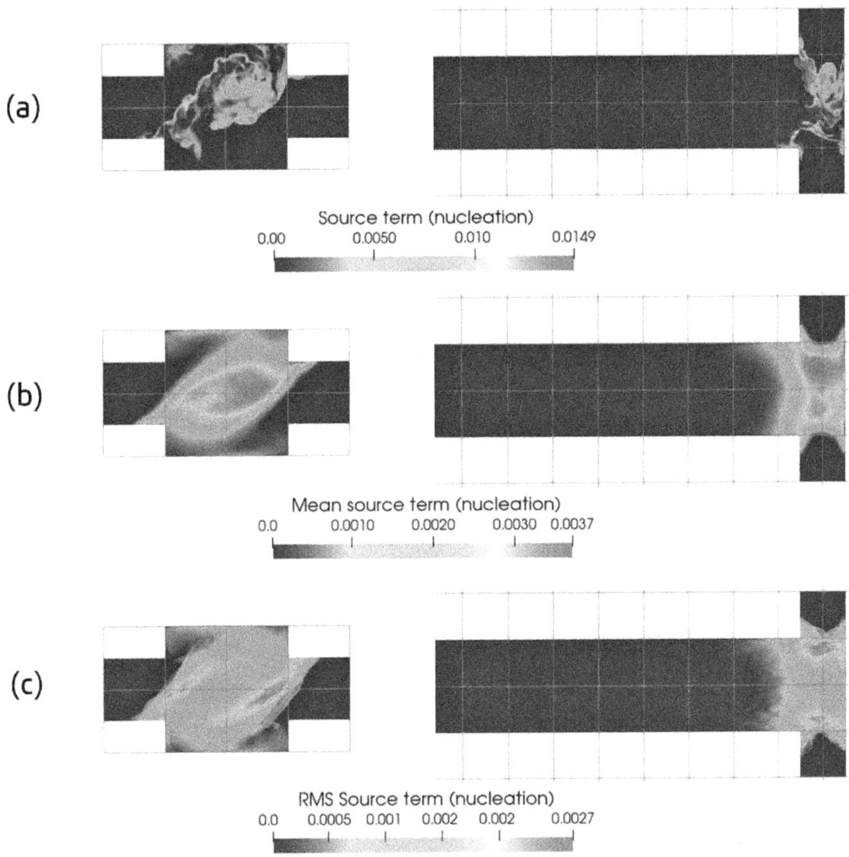

Figure 6.29 Nucleation consumption rate [kmol $m^{-3}s^{-1}$] (instantaneous (a), mean (b) and RMS(c)). Left: x-y plane at z=0, right: x-z plane at y=0 (only the first 40% of the channel is shown). Reproduced from Tang et al. (2022) under CC-BY 4.0 license.

particularly nucleation, which is a highly nonlinear function of supersaturation (Fig. 6.21). In order to better understand the dynamics of precipitation, we employ two timescales, one for nucleation, τ_{N}, and one for growth, τ_G, adopting the definitions proposed by Bałdyga and Bourne (1999):

$$\tau_{\mathrm{N}} = \frac{\int_0^\infty n dv}{B_N}, \tag{6.49}$$

$$\tau_G = \frac{3k_v^{2/3} \int_0^\infty v n dv}{k_a \int_0^\infty G v^{-1/3} n dv}. \tag{6.50}$$

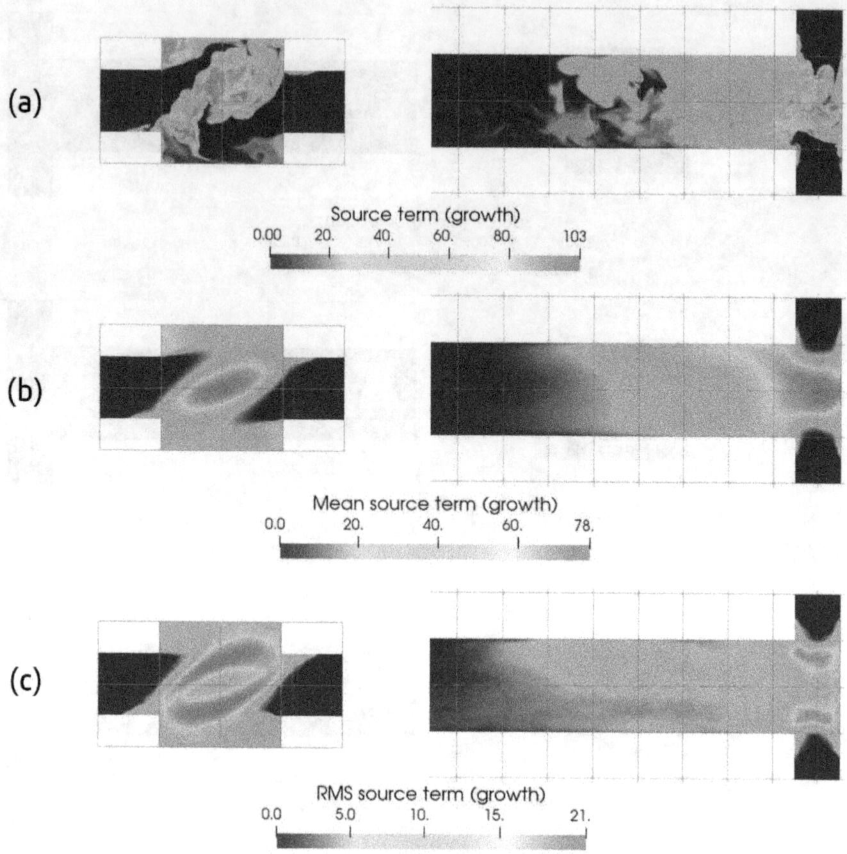

Figure 6.30 Growth consumption rate [kmol $m^{-3}s^{-1}$] (instantaneous (a), mean (b) and RMS (c)). Left: x-y plane at z=0, right: x-z plane at y=0 (only the first 40% of the channel is shown). Reproduced from Tang et al. (2022) under CC-BY 4.0 license.

The nucleation timescale is thus defined as the ratio of the total number of particles to the nucleation rate, while the growth timescale is defined as the ratio of the total particle volume to the total volumetric growth rate (all quantities are per unit volume of fluid).

The nucleation and growth timescales allow us to quantify the competing effects of these two kinetic and their interaction with flow. However, due to the high intermittency, the time-averaging of timescales does not provide useful information. Filtering is thus performed on the time signals to remove instances of extremely slow rates. The cut-off value for this filtering is the flow-through time in the mixing channel because most of the high reaction rates occur at the impingement zone where the residence time is typically far lower, and also because timescales

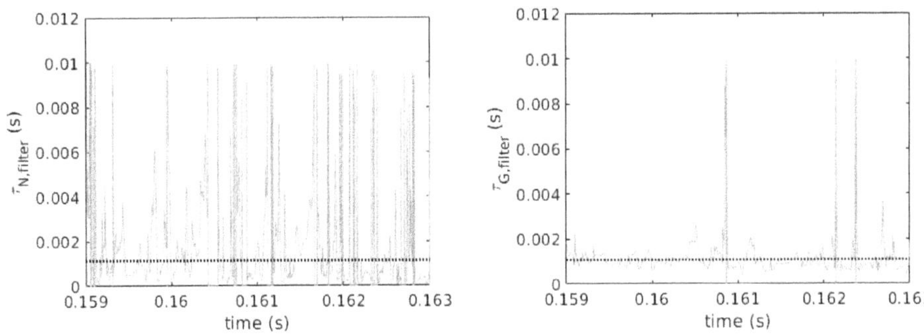

Figure 6.31 Time series of the filtered nucleation (left) and growth (right) timescales at the impingement centre. The dotted lines indicate the filtered means. Reproduced from Tang et al. (2022) under CC-BY 4.0 license.

longer than the flow-through time are indicative of non-reaction as the reactants would have been flushed away. The filtered timescales are shown in Fig. 6.31. It can be immediately seen that nucleation is highly intermittent, featuring sharp spikes, while growth exhibits a much smoother behaviour. This is due to the highly nonlinear form of the nucleation kinetics and implies strong effect of mixing on nucleation. The filtered timescales will also be employed for the Damköhler number analysis in the next section.

The time-averaged filtered timescales of nucleation and growth are juxtaposed in Fig. 6.32. By comparing this figure with Fig. 6.28, we can see that regions of high supersaturation feature smaller timescales that indicate rapid kinetics and strong influence on the evolution of the PSD. At the impingement, where precipitation is most intense, both nucleation and growth have small magnitude, yet nucleation is in general smaller and occurs within hot spots, as evidenced by Fig. 6.29.

It is also instructive to consider the thickness of the nucleation zones, as it is highly relevant to the application of mixing models. Nucleation is very localised and often takes place via bursts within thin zones. The thickness of these zones compared to the Kolmogorov scale indicates whether scale separation models that decouple chemistry and flow can be applied; for example, in combustion, if the reaction zones are much smaller than the Kolmogorov scale, flamelet-type models are applicable (see e.g. Peters, 2000, Veynante and Vervisch, 2002 or Poinsot and Veynante, 2005), where a locally laminar diffusion-reaction model is applied within the zone; otherwise, the Kolmogorov eddies penetrate the reaction zones and result in a distributed reaction field.

Fig. 6.33 shows an instantaneous contour of the nucleation source term in the impingement zone, as well as a graph of the source term along a line probe, where

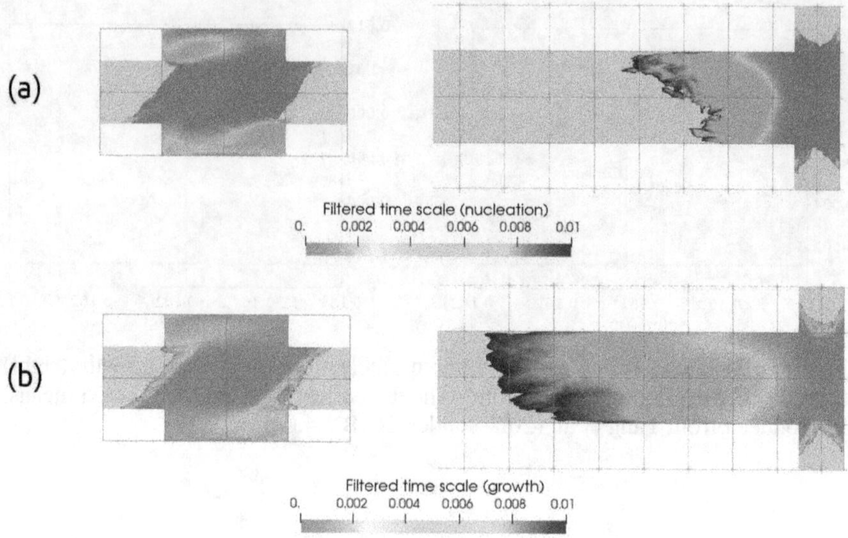

Figure 6.32 Contours of the time-average of filtered timescales [s]: nucleation (a); growth (b). The left-hand side of the legend bar indicates small values, that is, fast reaction rates. Non-reactive regions are shown in light grey. Left: x-y plane at z=0, right: x-z plane at y=0 (only the first 40% of the channel is shown). Reproduced from Tang et al. (2022) under CC-BY 4.0 license.

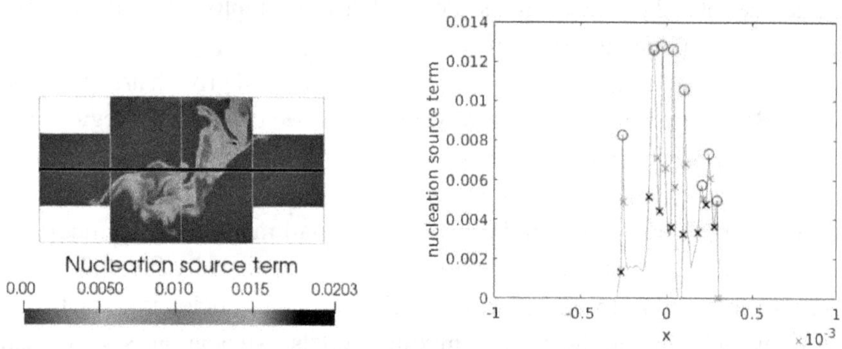

Figure 6.33 Left: a contour plot of an instance of the nucleation source term [$kmol\ m^{-3}s^{-1}$] at the impingement (x-y plane, z = 0) with a horizontal probe indicated by a line. Right: a graph of the nucleation source term along the probe, with circles indicating peaks and crosses indicating peaks and troughs in the first derivative around each source peak. Reproduced from Tang et al. (2022) under CC-BY 4.0 license.

the thickness of the zones is clearly visible. To quantify these results, spatial data of the nucleation zone thickness, l_n, are obtained at 30 uncorrelated time instants and 239 peaks are identified, giving a mean, standard deviation, maximum and minimum thickness as $2.9 \cdot 10^{-5}$ m, $1.2 \cdot 10^{-5}$ m, $9.6 \cdot 10^{-5}$ m and $1.6 \cdot 10^{-5}$ m, respectively (the PDF of the nucleation zone thickness is also shown in Tang et al., 2022). The minimum thickness is higher than the grid resolution and the mean thickness is resolved with 5 cells, so the grid resolution can be considered as sufficient. It must be remembered that this simulation is conducted with $Sc = 1$, and the nucleation zones can be even thinner if the Batchelor scale is resolved, as we will see in Section 6.3.7. The ratio of nucleation zone thickness to Kolmogorov scale, l_n/η, yields a mean, maximum and minimum value of 6.1, 19.1 and 1.2, respectively. Therefore, the Kolmogorov scale does penetrate the nucleation zones, a fact with important repercussions for precipitation modelling. Furthermore, since the reactant consumption due to nucleation is very small even at high nucleation rates (cf. Fig. 6.29), the high concentration in these zones is broken mainly by turbulent transport.

Interaction between Mixing and Precipitation

In order to probe into the interaction of mixing and precipitation, we define two Damköhler numbers (Da), one for nucleation and one for growth. Each Damköhler number is a ratio of a flow timescale to a kinetic process one; a high value thus indicates that the problem is mixing controlled, while a low value indicates the opposite. The two Da for nucleation and growth, respectively, are defined as follows:

$$Da_N = \frac{\tau_f}{\bar{\tau}_{N,filter}}, \tag{6.51}$$

$$Da_G = \frac{\tau_f}{\bar{\tau}_{G,filter}}. \tag{6.52}$$

Note that the filtered timescales are employed for the kinetic processes. Regarding the flow timescale, we employ the engulfment time, τ_E (Bałdyga, 2016; see also Bałdyga and Bourne, 1989, 1999):

$$\tau_E = 17.3\sqrt{\frac{\nu}{\epsilon}} = 17.3\tau_\eta, \tag{6.53}$$

where τ_η is the Kolmogorov timescale. The engulfment time is of relevance to the viscous convective scales, at which precipitation is occurring. The factor 17.3 comes from the derivation of the engulfment rate as $E = \ln 2/\tau_w$ (Bałdyga and Bourne, 1989), which is the reciprocal of the engulfment timescale and where the hydrodynamic lifetime, $\tau_w = 12(\nu/\epsilon)^{1/2}$, is taken to be the shortest mean lifetime of an eddy (Bałdyga and Bourne, 1984).

Contour plots of the two Damköhler numbers are shown in Fig. 6.34. Da_N is around unity at the impingement zone and subsequently vanishes at the channel where nucleation is mostly absent (cf. Fig. 6.29). Da_G is around unity at most

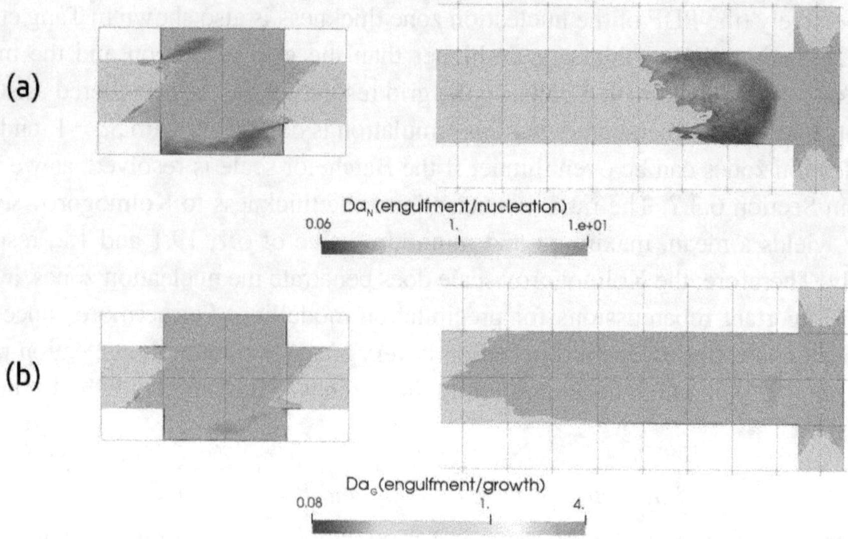

Figure 6.34 Damköhler number map based on the engulfment time: nucleation (a); growth (b). Non-reactive regions are shown in light grey. Left: x-y plane at z = 0, right: x-z plane at y = 0 (only the first 40% of the channel is shown). Reproduced from Tang et al. (2022) under CC-BY 4.0 license.

of the domain. These observations indicate that the process is neither mixing nor kinetics-controlled. High local reactant concentrations leads to nucleation bursts, which would be maintained for a long time if it were not for mixing, since reactant consumption due to nucleation is very low; however, mixing rapidly disperses such hot spots. Therefore, a model that decouples flow and kinetics is in risk of missing out on such effects.

Effect of Turbulent Fluctuations on Precipitation

We now turn our attention on the effect of turbulent fluctuations on precipitation, an issue of relevance to model development. The closure problem for the PBE in turbulent flow was discussed in detail in Chapter 5, where the various unclosed terms that result were identified. To facilitate the following discussion, we show here the averaged PBE for the present problem:

$$\frac{\partial \langle n\rangle}{\partial t} + \frac{\partial \left(\langle u_i\rangle\langle n\rangle\right)}{\partial x_i} + \frac{\partial \langle u_i' n'\rangle}{\partial x_i} + \frac{\partial \left(\langle G\rangle\langle n\rangle\right)}{\partial v} + \frac{\partial \langle G' n'\rangle}{\partial v} = D_p \frac{\partial^2 \langle n\rangle}{\partial x_i \partial x_i} + \langle B\rangle \delta(v - v_0). \tag{6.54}$$

The unclosed terms here are those involving $\langle u_i' n'\rangle$, $\langle G' n'\rangle$, $\langle B\rangle$ and $\langle G\rangle$. The first of these terms represents turbulent transport of particles and can be addressed with standard models, so we focus our attention on the remaining ones. It is worth also to write the averaged species transport equation:

$$\frac{\partial \langle C_\alpha \rangle}{\partial t} + \frac{\partial (\langle u_i \rangle \langle C_\alpha \rangle)}{\partial x_i} + \frac{\partial \langle u_i' C_\alpha' \rangle}{\partial x_i} = D_\alpha \frac{\partial^2 \langle C_\alpha \rangle}{\partial x_i \partial x_i} + \langle R_\alpha \rangle, \tag{6.55}$$

where again $\langle u_i' C_\alpha' \rangle$ represents turbulent transport of species, while the averaged source term can be written as:

$$\langle R_\alpha \rangle = \frac{\gamma_\alpha \rho_c}{MW_c} \left[\langle B \rangle v_0 - \int v \left[\frac{\partial (\langle G \rangle \langle n \rangle)}{\partial v} + \frac{\partial \langle G' n' \rangle}{\partial v} \right] dv \right]. \tag{6.56}$$

In the soot formation case in Section 6.2, the reaction source term was unclosed due to the complex reaction system that involved many intermediate steps; this is not the case here, as the ionic reactions are instantaneous. However, the source term is still unclosed due to the presence of $\langle B \rangle$ and $\langle G' n' \rangle$ that are involved in the consumption of chemical species due to precipitation.

The aim of the present section is to quantify the importance of the unclosed terms in the T-mixer case under consideration with the aid of the DNS-PBE simulation. To this aim, we compute the supersaturation, nucleation and growth rates using the mean concentrations and compare them with the mean values as obtained from DNS-PBE. We also investigate the term $\langle G' n' \rangle$ that appears in both Eqs. 6.54 and 6.55. It should be mentioned that, in a RANS model, the mean concentrations would be subject to the assumptions of the model rather than obtained from DNS, hence further discrepancies would result. Also, similar unclosed terms appear in an LES context, although their magnitude is expected to be smaller in that case.

Fig. 6.35 shows the spatial distribution of the mean supersaturation as computed from the DNS-PBE and from the average (i.e. $\langle S \rangle$ and $S(\langle C \rangle)$). It should be reminded that supersaturation involves the product of concentrations, which is why the two values differ. Overall, when computed with the mean values, the region of high supersaturation covers a wider area and the maximum is overpredicted by 14%. This is bound to influence both nucleation and growth rates but in different ways, as nucleation is a nonlinear function of supersaturation while growth is a linear one.

Fig. 6.36 shows the percentage difference in nucleation and growth[7] rates as computed from DNS-PBE and from the mean values. At the impingement zone, the differences are located mainly around the two ends, where the nucleation rate features both underpredictions and overpredictions by the mean-based model, while the growth rate is overpredicted. The level of overshoot or undershoot is quite high at the ends (100% or more), while it is very low at the centre of the zone. The ends are indicative of the impingement of the two jets and the onset of the formation

[7] As mentioned in Section 6.3.3, the volumetric growth rate is a linear function of the particle diameter, hence the difference is the same for all diameters.

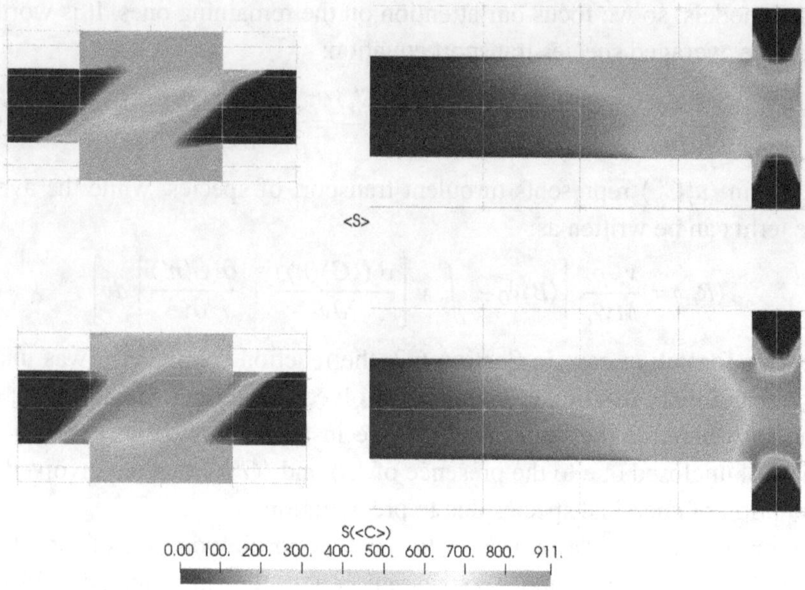

Figure 6.35 Mean supersaturation (top) and supersaturation obtained from mean concentration (bottom). Reproduced from Tang et al. (2023) under CC-BY 4.0 license.

of the helical vortex, and it can also be seen from Fig. 6.35 that supersaturation exhibits a large discrepancy when computed via the means in that region. A difference between the behaviours of nucleation and growth is also observed along the z-direction, both in the impingement region and in the main channel.

Tang et al. (2023) also examined the particle number density – growth correlations, $\langle G'n' \rangle$, which appears in Eqs. 6.54 and 6.55 (as shown by Eq. 6.56). Since this correlation is a function of particle size, they considered the correlation in the discretised equation, $\langle G'_k n'_k \rangle$, for three representative particle size intervals, namely 10 nm, 50 nm and 80 nm, indicative of the low, medium and high particle size ranges, respectively, and compared it with $\langle G \rangle \langle n \rangle$ and $\langle Gn \rangle$. The correlations are strongest at the stream boundaries, and their contribution to $\langle Gn \rangle$ in the impingement zone is significant, with typical values of 50–70% at 10 nm and 20–40% at the other two intervals. Since the number density-growth rate correlations actually appear in the averaged PBE (Eq. 6.54) as a gradient in the particle volume domain, it is instructive to examine also the magnitude of this gradient. The contribution of the fluctuating part is more significant at the smaller end of the size spectrum, similar to the trend in $\langle G'n' \rangle$. It can be concluded that the growth term can be considerably underestimated if fluctuations are neglected, with two consequences: first, due to the imbalance in the underestimation along the size spectrum, the number density at the lower end of it would be overpredicted, and second, the

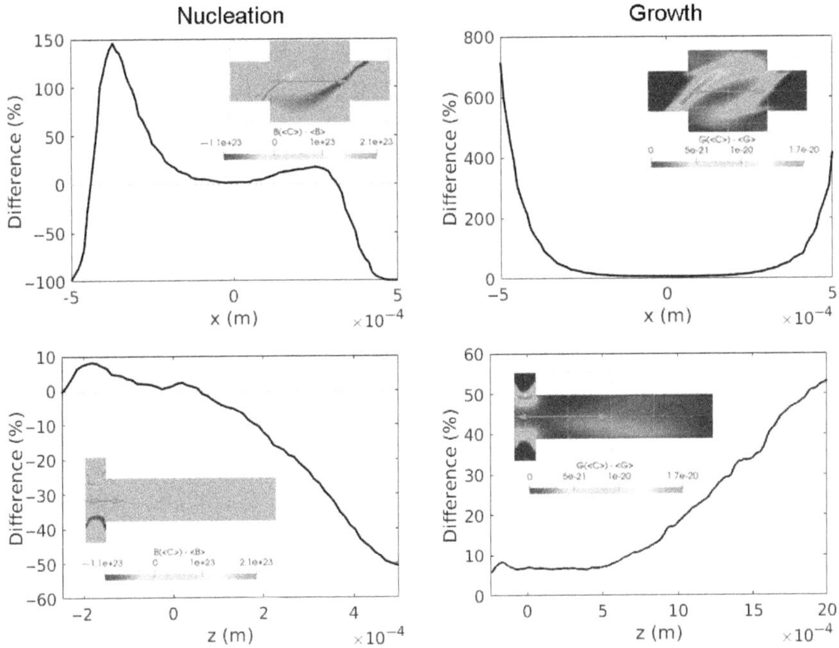

Figure 6.36 Percentage difference between the rates computed with mean values and the DNS means across the centreline of the impingement zone in the x-direction (top) and along the centreline of the mixing channel in the z-direction (bottom). The differences are calculated as $(B(\langle C\rangle) - \langle B\rangle)/\langle B\rangle$ and $(G(\langle C\rangle) - \langle G\rangle)/\langle G\rangle$, respectively. Reproduced from Tang et al. (2023) under CC-BY 4.0 license.

supersaturation consumption would also be underpredicted, resulting in an overprediction of supersaturation that would favour nucleation over growth, which would also bias the PSD towards smaller sizes.

Tang et al. (2023) considered also the contribution of the fluctuating term to the consumption term in the scalar transport equation (Eq. 6.56). In contrast with the PBE, in this case, the contribution turned out to be small. To explain this, the integral was split into contributions from each size interval. It was shown that both positive and negative contributions were present at different sizes, hence the integral over all sizes turned out to be small.

6.3.7 Scaled-Down T-Mixer: Effect of Sub-Kolmogorov Scales on Precipitation

Problem Description and Preliminary Considerations

We now turn our attention to the mixing at scales below the Kolmogorov scale. This issue was first mentioned in Section 4.4.5 in association with the diffusivity of particles. However, it is also an issue with the diffusivity of chemical species in liquid-phase flows where the Schmidt number can be very high. This is indeed the case in the present study, and the diffusivity of the ionic species is actually of greater concern than that of the particles. This is because, as the results have so far shown, the precipitation rates (particularly nucleation) are sensitive to variations in local supersaturation. Therefore, if the spatial distribution of supersaturation features thinner zones, this could affect the kinetic processes and ultimately the PSD. For example, for $Sc = 1{,}000$, the Batchelor scale, η_c, is $\sqrt{1{,}000} \approx 31$ times smaller than the Kolmogorov scale. In the context of the simulation of the Schwarzer et al. (2006) case shown in the previous section, to maintain a ratio $(\delta x/\eta_c)$ of about 2, the grid spacing δx should be around 30 times finer in each direction, which would mean 570 billion cells.

The results in the previous section were obtained with $Sc = 1$, based on the hypothesis that the excessive diffusion given by $Sc = 1$ to some extend compensates for the effect of the unresolved sub-Kolmogorov scales, while the use of high Sc without resolving these scales would require a TVD scheme to suppress the resulting numerical oscillations, which would amount to adding artificial diffusion in a manner similar to the use of $Sc = 1$. The comparison with experiments showed that the exit PSD exhibits good agreement, but the validity of the hypothesis should nevertheless be examined.

Therefore, in this section, we show results from a new T-mixer case with both $Sc = 1$ and 10. The new case is almost identical to the T-mixer in the previous section but scaled down, such that at $Sc = 10$ and $Re = 400$ – the lowest Re for which turbulent flow could be achieved in a T-mixer (Telib et al., 2004) – it is possible to resolve the Batchelor scale with a maximum cell-size-to-η_c ratio of 2. As it is still not possible to resolve the Batchelor scale at a Schmidt number of the order of 1,000, this can only be regarded as a preliminary attempt to study this open question. Nevertheless, the comparison between the $Sc = 1$ and $Sc = 10$ cases allows investigating the following research questions:

- How does the structure of the reaction zones change at higher Sc?
- What is the effect of the sub-Kolmogorov scales on the precipitation rates and the PSD?

The T-mixer under investigation is shown in Fig. 6.37. As mentioned, it is similar to the geometry simulated in the previous section but scaled down and with a few changes in order to make the simulations feasible. The length of the channel is cut down to 30% of the one considered in Section 6.3.6, as this is the region where most of the precipitation has been found to take place and, furthermore, the inlets have the shape of a square duct for better conformality with a Cartesian grid. Two grids are employed, to be referred to as ‘fine’ and ‘medium’ and featuring a smallest

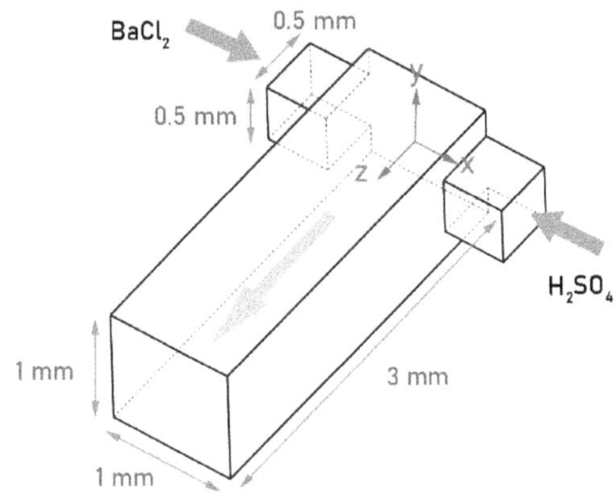

Figure 6.37 Illustration on the process setup for the Re = 400 case. Reproduced from Tang et al. (2023) under CC-BY 4.0 license.

cell size, δx, of $4 \cdot 10^{-6}$ m and $8 \cdot 10^{-6}$ m respectively. While both grids are able to resolve the Kolmogorov scale, the fine grid also resolves the Batchelor scale at $Sc = 10$. In order to achieve a finer resolution in the impingement zone, stretching is applied along the downstream of the mixing channel and the inlets, thus resulting in 15 and 6 million of cells in the fine and medium grid, respectively. As with the previous simulations, the reactants are injected once a statistically steady flow has been established and the mean PSD is extracted after 15 flow-through times.

Two grids were prepared for this investigation; they are called 'fine' and 'medium' resolution in the present section and feature a smallest cell size, δx, of $4 \cdot 10^{-6}$ m and $8 \cdot 10^{-6}$ m respectively. This way, both grids have adequate resolution to resolve the Kolmogorov scale, but the fine resolution grid is also able to resolve the Batchelor scale at $Sc = 10$. In order to reduce the number of cells in the grid, stretching is applied along the downstream of the mixing channel and the inlets, making the resolution finer in the impingement zone. This amounts to 15 and 6 millions of cells in the fine and medium resolution grids, respectively. Regarding resolution of the Kolmogorov scale, the two grids have $\delta x/\eta$ ratios of 1.36 (medium) and 0.68 (fine), while for the Batchelor scale, the corresponding ratios are 4.30 and 2.14, respectively. Three simulations have been performed, with $Sc = 1$ at medium resolution and $Sc = 10$ at both medium and fine resolutions. The $Sc = 1$ case at medium resolution resolves the Kolmogorov scale, which is also the smallest scale of mixing under this assumption. At $Sc = 10$, sub-Kolmogorov scales appear with respect to scalar mixing and the simulation is under-resolved at medium resolution and fully resolved at fine resolution (with respect to the Batchelor scale).

Table 6.7 *Mean iso-surface area of the nucleation consumption rate (dimensionless, normalised by the mixing channel cross-sectional area).*

Case / percentage	20% max	50% max	70% max
Medium, Sc = 1	4.99	3.78	3.20
Medium, Sc = 10	7.63	5.84	5.17
Fine, Sc = 10	9.86	7.67	6.81

Reproduced from Tang et al. (2023) under CC-BY 4.0 license.

Table 6.8 *Mean iso-surface area of the growth consumption rate (dimensionless, normalised by the mixing channel cross-sectional area).*

Case / percentage	20% max	50% max	70% max
Medium, Sc = 1	17.3	11.7	8.45
Medium, Sc = 10	42.8	32.8	15.6
Fine, Sc = 10	37.6	29.5	23.6

Reproduced from Tang et al. (2023) under CC-BY 4.0 license.

Structure of Reaction Zones

At higher Sc, it is expected that the reaction zones will be thinner, a fact that may affect the reaction rates. To examine this effect, Tang et al. (2023) extracted iso-surfaces of the nucleation and growth consumption rates at 30 uncorrelated time instants and at 20%, 50% and 70% of their maximum time-averaged consumption rate. The time-averaged results for the iso-surface area, normalised by the mixing channel cross-sectional area, are shown in Tables 6.7 and 6.8, while an indicative picture is provided in Fig. 6.38 showing instantaneous snapshots for the 70% case. The $Sc = 10$ (fine resolution) case exhibits finer structures and higher iso-surface area for both nucleation and growth, while the changes in the growth iso-surfaces are more noticeable because growth covers a more extensive region. In the $Sc = 10$ but the sub-Kolmogorov scales are not resolved (the medium grid case), the iso-surface areas have an intermediate value between the $Sc = 1$ and $Sc = 10$ fully resolved case. We also notice that the extent of the reaction zones decreases at increasing percentage; at the highest percentage, only the impingement zone features high rates, and the changes are more noticeable in growth than in nucleation because the former extends further into the mixing channel.

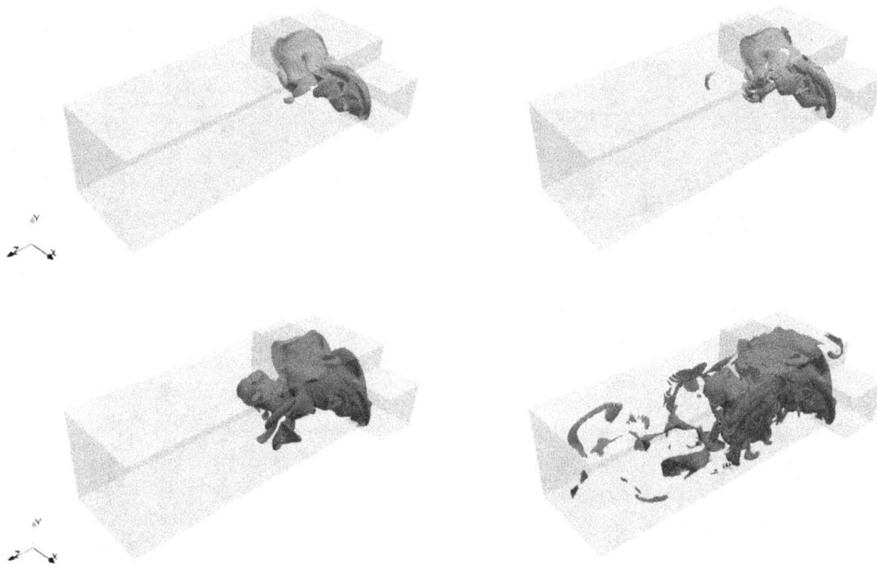

Figure 6.38 Iso-surface of instantaneous nucleation (top) and growth (bottom) consumption at 70% of the maximum at Schmidt number 1 (left) and 10 in the fine resolution (right). Reproduced from Tang et al. (2023) under CC-BY 4.0 license.

A comparison of the instantaneous contours of supersaturation, nucleation consumption rate and growth consumption rate at *Sc* of 1 and 10 (fine resolution) is shown in Fig. 6.39. The thinner zones in the $Sc = 10$ case are clearly observed, but the maximum rates in the $Sc = 1$ case are also lower compared with $Sc = 10$. In other words, the coverage of reaction zones competes with the magnitude of the local reaction rate. Thus, while resolving the Batchelor scale seems necessary in order to get the correct small-scale structure at high *Sc*, the competition between these two mechanisms may result in a small net effect on the important variables for the process outcome, namely the precipitation rates and the PSD, which are examined in the next section.

Precipitation Rates and Particle Size Distribution

In order to quantify the effect of thinner reaction zones at high *Sc* on the process outcome, we first examine the volume-averaged mean reaction rates in the entire domain, obtained by integrating the local mean nucleation and growth consumption rates (Table 6.9). The differences are rather small, regardless of the Schmidt number and grid resolution. This is contrasted with the considerable difference in the iso-surface area observed in the previous section, but can be explained by the fact that, while reactions are confined to thinner zones with larger iso-surface area,

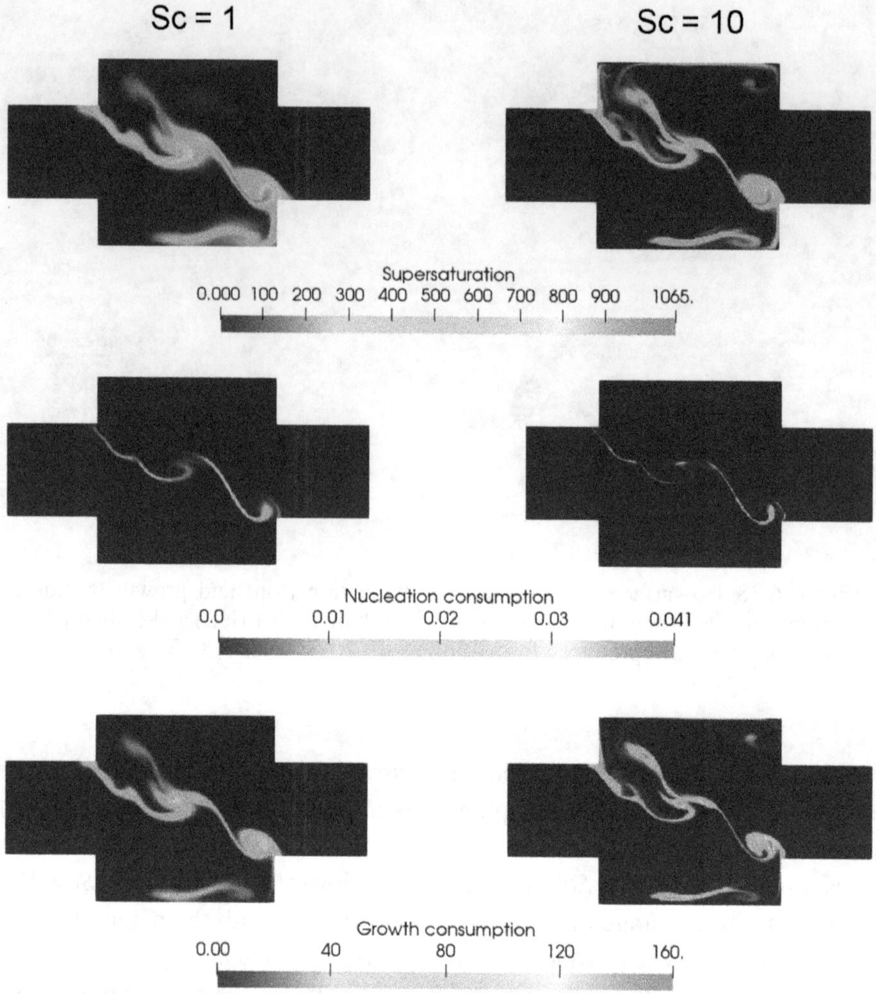

Figure 6.39 Instantaneous supersaturation (top), nucleation consumption rate (middle) and growth consumption rate (bottom) contours at $Sc = 1$ (left) and $Sc = 10$ in the fine resolution(right). Reproduced from Tang et al. (2023) under CC-BY 4.0 license.

greater local rates of consumption and kinetics arise within these structures, as evidenced in Fig. 6.39.

We now turn our attention to the difference in the exit plane-averaged PSD between the three cases, shown in Fig. 6.40. No major differences are evident between all cases. The two $Sc = 10$ cases, in particular, provide almost identical results, while the $Sc = 1$ case exhibits a small underprediction around the peak and a small skew towards larger sizes. Tang et al. (2023) also examined the evolution

Table 6.9 *Volume-averaged mean consumption rate (kmol/(m³ · s)).*

	Nucleation	Growth
Medium, Sc = 1	$1.692 \cdot 10^{-4}$	16.3763
Medium, Sc = 10	$1.752 \cdot 10^{-4}$	15.8275
Fine, Sc = 10	$1.673 \cdot 10^{-4}$	15.9995

Reproduced from Tang et al. (2023) under CC-BY 4.0 license.

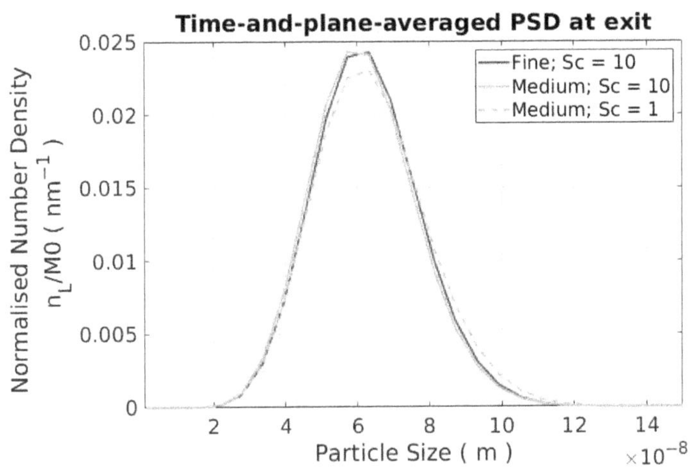

Figure 6.40 Plane-averaged particle size distribution. Reproduced from Tang et al. (2023) under CC-BY 4.0 license.

of the zeroth and first moment along the mixing channel and found that the differences are relatively minor, though the $Sc = 1$ case results in an underprediction of the zeroth moment and an overprediction in the first one. This suggests that higher Sc slightly favours nucleation over growth due to the more confined high supersaturation zones. It must be noted that the comparison of the exit PSD with experiments (Fig. 6.27) also exhibited a minor underprediction of the peak and a slight skew towards larger sizes. However, it is not possible to ascribe this with confidence to the effect of Sc, as uncertainties in kinetics and possibly experimental measurements are likely to be more significant. We should also note that the TVD scheme introduces numerical diffusion at high Sc, hence further increase in Sc may not improve the solution.

6.3.8 Concluding Remarks

The third case study involved a liquid-phase flow with particle formation, where there were only two kinetic processes present, nucleation and growth. While the problem involved turbulent flow, a DNS approach was adopted that circumvents the need for modelling turbulence–chemistry and turbulence–PBE interaction. The purpose of this case study was to examine how fluid dynamics interacts with the kinetic processes in the PBE, how this interaction shapes the evolution of the PSD and what are the implications for modelling with RANS or LES.

The first problem that was simulated was an experiment of $BaSO_4$ precipitation in a T-mixer reported by Schwarzer et al. (2006). While DNS resolves the flow field, the DNS-PBE simulation involves kinetic models. Furthermore, a problem particular to liquid mixing is the high Schmidt number and the sub-Kolmogorov scales that result; these scales cannot be resolved with present-day computational resources, and $Sc = 1$ was used in this simulation. Therefore, the results must be validated with experiments. The exit PSD in the T-mixer was predicted to a good level of accuracy, indicating that the kinetic processes were properly accounted for by the DNS-PBE approach.

Afterwards, an investigation was conducted about the interaction of turbulence with nucleation and growth. The balance between these processes determines the evolution of the PSD, as nucleation gives rise to a large number of small crystals, but also features a highly nonlinear dependence on supersaturation which makes it very intermittent. It was found that nucleation takes place in bursts within thin zones. The Kolmogorov scale is smaller than these zones but comparable, which implies that chemistry and flow cannot be decoupled in precipitation models. The interaction between mixing and precipitation was also analysed by defining separate Damköhler numbers for nucleation and growth. It was found that, on the whole, the process is neither mixing nor kinetics controlled.

The DNS-PBE approach was also used to shed light on the closure problem associated with the fluctuating nucleation and growth rates. The kinetic rates approximated with mean concentration were compared with the mean rates as evaluated by DNS-PBE and significant differences were found, especially in intense precipitation regions. The number density-growth correlations were investigated, and it was found that the growth term can be considerably underestimated if fluctuations are neglected, resulting in an underprediction of the number density at the lower end of the size spectrum and an overprediction of supersaturation; the latter would also bias the PSD towards smaller sizes by favouring nucleation over growth. On the other hand, the effect of fluctuations on the reactant consumption term was found to be minimal, due to a compensation between positive and negative fluctuation terms.

Subsequently, the number density-growth correlations were investigated. It was shown that negative correlations are found in areas of high mean growth rate since particle growth inherently relocates the number densities from lower intervals to higher intervals, causing number density reduction and hence negative correlation terms. On the other hand, nucleation generates particles and results in positive correlations at small size intervals. The gradient of these correlations in the particle size domain provides the fluctuation contribution in the PBE source term, whose magnitude was found to be significant for the PSD evolution. However, the effect of fluctuations on the reactant consumption term was found to be minimal, due to a compensation between positive and negative fluctuation terms.

The problem of high Schmidt number has been a well known challenge in the modelling of mixing in liquid phase flow and is highly relevant to many chemical engineering applications. In the present study, an attempt was made to shed light into this problem by performing simulations of precipitation in a T-mixer at $Re = 400$ (the lowest possible value to achieve turbulent flow in a T-mixer) at $Sc = 1$ and $Sc = 10$. The results suggest that the sub-Kolmogorov scales have two competing effects. The particle formation zones are more confined at high Schmidt number, but also feature more intense precipitation rates, and these two effects compete with each other, resulting in a very small effect on the PSD. Higher Sc slightly favours nucleation over growth due to the more confined high supersaturation zones, which might explain the small discrepancy between simulations and experiments in the Schwarzer et al. (2006), but other uncertainties (kinetics and possibly experimental measurements) are likely to be more significant. The problem of high Sc flows requires more research in the future, as better computational resources become available.

Overall, this last case study highlighted the interactions of flow and kinetic processes in turbulent flow and tackled the issue of high Sc which is particular to liquid-phase flows.

Appendix A
Coupling of PBE with Fluid Flow, Heat and Mass Transfer

In many problems, the population balance must be coupled with the equations of fluid dynamics, thermodynamics and transport phenomena. In the present appendix, we summarise the main transport equations that are used throughout the book and show how they should be augmented for the coupling with the PBE. Particular emphasis is placed on the equations for multicomponent reacting flows, as they appear in numerous population balance problems such as nanoparticle synthesis, soot formation and precipitation, examples of which are shown in Chapter 6.

The transport equations are based on the physical principles of conservation of mass, momentum and energy. When these equations are coupled with the population balance, additional terms may need to be included. In the problems considered in the present book, the flows are dilute and the exchange of momentum between the particulate phase and the fluid can be neglected. The main additional terms required are thus source terms in the species transport equations to account for the exchange of mass between species and particles. In soot and gas-phase nanoparticle synthesis, radiative heat transfer from the particles should also be included, while there may be an enthalpy change associated with the particle formation or oxidation.

A.1 Mass Transport

In single-component flows, conservation of mass is expressed by the continuity equation. In multicomponent flows, individual conservation equations can be written for each species, which include diffusion and possibly source terms. Constitutive equations must be provided for the diffusion fluxes. The source terms account for chemical reactions and population balance kinetic processes that consume or release species (i.e. nucleation and growth). Examples of this linking between species transport and population balance equations can be found in all the case studies in Chapter 6.

A.1.1 Continuity Equation

The principle of conservation of mass for a single-component flow or a mixture of components is expressed by the continuity equation. For variable density flows, the appropriate form is:

$$\frac{\partial \rho}{\partial t} + \frac{\partial (\rho u_i)}{\partial x_i} = 0. \tag{A.1}$$

The derivation of this equation can be found in many sources, such as Bird et al. (1960). For constant density flows, the continuity equation is further simplified as follows:

$$\frac{\partial u_i}{\partial x_i} = 0. \tag{A.2}$$

A.1.2 Multicomponent Mixtures

In a multicomponent mixture, mass transport can be written for each individual species, the sum of which yields the continuity equation. Mass transport equations may be written in terms of mole-based or mass-based quantities. Mole-based quantities are more naturally suited to chemical kinetics, but mass-based quantities are preferable when dealing with variable density flows. The mass fraction of a species, Y_α, is related to its concentration, C_α, in the following way:

$$Y_\alpha = \frac{m_\alpha}{m_{mix}} = \frac{C_\alpha MW_\alpha}{\rho}, \tag{A.3}$$

where MW_α is the molecular weight of species a. The mass fractions must satisfy the following condition:

$$\sum_{a=1}^{N} Y_\alpha = 1. \tag{A.4}$$

Mass transport equations can be straightforwardly written for the species mass density, defined as:

$$\rho_\alpha = \rho Y_\alpha. \tag{A.5}$$

The sum of all mass densities is equal to the mixture density. The conservation of mass equation for an individual species can now be written as follows:

$$\frac{\partial \rho_\alpha}{\partial t} + \frac{\partial (\rho_\alpha u_{a,i})}{\partial x_i} = \dot{w}_{a,r} + \dot{w}_{a,PBE}, \tag{A.6}$$

or

$$\frac{\partial (\rho Y_\alpha)}{\partial t} + \frac{\partial (\rho Y_\alpha u_{a,i})}{\partial x_i} = \dot{w}_{a,r} + \dot{w}_{a,PBE}, \tag{A.7}$$

where $\dot{w}_{a,r}$ and $\dot{w}_{a,PBE}$ are source terms (in mass per unit volume) due to chemical reactions and population balance, respectively. The velocity $u_{a,i}$ is the sum of the mixture velocity and the *diffusion velocity*, $V_{a,i}$, which is the velocity with which an individual species moves relative to the mixture due to diffusion:

$$u_{a,i} = u_i + V_{a,i}. \tag{A.8}$$

The species transport equation may now be written as:

$$\frac{\partial(\rho Y_\alpha)}{\partial t} + \frac{\partial(\rho u_i Y_\alpha)}{\partial x_i} = -\frac{\partial j_{a,i}}{\partial x_i} + \dot{w}_{a,r} + \dot{w}_{a,PBE}, \tag{A.9}$$

where $j_{a,i}$ is the diffusive mass flux, which can be expressed in terms of the diffusion velocity as follows:

$$j_{a,i} = \rho V_{a,i} Y_\alpha. \tag{A.10}$$

In the absence of the PBE source term, the sum of all species transport equations must yield the continuity equation for the mixture. This can be readily verified for the accumulation and convection terms. Additionally, the diffusion velocities must satisfy:

$$\sum_{a=1}^{N} (Y_\alpha V_{a,i}) = 0, \tag{A.11}$$

which expresses the fact that diffusion should not result in bulk motion and is satisfied when the most rigorous diffusion laws are used, but remedies may have to be sought for simplified expressions, as we will see in the next section. The reaction source terms must sum to zero for a valid reaction stoichiometry:

$$\sum_{a=1}^{N} \dot{w}_{a,r} = 0. \tag{A.12}$$

It must be noted that $\dot{w}_{a,r}$ does not include the reactions that result in mass exchange between continuous and particulate phase (nucleation and growth).

In the presence of the PBE source term, there is an exchange of mass between the species and particulate phase and therefore the sum of the species conservation equations does not yield the mixture continuity equation. However, the overall mass balance must still be satisfied when considering the change in the moment of the distribution that is indicative of particle volume and the PBE source terms in the species transport equations.

The mass-based conservation of mass equation, Eq. A.9, is used for the variable density problems in Section 6.2. For constant density problems, such as the precipitation process in Section 6.3, the following form of species transport equation in terms of concentrations is employed:

$$\frac{\partial C_\alpha}{\partial t} + \frac{\partial (u_i C_\alpha)}{\partial x_i} = -\frac{\partial J^*_{a,i}}{\partial x_i} + \dot{R}_{a,r} + \dot{R}_{a,PBE}, \tag{A.13}$$

where $J^*_{a,i}$, $\dot{R}_{a,r}$ and $\dot{R}_{a,PBE}$ are the diffusion fluxes,[1] reaction source terms and PBE source terms, respectively, written in terms of concentrations.

A.1.3 The Diffusion Flux

Expressions for the diffusion flux in multicomponent mixtures can take rather complicated forms. In the following, we will discuss the basis for the common forms encountered in population balance – species transport problems such as the ones considered in Chapter 6. More details on this subject can be found in Bird et al. (1960), Hirschfelder et al. (1964) or Kee et al. (2003).

The simplest case is that of diffusion in a binary system, where the diffusion flux can be described by Fick's law. If the mole fractions of the two species are X_A and X_B, the flux is:

$$j_{A,i} = -C_{tot} D_{AB} \frac{\partial X_A}{\partial x_i}, \tag{A.14}$$

where D_{AB} is the diffusion coefficient for the mixture of species A and B and C_{tot} is the total concentration. In terms of the mass fractions Y_A and Y_B, we can write:

$$j_{A,i} = -\rho D_{AB} \frac{\partial Y_A}{\partial x_i}. \tag{A.15}$$

Similar equations can be written for species B.

In the case of multicomponent mixtures, separate diffusion coefficients must be defined for each pair of species. Differential diffusion (i.e. different diffusion rates for different species) is more pronounced when the species have widely varying molecular weights. The most rigorous way of accounting for diffusion in multicomponent mixtures is via the Stefan–Maxwell equations (Bird et al., 1960; Kee et al., 2003). Implementation of the Stefan–Maxwell equations in CFD codes is difficult, however, because they require the solution of a system of linear equations. An approximation often used in the computation of laminar flames is the Hirschfelder and Curtiss approximation (Hirschfelder et al., 1964), which yields the diffusion velocities via a form similar to Fick's law. The condition that the sum of diffusion fluxes must be equal to zero is not satisfied when approximate forms are used, however. There are several ways to remedy this issue, such as by applying a velocity correction (Poinsot and Veynante, 2005). There are also further physical

[1] The notation indicates that the diffusion velocity is expressed in terms of the molar average velocity; see Bird et al. (1960) for more details.

effects that have not been considered here, such as pressure effects, body forces and diffusion due to temperature gradients (Soret effect).

In general, the choice of diffusion law depends on its relative importance compared to other aspects of the problem and the type of flow considered. In turbulent flows, molecular diffusion is not a dominant effect and Fick's law with a single diffusion coefficient is usually employed. This assumption can be based on the fact that a species is often present in excess (e.g. N_2 in the case of combustion), and the species are assumed to diffuse into the mixture with a single diffusion coefficient. Laminar flows require more detailed diffusion laws, and the difference in the diffusivities of species such as hydrogen (which has a very small molecular weight) can be important.

In the problems considered in the present book, multicomponent diffusion is modelled with Fick's law, written as[2]:

$$j_\alpha = -\rho D_\alpha \frac{\partial Y_\alpha}{\partial x_i}. \tag{A.16}$$

By substituting Eq. A.16 into Eq. A.9, we obtain:

$$\frac{\partial(\rho Y_\alpha)}{\partial t} + \frac{\partial(\rho u_i Y_\alpha)}{\partial x_i} = \frac{\partial}{\partial x_i}\left(\rho D_\alpha \frac{\partial Y_\alpha}{\partial x_i}\right) + \dot{w}_{a,r} + \dot{w}_{a,PBE}, \tag{A.17}$$

which is the species transport equation for a binary mixture. Note that diffusivities depend on temperature and may thus vary with space. For constant density flows, the following equation in terms of concentration may be preferable:

$$\frac{\partial C_\alpha}{\partial t} + \frac{\partial(u_i C_\alpha)}{\partial x_i} = \frac{\partial}{\partial x_i}\left(D_\alpha \frac{\partial C_\alpha}{\partial x_i}\right) + \dot{R}_{a,r} + \dot{R}_{a,PBE}. \tag{A.18}$$

A.1.4 The Reaction Source Term

For reacting flows, the chemical reaction source can be written as follows:

$$\dot{R}_{a,r} = \sum_{b=1}^{M} \nu_{ab} r_b(\mathbf{C}, T), \tag{A.19}$$

where M is the number of reactions, ν_{ab} is the stoichiometric coefficient of species a in reaction b and r_b is the reaction rate of reaction b, which depends on the concentrations and temperature. Reaction kinetics will not be covered here; one may consult e.g. Kee et al. (2003) or Froment et al. (2011) for more details.

[2] Note that in Eq. A.16 and elsewhere, no summation on the species index is implied.

When the species transport equations are formulated in terms of mass fractions, the species mass production rate can be written as follows:

$$\dot{w}_\alpha = MW_\alpha \sum_{b=1}^{M} \nu_{ab} r_b(\mathbf{Y}, T), \tag{A.20}$$

where the reaction rate has been expressed in terms of the mass fractions, which are related to the concentrations according to Eq. A.3.

A.1.5 The Population Balance Kinetics Source Term

The source due to the population balance is associated with the nucleation and growth terms, since aggregation and breakage do not involve transfer of mass between the continuous and particulate phases.

Consider a reaction of a species, A, possibly with other species, leading to the formation of a particulate phase, C_s, via nucleation and/or growth. This reaction can be written in the following way:

$$\nu_{a,PBE} A + \ldots \rightarrow C_s,$$

where $\nu_{a,PBE}$ is the stoichiometric coefficient of A in this reaction. The PBE source term for species A can thus be formulated based on the nucleation and growth rates, as follows:

$$R_{\alpha,PBE} = \frac{\nu_{a,PBE}\rho_p}{MW_p} \left[B v_0 + \int_0^\infty v \frac{\partial (Gn)}{\partial v} dv \right], \tag{A.21}$$

where ρ_c and MW_p are the volume and molecular weight of the particles, respectively. The first term in the brackets is the total volume of the nuclei produced, while the second term is the volume of particulate phase produced by growth.

A.2 Momentum Transport

The flows in the population balance problems considered in the present book are viscous flows, often with variable density. The appropriate form of momentum transport equations are therefore the Navier–Stokes equations:

$$\frac{\partial (\rho u_i)}{\partial t} + \frac{\partial (\rho u_j u_i)}{\partial x_j} = -\frac{\partial p}{\partial x_i} + \frac{\partial \tau_{ij}}{\partial x_j} + \rho g_i, \tag{A.22}$$

where g_i is acceleration of gravity and τ_{ij} is the stress tensor:

$$\tau_{ij} = \mu \left(\frac{\partial u_i}{\partial x_j} + \frac{\partial u_j}{\partial x_i} \right) - \frac{2}{3} \mu \frac{\partial u_k}{\partial x_k} \delta_{ij}, \tag{A.23}$$

where μ is the viscosity of the fluid. A second parameter, the bulk viscosity, has been omitted as it is usually negligible in the problems considered in the present book. The derivation of the Navier–Stokes equations can be found in numerous references, such as Schlichting (1979) and White (2006).

The Navier–Stokes equations can also be written in the following form:

$$\rho\frac{\partial u_i}{\partial t} + \rho u_j\frac{\partial u_i}{\partial x_j} = -\frac{\partial p}{\partial x_i} + \frac{\partial \tau_{ij}}{\partial x_j} + \rho g_i. \tag{A.24}$$

This form emphasizes the point of view of a moving fluid element and is called the *non-conservation form*, as opposed to the *conservation form* of Eq. A.22. It must be stressed that both forms are equivalent, as can be shown by substituting the continuity equation into Eq. A.22.

In variable density nonisothermal flows, the Navier–Stokes equations are coupled with the energy transport equation via the density, which changes with heat release, and the viscosity, which is temperature dependent. Furthermore, the pressure is related to temperature and density via an equation of state, usually the ideal gas law.

For constant density and viscosity, the following simplified form is obtained:

$$\frac{\partial u_i}{\partial t} + \frac{\partial (u_j u_i)}{\partial x_j} = -\frac{1}{\rho}\frac{\partial p}{\partial x_i} + \nu\frac{\partial^2 u_i}{\partial x_j \partial x_j} + g_i, \tag{A.25}$$

where $\nu = \mu/\rho$ is the kinematic viscosity. The equivalent non-conservation form is:

$$\frac{\partial u_i}{\partial t} + u_j\frac{\partial u_i}{\partial x_j} = -\frac{1}{\rho}\frac{\partial p}{\partial x_i} + \nu\frac{\partial^2 u_i}{\partial x_j \partial x_j} + g_i. \tag{A.26}$$

In constant density flows, there is no equation of state relating pressure with density and temperature, hence the pressure must be implicitly calculated by the coupled continuity and Navier–Stokes equations.

A.3 Energy Transport

The energy transport equation is derived from the first law of thermodynamics and can take various forms. It is required for non-isothermal flows. Several of the problems in the present book involve reacting flows with multicomponent mixtures and heat release due to chemical reactions, hence the appropriate forms of the equations will be briefly discussed below. More details can be found in Bird et al. (1960), Kee et al. (2003) and Poinsot and Veynante (2005).

Before formulating an energy transport equation, it must first be decided what will be the transported quantity, as several forms of energy may enter the equation as either parts of the transported quantity or source terms. In the present section, the quantity to be considered is the sum of chemical and thermal energies, which will be called the *absolute* energy and denoted as e for specific internal energy or h for specific enthalpy. The consequence of this choice is that the heat of reaction does not appear as a source term.

The main source terms in the energy transport equation that are associated with the population balance are radiation from particles (in problems such as soot formation and gas-phase nanoparticle synthesis) and enthalpy of condensing or oxidising species.

A.3.1 The First Law of Thermodynamics

The starting point for the derivation of the energy transport equation is the first law of thermodynamics for a fluid flow, which can be written as follows:

$$\rho \frac{\partial}{\partial t}\left(e + \frac{u_i^2}{2}\right) + \rho u_j \frac{\partial}{\partial x_j}\left(e + \frac{u_i^2}{2}\right) = \dot{q} - \dot{w}. \tag{A.27}$$

In this form, the transported quantity is the sum of the specific internal (which includes chemical and thermal) and kinetic energy. Other forms of energy as well such as potential, electromagnetic or nuclear energy may need to be considered in some problems and may be included either as transported variables or as heat or work sources.

The heat sources in Eq. A.27 usually include heat conduction and radiation. The work term includes the work by surface forces (pressure and viscous forces) and by body forces, such as gravity. Heat release due to chemical reactions does not appear as a source since the chemical energy is included in the specific internal energy; reaction simply converts energy from chemical to thermal. Note, however, that it would appear as a source if a transport equation for thermal energy alone was to be written. By introducing expressions for the aforementioned sources, the transport equation becomes:

$$\rho \frac{\partial}{\partial t}\left(e + \frac{u_i^2}{2}\right) + \rho u_j \frac{\partial}{\partial x_j}\left(e + \frac{u_i^2}{2}\right) = -\frac{\partial q_j}{\partial x_j} + \dot{q}_{rad} - \frac{\partial (p u_j)}{\partial x_j} + \frac{\partial (\tau_{ij} u_i)}{\partial x_j} + \rho g_j u_j, \tag{A.28}$$

where q_j is the heat conduction flux vector and $\dot{q}_{rad}$ is the source due to radiative heat transfer. The last three terms represent the work by pressure, shear and gravitational forces, respectively.

A.3.2 The Transport Equation for Chemical and Thermal Energy

We now wish to write a transport equation for specific energy that includes chemical and thermal but not kinetic energy. To this end, we first derive a transport equation for specific kinetic energy by multiplying the momentum equation, Eq. A.24, with the velocity:

$$\rho\frac{\partial}{\partial t}\left(\frac{u_i^2}{2}\right)+\rho u_j\frac{\partial}{\partial x_j}\left(\frac{u_i^2}{2}\right)=-u_i\frac{\partial p}{\partial x_i}+u_i\frac{\partial \tau_{ij}}{\partial x_j}+u_i\rho g_i. \tag{A.29}$$

By subtracting the last equation from Eq. A.28, we arrive at the transport equation for specific internal energy:

$$\rho\frac{\partial e}{\partial t}+\rho u_j\frac{\partial e}{\partial x_j}=-\frac{\partial q_j}{\partial x_j}+\dot{q}_{rad}-p\frac{\partial u_j}{\partial x_j}-\tau_{ij}\frac{\partial u_i}{\partial x_j}. \tag{A.30}$$

The penultimate term on the right-hand side denotes the reversible thermodynamic work due to expansion or compression of the fluid element, while the last term represents irreversible dissipation of mechanical energy into heat by viscous forces.

It is often more convenient to employ an equation for the specific enthalpy, defined as:

$$h=e+\frac{p}{\rho}. \tag{A.31}$$

From this expression, we have:

$$\rho\frac{\partial h}{\partial t}+\rho u_j\frac{\partial h}{\partial x_j}=\rho\frac{\partial e}{\partial t}+\rho u_j\frac{\partial e}{\partial x_j}+\frac{\partial p}{\partial t}-\frac{p}{\rho}\frac{\partial \rho}{\partial t}+u_j\frac{\partial p}{\partial x_j}-\frac{pu_j}{\rho}\frac{\partial \rho}{\partial x_j}. \tag{A.32}$$

Using the continuity equation (Eq. A.1), we have:

$$\rho\frac{\partial h}{\partial t}+\rho u_j\frac{\partial h}{\partial x_j}=\rho\frac{\partial e}{\partial t}+\rho u_j\frac{\partial e}{\partial x_j}+\frac{\partial p}{\partial t}+\frac{\partial (pu_j)}{\partial x_j}. \tag{A.33}$$

By combining this equation with Eq. A.30, we now obtain:

$$\rho\frac{\partial h}{\partial t}+\rho u_j\frac{\partial h}{\partial x_j}=-\frac{\partial q_j}{\partial x_j}+\dot{q}_{rad}+\frac{\partial p}{\partial t}+u_j\frac{\partial p}{\partial x_j}-\frac{\partial (\tau_{ij}u_i)}{\partial x_j}+\rho g_j u_j. \tag{A.34}$$

The pressure gradients in the right-hand side can be important in high-speed flows and internal flows, but they can be neglected in open flames where the pressure is roughly constant, such as the problems treated in Chapter 6. If the viscous dissipation and body forces are also neglected, we arrive at the following simplified form:

$$\rho\frac{\partial h}{\partial t}+\rho u_j\frac{\partial h}{\partial x_j}=-\frac{\partial q_j}{\partial x_j}+\dot{q}_{rad}. \tag{A.35}$$

A.3.3 Heat Conduction

Heat conduction can be described by Fourier's law, so the corresponding flux is:

$$q_j = -k\frac{\partial T}{\partial x_j}, \tag{A.36}$$

where k is the thermal conductivity. This flux can be expressed in terms of internal energy or enthalpy as follows:

$$q_j = -\frac{k}{\rho c_v}\frac{\partial e}{\partial x_j} = -\frac{k}{\rho c_p}\frac{\partial h}{\partial x_j}, \tag{A.37}$$

where c_v and c_p are the specific heat capacities under constant volume and pressure, respectively.

A.3.4 Radiative Heat Transfer

In high temperature gas flows with particles, such as the flames considered in Section 6.2, the radiation from particles is important and also provides for a heat exchange mechanism between the species and the particles. Models of radiative heat transfer will not be discussed here in detail; the reader may consult Modest and Mazumder (2022) for a comprehensive exposition of radiative heat transfer or Coelho (2007) for a focus on the models used in turbulent reacting flows. We will only show here the optically thin model, as it is very commonly used in flame simulations and is employed in Section 6.2.

For species, it is sufficient to consider the most abundant ones such as H_2O, CO_2, CH_4 and CO. The equation is:

$$\dot{q}_{rad} = -\frac{4\sigma}{\rho}(T^4 - {T_b}^4)\sum_i a_{p,i}p_i, \tag{A.38}$$

where σ is the Stefan–Boltzmann constant, T_b is the ambient background temperature, p_i is the partial pressure of species i and $a_{p,i}$ are the Planck mean absorption coefficients expressed as polynomials in temperature (Grosshandler, 1993).

A similar equation is used for radiation from particles:

$$\dot{q}_{rad,p} = -\frac{4\sigma}{\rho}C_s f_v(T^5 - T^5_{amb}), \tag{A.39}$$

where C_s is a proportionality constant.

A.3.5 Multicomponent and Reacting Flows

In multicomponent and reacting flows, such as those considered in Section 6.2, enthalpy changes are brought by diffusion and chemical reaction. As mentioned, the energy transport equation has been formulated here in terms of absolute energy including the standard enthalpy of formation, hence the enthalpy of reaction does not appear as a source of heat. However, transport of species with different standard enthalpies of formation, as well as different specific heat capacities, must be accounted for.

The specific enthalpy of a chemical species, h_α, can be written as a sum of its standard enthalpy of formation, h^o_α, and its sensible enthalpy as follows:

$$h_\alpha = h^o_\alpha + c_{p,\alpha} T, \tag{A.40}$$

where $c_{p,\alpha}$ is the specific heat capacity of species a, such that:

$$c_p = \sum_{a=1}^{N} Y_\alpha c_{p,\alpha}. \tag{A.41}$$

We can thus write the following expression for the enthalpy of the mixture:

$$h = \sum_{a=1}^{N} Y_\alpha \left(h^o_\alpha + c_{p,\alpha} T\right). \tag{A.42}$$

Assuming diffusion to be described by Fick's law (Eq. A.16), we have an additional heat flux term arising from the diffusion of species with different specific enthalpies. Therefore, the total heat flux term can be written as:

$$q_{a,j} = -\frac{Y_\alpha k}{\rho_\alpha c_p} \frac{\partial h_\alpha}{\partial x_j} - h_\alpha \rho D_\alpha \frac{\partial Y_\alpha}{\partial x_j}. \tag{A.43}$$

We now introduce this flux into Eq. A.35, under the same assumptions that apply to that equation, to obtain:

$$\rho \frac{\partial h}{\partial t} + \rho u_j \frac{\partial h}{\partial x_j} = \frac{\partial}{\partial x_j}\left(\frac{Y_\alpha k}{\rho_\alpha c_p} \frac{\partial h_\alpha}{\partial x_j} + h_\alpha \rho D_\alpha \frac{\partial Y_\alpha}{\partial x_j}\right) + \dot{q}_{rad}. \tag{A.44}$$

Appendix B

Implementation of the Conservative Finite Volume Discretisation Method

The present appendix provides some more details regarding the implementation of the conservative finite volume method for the solution of the population balance equation, the main elements of which were discussed in Section 4.2.8. This method is used in the case studies in Chapter 6. Below, we show the details of the construction of the aggregation and fragmentation maps, the time advancement and the evaluation of integrals.

B.1 Construction of Aggregation and Fragmentation Map

B.1.1 Aggregation Map

The starting point for obtaining the final discretised equations is Eq. 4.73, where the aggregation birth term was decomposed into two terms that have the same form as the death term. The treatment of the death term is straightforward, as it may be written as a sum according to 4.74. Thus, Eq. 4.73 may be rewritten as follows:

$$\begin{aligned}\hat{v}_i \frac{dn_i(t)}{dt} = & \int_{v_{i-1}-w}^{v_i-w} \int_{v_{min}}^{\frac{u+w}{2}} u\beta(w',u)n(w',t)n(u,t)dw'du \\ & + \int_{v_{i-1}-w}^{v_i-w} \int_{v_{min}}^{\frac{u+w}{2}} w'\beta(w',u)n(w',t)n(u,t)dw'du \\ & - n_i(t) \sum_{j=1}^{m} \left(n_j(t) \int_{v_{i-1}}^{v_i} \int_{w_{j-1}}^{w_j} v\beta(v,w)dwdv \right). \end{aligned} \tag{B.1}$$

The task now is to obtain the discretised forms of the two aggregation birth integrals. Fig. B.1 shows their domain of integration (which is the same for both of them) in *w*-*u* space. These two integrals correspond to the birth of particles within interval Δv_i in the *v*-space of daughter particles.

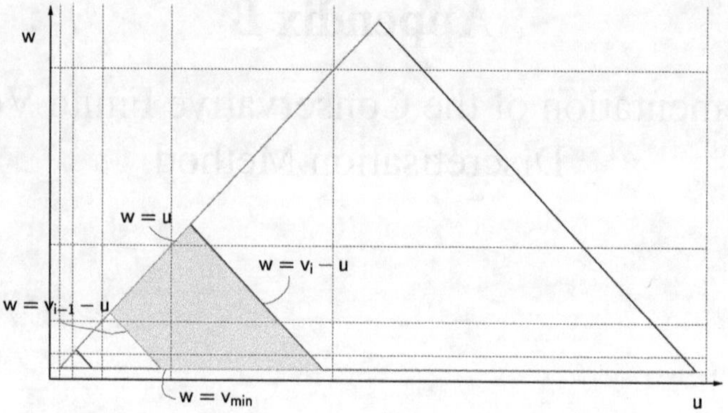

Figure B.1 Domain of integration (shaded) for the birth integrals. The dotted lines represent the grid. Reproduced from O'Sullivan and Rigopoulos (2022) under CC-BY 4.0 license.

The main difficulty in constructing the aggregation map for an arbitrary grid is that, as can be seen from Fig. B.1 the boundaries of the integration domain do not coincide with the grid nodes. Therefore, the integration domain is going to be further partitioned into subdomains over which the number density products are constant. We introduce an *auxiliary grid* of *complementary nodes* defined as follows:

$$U_1 = v_{i-1} - w_j, \tag{B.2a}$$

$$U_2 = v_{i-1} - w_{j-1}, \tag{B.2b}$$

$$U_3 = v_i - w_j, \tag{B.2c}$$

$$U_4 = v_i - w_{j-1}. \tag{B.2d}$$

The auxiliary grid is depicted in Fig. B.2. Its nodes indicate the volumes of complementary particles in the u-axis that combine with particles in the w-axis to form daughter particles at the boundaries of Δv_i. Since a unique set of complementary nodes is required for every combination of Δv_i and Δw_j, the complementary nodes can be arranged into a two-dimensional array, $U_k(\Delta v_i, \Delta w_j)$. Fig. B.2 shows the domain of integration for a single combination of Δv_i and Δw_j, as well as the auxiliary grid corresponding to this particular combination. Finally, Fig. B.3 shows the fully partitioned aggregation map.

The partitioning of the birth space, together with the approximation of the distribution via pairwise constant functions, makes it now possible to split the birth term into a sum of terms involving products of constant number densities. To illustrate the procedure, we consider the grid shown in Fig. B.4. Three shapes can be identified, indicated as A, B_1 and B_2, and the birth term integral over subdomain A takes the following form:

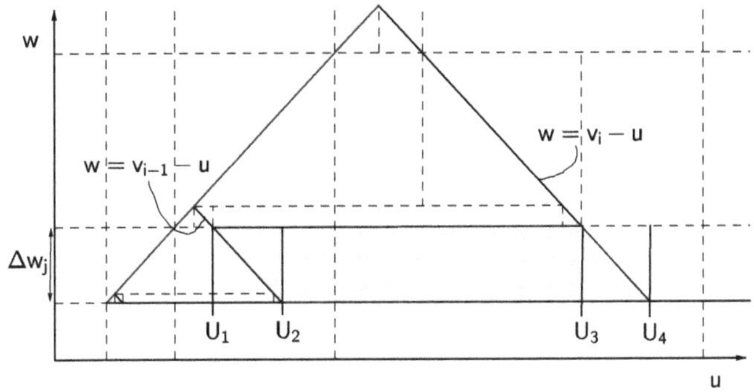

Figure B.2 A single combination of Δv_i and Δw_j leading to an auxiliary grid $U_k(\Delta v_i, \Delta w_j)$. The shaded region indicates the domain of integration for the given combination of intervals, which is part of the domain of integration for the birth term in Δv_i. Reproduced from O'Sullivan and Rigopoulos (2022) under CC-BY 4.0 license.

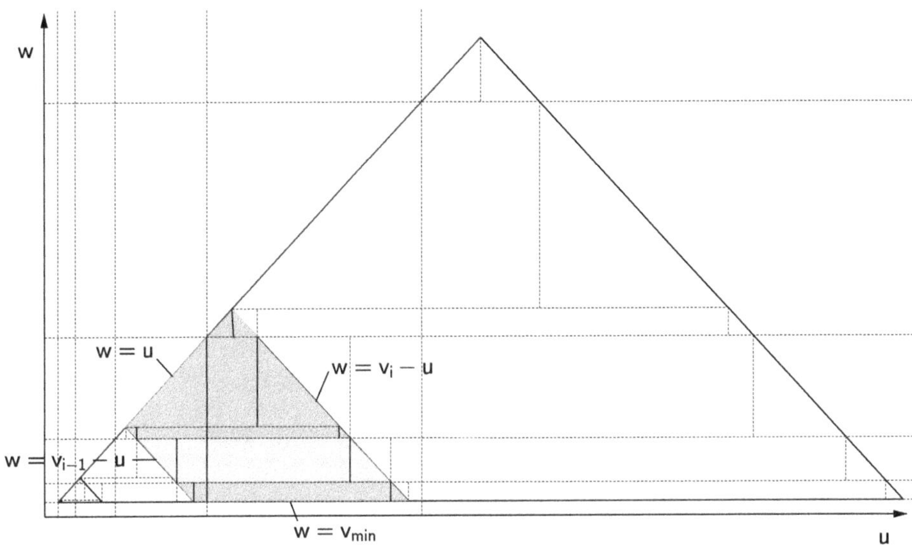

Figure B.3 The fully partitioned aggregation map, showing all possible daughter intervals. The entire shaded region indicates the domain of integration depicted in Fig. B.1, while the lighter shaded region corresponds to the domain illustrated in Fig. B.2. Reproduced from O'Sullivan and Rigopoulos (2022) under CC-BY 4.0 license.

$$I_A = n_j(t)n_k(t)\int_{w_{j-1}}^{w_j}\int_{v_{i-1}-w'}^{U_2} u\beta(w',u)dudw'$$
$$+ n_j(t)n_k(t)\int_{w_{j-1}}^{w_j}\int_{v_{i-1}-w'}^{U_2} w'\beta(w',u)dudw'. \tag{B.3}$$

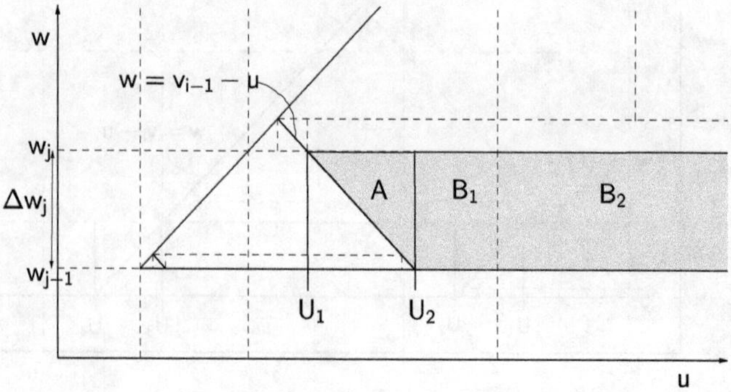

Figure B.4 Close view of an individual combination of parent intervals and associated sub-partitioning of the rectangular middle section. The vertical dotted line separating B_1 and B_2 corresponds to u_k, which is a further sub-partitioning node used in Eq. B.4b. Reproduced from O'Sullivan and Rigopoulos (2022) under CC-BY 4.0 license.

The rest of the shaded region in Fig. B.4 is crossed by a grid node boundary. Therefore, that part of the domain is split into the two subdomains B_1 and B_2, each corresponding to a different product of number densities:

$$\begin{aligned} I_{B_1} &= n_j(t)n_k(t)\int_{w_{j-1}}^{w_j}\int_{U_2}^{u_k} u\beta(w',u)dudw' \\ &+ n_j(t)n_k(t)\int_{w_{j-1}}^{w_j}\int_{U_2}^{u_k} w'\beta(w',u)dudw', \end{aligned} \tag{B.4a}$$

$$\begin{aligned} I_{B_2} &= n_j(t)n_{k+1}(t)\int_{w_{j-1}}^{w_j}\int_{u_k}^{U_3} u\beta(w',u)dudw' \\ &+ n_j(t)n_{k+1}(t)\int_{w_{j-1}}^{w_j}\int_{u_k}^{U_3} w'\beta(w',u)dudw'. \end{aligned} \tag{B.4b}$$

In order to generalise this procedure, let S represent the domain of integration in the aggregation birth term for interval i. This domain is partitioned into $P(i)$ subdomains, each of them denoted by S_p where $p \in (1,2,3,\ldots,P(i))$, such that the number density products are constant over each subdomain. The complete discretised equation can now be written in the following form:

$$\begin{aligned} \hat{v}_i\frac{dn_i}{dt} &= \sum_{p=1}^{P(i)}\left(n_{j(p)}(t)n_{k(p)}(t)\underbrace{\int_{S_p} u\beta(w',u)dw'du}_{I_k(p)} + n_{j(p)}(t)n_{k(p)}(t)\underbrace{\int_{S_p} w'\beta(w',u)dw'du}_{I_j(p)}\right) \\ &- \sum_{j=1}^{m}\left(n_i(t)n_j(t)\int_{v_{i-1}}^{v_i}\int_{w_{j-1}}^{w_j} v\beta(v,w)dwdv\right), \end{aligned} \tag{B.5}$$

where $P(i)$ is the total number of subdomains that contribute to the birth term in daughter interval Δv_i. The construction of the aggregation map, therefore, involves locating and classifying the subdomains of integration. As mentioned in Section 4.2.8, the construction of the map is carried out once in the beginning of the computation and tabulated. The procedure for the construction of the map can be summarised as follows:

- Generate a grid, $v_i = (v_{min}, v_{max})$, consisting of m intervals $\Delta v_i = v_i - v_{i-1}$.
- Perform a nested loop over all daughter and parent intervals comprising the following steps:
  - Establish the auxiliary grid for each combination of daughter and parent interval, $U_K(\Delta v_i, \Delta w_j)$.
  - Locate and classify the subdomains S_p corresponding to products of constant number densities.
  - Calculate the integrals $I_{j(p)}$ and $I_{k(p)}$.
  - Store the aggregation integrals. For convenience, all unique combinations (i,j,k,I_j,I_k) may be stored into a single array and indexed by a global pointer.

We now detail the algorithm for identifying all integration subdomains in the aggregation map, for an arbitrary non-uniform grid. It is necessary to distinguish between three main cases: an expanding, a uniform and a contracting grid.

- Calculate auxiliary grid, $U_k(\Delta v_i, \Delta w_j)$.
- Expanding grid: $\Delta v_i > \Delta w_j$
  - Three classes of shapes appear: A *left-side triangle*, a *rectangular middle section* and a *right-side triangle*. Note how the auxiliary grid aligns with the boundaries of each shape (Fig. B.2).
  - Divide these shapes into subdomains such that the product of number densities is constant over each of them.

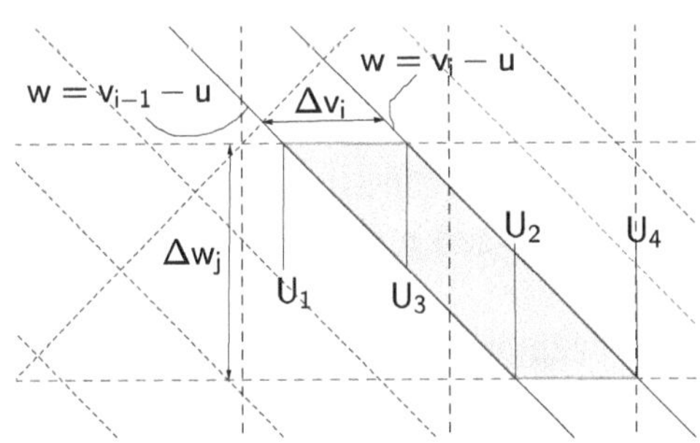

Figure B.5 Example of an interval on a contracting grid, used to highlight the left triangle, right triangle and parallelogram found in the domain of integration. Reproduced from O'Sullivan and Rigopoulos (2022) under CC-BY 4.0 license.

- Uniform grid: $\Delta v_i = \Delta w_j$
  - Since $\Delta v_i = \Delta w_j \rightarrow U_3 = U_2$. Thus, the central rectangular subdomain found in the expanding case collapses and just two triangular subdomains are left to be evaluated: a left-side triangle and a right-side triangle.
- Contracting grid: $\Delta v_i < \Delta w_j$
  - Since $\Delta v_i < \Delta w_j \rightarrow U_3 < U_2$. Thus, U_3 is now to left of U_2 by definition and the remaining overall shape can be viewed as a left-side triangle, a right-side triangle and a parallelogram in between, as shown in Fig. B.5.

B.1.2 Fragmentation Map

The fragmentation map is considerably simpler than the aggregation one. Only rectangular and triangular shapes arise, as can be seen in Fig. B.6. Consider any triangular domain along the line $z = v$. This class of shapes represents the contribution of parent particles from interval Δz_q fragmenting into daughter particles into the same interval and can, therefore, be ignored. The remaining rectangular domains represent the birth of particles into interval Δv_i as a result of the death of particles from parent intervals Δz_q. For an individual parent–daughter interval pair, the integral that needs to be evaluated numerically is shown in Eq. B.6, where S_r represents the rectangular domain with integration limits unique to index r, while q is the parent interval that contributes particles to daughter interval Δv_i.

$$F_{q(r)} = n_{q(r)} \underbrace{\int_{S_r} v\gamma(z)b(v,z)dz\,dv}_{I_{q(r)}} . \tag{B.6}$$

Further sub-partitioning of the domains into subdomains, as was done for aggregation, is not required for fragmentation, as the number density is constant over every rectangular domain.

The procedure for the construction of the fragmentation map can be summarised as follows:

- Generate a grid, $v_i = (v_{min}, v_{max})$, consisting of m intervals $\Delta v_i = v_i - v_{i-1}$.
- Locate the triangular shapes and remove them from the set of fragmentation domains to be integrated.
- Locate the rectangular integration domains using a nested loop – one loop over all daughter intervals and one over all possible parent intervals:
  - Calculate the integrals $I_{q(r)}$.
  - Create a unique pointer and store the fragmentation triplet, $\{i, q, I_{q(r)}\}$, for each rectangular domain.

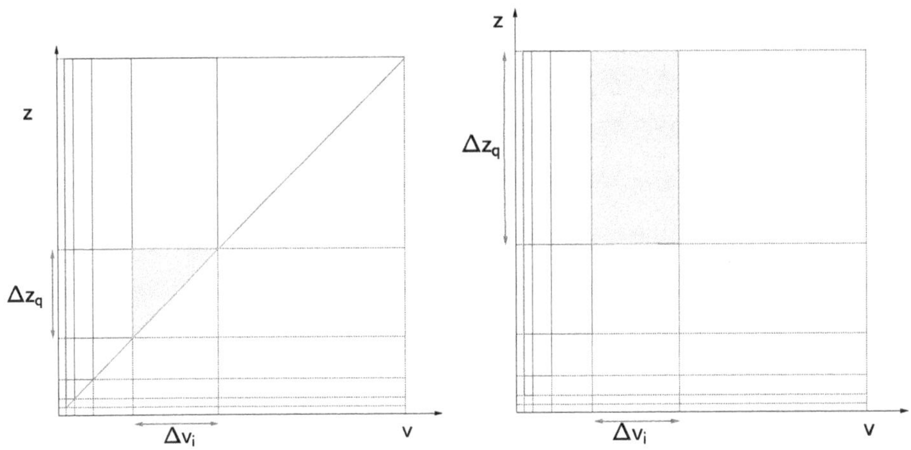

Figure B.6 Left: Initial fragmentation map, showing all possible domains of integration. The shaded region indicates a triangular domain along the line $z = v$, whose contribution can be ignored. Right: Fragmentation map after removal of triangular domains. The shaded region indicates one of the rectangular domains that should be considered. Reproduced from O'Sullivan and Rigopoulos (2022) under CC-BY 4.0 license.

B.2 Time Advancement

Any temporal discretisation method can be employed for the time advancement of the discretised equations. An explicit Euler scheme is shown below, corresponding to the implementation in O'Sullivan and Rigopoulos (2022), which has been found to be sufficient as the equations are not stiff. The procedure to be carried out at each time step is summarised below:

1. At time step n, for every interval, the death terms $D^n_{j(p)}$ and $D^n_{k(p)}$ over the whole domain that contribute to the birth term in that interval are calculated as follows:

$$D^n_{j(p)} = n^n_{j(p)} n^n_{k(p)} I_{j(p)}, \tag{B.7a}$$

$$D^n_{k(p)} = n^n_{j(p)} n^n_{k(p)} I_{k(p)}. \tag{B.7b}$$

2. The birth terms are now computed by summing the respective death terms:

$$B^n_{i(p)} = D^n_{j(p)} + D^n_{k(p)}. \tag{B.8}$$

3. The number densities at time step $n + 1$ are now updated into temporary values (denoted by *) as follows:

$$n_i^{n+1,*} = n_i^n + \frac{B_{i(p)}^n}{\hat{v}_{i(p)}}\delta t, \tag{B.9a}$$

$$n_{j(p)}^{n+1,*} = n_{j(p)}^n - \frac{D_{j(p)}^n}{\hat{v}_{j(p)}}\delta t, \tag{B.9b}$$

$$n_{k(p)}^{n+1,*} = n_{k(p)}^n - \frac{D_{k(p)}^n}{\hat{v}_{k(p)}}\delta t, \tag{B.9c}$$

where $\hat{v}_{i(p)}$, $\hat{v}_{j(p)}$ and $\hat{v}_{k(p)}$ are defined as in Eq. 4.71.

This process is repeated until all contributions have been accounted for, so that, in the end, the number density at time step $n+1$ has been calculated as follows:

$$n_i^{n+1} = n_i^n + \frac{1}{\hat{v}_i}\sum_{p=1}^{P(i)} B_{i(p)}^n \delta t - \frac{n_i^n}{\hat{v}_i}\sum_{j=1}^{m} n_j^n \int_{v_{i-1}}^{v_i}\int_{w_{j-1}}^{w_j} v\beta(v,w)dwdv\delta t. \tag{B.10}$$

A similar algorithm is employed for fragmentation:

1. At time step n, for every interval, the fragmentation birth term, $F_{q(r)}^n$, is calculated using the tabulated integral $I_{q(r)}$:

$$F_{q(r)}^n = n_{q(r)}^n I_{q(r)}. \tag{B.11}$$

2. The number densities at time step $n+1$ are now updated as follows:

$$n_i^{n+1,*} = n_i^n + \frac{F_{q(r)}^n}{\hat{v}_i}\delta t, \tag{B.12a}$$

$$n_q^{n+1,*} = n_q^n - \frac{F_{q(r)}^n}{\hat{v}_q}\delta t. \tag{B.12b}$$

At the end, the number density at step $n+1$ has been calculated as follows:

$$n_i^{n+1} = n_i^n + \frac{1}{\hat{v}_i}\left[\sum_{r=1}^{R(i)} F_{q(r)}^n\right]\delta t - \frac{1}{\hat{v}_i}\left[\sum_{s=1}^{S(i)} F_{q(s)}^n\right]\delta t, \tag{B.13}$$

where $R(i)$ represents the number of integration domains that contribute mass to interval Δv_i and $S(i)$ represents the number of domains where Δv_i is the parent interval.

Appendix C

Derivation of the PDF Transport Equation

In the present appendix, we show the derivation of the PDF transport equation. To keep the notation as simple as possible, the form that will be derived is for the PDF of a single scalar with a nonlinear source term, constant diffusivity and in a constant density flow (Eq. 5.66). Extensions relaxing the above limitations can be carried out in a straightforward manner. The derivation of PDF transport equations can be accomplished in several ways, including the use of the fine grained density (Lundgren, 1967; O'Brien, 1980) and the characteristic function (Kollmann, 1990; Dopazo, 1994). The method adopted here is that of the fine-grained density, following O'Brien (1980). Before proceeding with the derivation, we first introduce this concept and show some essential identities that will be employed. More details on the concepts involved in this derivation can be found in Pope (1985, 2000).

C.1 Random Variables and Stochastic Processes

In a turbulent flow, all quantities are regarded as *random* (or *stochastic*) *variables*. By this, we mean that they have a different value in every individual realisation of the turbulent flow, although each realisation is governed by a deterministic set of equations.

Consider a random variable X with PDF $f(x)$, where x is the corresponding sample-space variable. The average of any function of this random variable, $g(X)$, can be calculated as:

$$\langle g(X)\rangle = \int_{-\infty}^{\infty} g(x)f(x)dx. \tag{C.1}$$

For two random variables X and Y, their joint PDF, $f_{XY}(x,y)$, is the probability that $x < X < x+dx$ and $y < Y < y+dy$. The conditional expectation of X on Y is denoted by $\langle X|Y=y\rangle$ and is the average value of X when Y assumes the value y. It is related to the PDF of Y and joint PDF of X and Y via the following equation:

$$f(y)\langle X|Y=y\rangle = \int_{-\infty}^{\infty} x f_{XY}(x,y)dx. \tag{C.2}$$

A *stochastic process* is a set of random variables. We are interested in processes that are indexed by time, and such a process can be denoted by $X(t)$. In the following, we will be interested in spatially varying stochastic processes, denoted as $X(\mathbf{x},t)$. More information on random variables and stochastic processes can be found, for example, in Gardiner (1985), Kloeden and Platen (1992) or Stirzaker (2005).

C.2 Fine-Grained Density and Essential Identities

The following derivation relies heavily on the properties of delta functions, a concise summary of which can be found in Appendix C of Pope (2000). We note here (without proof) three identities that will be needed:

$$\delta(x-y) = \delta(y-x), \tag{C.3}$$

$$\int_{-\infty}^{\infty} f(x)\delta(x-x_0) = f(x_0), \tag{C.4}$$

$$\frac{\partial}{\partial x}\delta(x-y) = -\frac{\partial}{\partial y}\delta(y-x). \tag{C.5}$$

Let us now consider a random variable X having a PDF $f(x)$. We define the *fine-grained density* as follows:

$$\mathcal{F}(x) = \delta(x-X). \tag{C.6}$$

The fine-grained density can be physically interpreted as the PDF of the random variable in a single realisation. Since, in that realisation, the random variable can have only one value, its PDF (i.e. $\mathcal{F}$) approaches infinity for that value and is zero elsewhere, hence the use of the delta function (which has the properties of a PDF, such as normalisation to 1). The PDF can now be written as an ensemble average over the individual realisations via the following identity:

$$f(x) = \langle \mathcal{F}(x) \rangle. \tag{C.7}$$

This identity can be shown using Eqs. C.1–C.4:

$$\langle \mathcal{F}(x) \rangle = \langle \delta(x-X) \rangle = \int_{-\infty}^{\infty} \delta(x-\xi)f(\xi)d\xi = \int_{-\infty}^{\infty} \delta(\xi-x)f(\xi)d\xi = f(x),$$

where ξ, x are both variables representing the sample space of the random variable X.

Let X, Y be two random variables. Another identity that will be useful is:

$$f(y)\langle X|Y=y\rangle = \langle X\delta(y-Y)\rangle. \tag{C.8}$$

This is proven as follows:

$$\begin{aligned}\langle X\delta(y-Y)\rangle &= \int_{-\infty}^{\infty}\int_{-\infty}^{\infty} x\delta(y-\psi)f_{XY}(x,\psi)d\psi dx \\ &= \int_{-\infty}^{\infty} xf_{XY}(x,y)dx \\ &= f(y)\langle X|Y=y\rangle,\end{aligned}$$

where y, ψ are both variables representing the sample space of the random variable Y. The last step follows by application of Eq. C.2.

Now let $X(t)$ be a stochastic process. Then:

$$\frac{\partial}{\partial t}\delta(x-X(t)) = -\frac{\partial}{\partial x}\left[\frac{dX(t)}{dt}\delta(x-X(t))\right]. \tag{C.9}$$

This identity is proven by first applying the chain rule:

$$\frac{\partial}{\partial t}\delta(x-X(t)) = \frac{\partial}{\partial X}\delta(x-X(t))\frac{dX(t)}{dt}.$$

We now use the identities C.5, C.3: and obtain:

$$\begin{aligned}\frac{\partial}{\partial X}\delta(x-X(t))\frac{dX(t)}{dt} &= -\frac{\partial}{\partial x}\delta(X(t)-x)\frac{dX(t)}{dt} \\ &= -\frac{\partial}{\partial x}\left[\frac{dX(t)}{dt}\delta(x-X(t))\right].\end{aligned}$$

Similarly, for a spatially varying stochastic process, $X(\mathbf{x},t)$, we have:

$$\frac{\partial}{\partial x_i}\delta(x-X(\mathbf{x},t)) = -\frac{\partial}{\partial x}\left[\frac{\partial X(\mathbf{x},t)}{\partial x_i}\delta(x-X(\mathbf{x},t))\right]. \tag{C.10}$$

Identities C.9 and C.10 are useful for the manipulation of time and space derivatives in the derivation of the PDF transport equation.

C.3 Derivation of the PDF Transport Equation for a Single Scalar

We will now derive the PDF transport equation for a single scalar, ϕ, with a nonlinear source term, $S(\phi)$, and constant diffusivity, in a constant density flow. The starting point will be Eq. 5.61, where the velocity has been decomposed as $u_i = \langle u_i\rangle + u_i'$, shown again here for reference:

$$\frac{\partial\phi}{\partial t} + \frac{\partial(\langle u_i\rangle\phi)}{\partial x_i} + \frac{\partial(u_i'\phi)}{\partial x_i} = -\frac{\partial J_{\phi,i}}{\partial x_i} + S(\phi), \tag{C.11}$$

where the mean velocity, $\langle u_i \rangle$, will be regarded as known (i.e. computed via the RANS equations and a turbulence model). The scalar, on the other hand, has not been decomposed. We now introduce the spatially dependent PDF, $f(\psi; \mathbf{x}, t)$, and use Identity C.7 to represent the PDF and obtain the equation for its temporal evolution, shown also in Chapter 5 (Eq. 5.65):

$$\frac{\partial f(\psi; \mathbf{x}, t)}{\partial t} = \frac{\partial}{\partial t} \langle \delta(\psi - \phi(\mathbf{x}, t)) \rangle. \tag{C.12}$$

We now employ Identity C.9 to obtain Eq. 5.65 of Chapter 5:

$$\frac{\partial f(\psi; \mathbf{x}, t)}{\partial t} = -\left\langle \frac{\partial}{\partial \psi} \left[\frac{\partial \phi(\mathbf{x}, t)}{\partial t} \delta(\psi - \phi(\mathbf{x}, t)) \right] \right\rangle. \tag{C.13}$$

In the following, the dependence of f and ϕ will be omitted for brevity. Introducing Eq. C.11, we obtain:

$$\frac{\partial f}{\partial t} = T_1 + T_2 + T_3 + T_4, \tag{C.14}$$

where:

$$T_1 = -\left\langle \frac{\partial}{\partial \psi} \left[-\frac{\partial \left(\langle u_i \rangle \phi \right)}{\partial x_i} \delta(\psi - \phi) \right] \right\rangle,$$

$$T_2 = -\left\langle \frac{\partial}{\partial \psi} \left[-\frac{\partial \left(u_i' \phi \right)}{\partial x_i} \delta(\psi - \phi) \right] \right\rangle,$$

$$T_3 = -\left\langle \frac{\partial}{\partial \psi} \left[-\frac{\partial J_{\phi,i}}{\partial x_i} \delta(\psi - \phi) \right] \right\rangle,$$

$$T_4 = -\left\langle \frac{\partial}{\partial \psi} \left[S(\phi) \delta(\psi - \phi) \right] \right\rangle.$$

The four terms on the right-hand side denote transport by the mean velocity, transport by the fluctuating velocity, diffusion and chemical reaction. Each of them will now be considered separately.

Using Identity C.10 and the continuity equation, $\partial \langle u_i \rangle / \partial x_i = 0$, the first term becomes:

$$T_1 = -\langle u_i \rangle \left\langle -\frac{\partial}{\partial \psi} \left[\frac{\partial \phi}{\partial x_i} \delta(\psi - \phi) \right] \right\rangle = -\langle u_i \rangle \left\langle \frac{\partial}{\partial x_i} \delta(\psi - \phi) \right\rangle,$$

or:

$$T_1 = -\langle u_i \rangle \frac{\partial f}{\partial x_i}. \tag{C.15}$$

For the second term, we use Identities C.8, C.5 and $\partial\langle u_i'\rangle/\partial x_i = 0$ (also resulting from continuity):

$$T_2 = \left\langle \frac{\partial}{\partial \psi}\left[u_i' \frac{\partial \phi}{\partial x_i}\delta(\psi - \phi)\right]\right\rangle = \left\langle u_i' \frac{\partial \phi}{\partial x_i}\frac{\partial}{\partial \psi}\delta(\psi - \phi)\right\rangle$$
$$= -\left\langle u_i' \frac{\partial \phi}{\partial x_i}\frac{\partial}{\partial \phi}\delta(\psi - \phi)\right\rangle = -\left\langle u_i' \frac{\partial}{\partial x_i}\delta(\psi - \phi)\right\rangle$$
$$= -\frac{\partial}{\partial x_i}\left\langle u_i'\delta(\psi - \phi)\right\rangle,$$

or:

$$T_2 = -\frac{\partial}{\partial x_i}\left(\left\langle u_i' | \phi = \psi\right\rangle f\right). \tag{C.16}$$

This term is unclosed because it involves the conditional expectation of fluctuating velocity on the scalar.

For the third term, by applying Identity C.8, we have:

$$T_3 = \frac{\partial}{\partial \psi}\left[\left\langle \left.\frac{\partial J_{\phi,i}}{\partial x_i}\right| \phi = \psi\right\rangle f\right]. \tag{C.17}$$

This term is also unclosed, as it introduces an additional variable (conditional expectation).

For the fourth term, we have:

$$T_4 = -\left\langle \frac{\partial}{\partial \psi}\left[S(\phi)\delta(\psi - \phi)\right]\right\rangle = -\frac{\partial}{\partial \psi}\left(\langle S(\phi)|\phi = \psi\rangle f\right),$$

or:

$$T_4 = -\frac{\partial}{\partial \psi}\left(S(\psi) f\right). \tag{C.18}$$

This term is thus closed, a fact that constitutes the main advantage of transported PDF methods.

Putting all these terms into Eq. C.11, we arrive at the PDF transport equation (Eq. 5.66 in Chapter 5):

$$\frac{\partial f}{\partial t} + \langle u_i\rangle \frac{\partial f}{\partial x_i} + \frac{\partial}{\partial x_i}\left(\left\langle u_i' | \phi = \psi\right\rangle f\right)$$
$$= \frac{\partial}{\partial \psi}\left[\left\langle \left.\frac{\partial J_{\phi,i}}{\partial x_i}\right| \phi = \psi\right\rangle f\right] - \frac{\partial}{\partial \psi}\left(S(\psi) f\right). \tag{C.19}$$

C.4 The Diffusion Term and Scalar Dissipation

By taking the diffusive flux to be given by Fick's law with constant diffusivity:

$$J_{\phi,i} = -D\frac{\partial \phi}{\partial x_i},$$

the term T_3 in the previous section can be elaborated as follows:

$$\begin{aligned}
T_3 &= -\left\langle \frac{\partial}{\partial \psi}\left[-\frac{\partial J_{\phi,i}}{\partial x_i}\delta(\psi-\phi)\right]\right\rangle \\
&= -\left\langle \frac{\partial}{\partial \psi}\left[D\frac{\partial^2 \phi}{\partial x_i \partial x_i}\delta(\psi-\phi)\right]\right\rangle \\
&= -\left\langle \left[D\frac{\partial^2 \phi}{\partial x_i \partial x_i}\frac{\partial}{\partial \psi}\delta(\psi-\phi)\right]\right\rangle \\
&= -\left\langle \frac{\partial}{\partial x_i}\left[D\frac{\partial \phi}{\partial x_i}\frac{\partial}{\partial \psi}\delta(\psi-\phi)\right]\right\rangle + \left\langle D\frac{\partial \phi}{\partial x_i}\frac{\partial}{\partial x_i}\left[\frac{\partial}{\partial \psi}\delta(\psi-\phi)\right]\right\rangle \\
&= -\left\langle \frac{\partial}{\partial x_i}\left[\frac{\partial}{\partial \psi}\left[D\frac{\partial \phi}{\partial x_i}\delta(\psi-\phi)\right]\right]\right\rangle + \left\langle D\frac{\partial \phi}{\partial x_i}\frac{\partial}{\partial \psi}\left[\frac{\partial}{\partial x_i}\delta(\psi-\phi)\right]\right\rangle \\
&= \left\langle \frac{\partial}{\partial x_i}\left[D\frac{\partial}{\partial x_i}\delta(\psi-\phi)\right]\right\rangle - \left\langle D\frac{\partial \phi}{\partial x_i}\frac{\partial^2}{\partial \psi^2}\left[\frac{\partial \phi}{\partial x_i}\delta(\psi-\phi)\right]\right\rangle \\
&= D\frac{\partial^2 f}{\partial x_i^2} - \frac{\partial^2}{\partial \psi^2}\left\langle D\frac{\partial \phi}{\partial x_i}\frac{\partial \phi}{\partial x_i}\delta(\psi-\phi)\right\rangle,
\end{aligned}$$

or:

$$T_3 = D\frac{\partial^2 f}{\partial x_i^2} - \frac{\partial^2}{\partial \psi^2}\left[\left\langle \left(D\frac{\partial \phi}{\partial x_i}\frac{\partial \phi}{\partial x_i}\right)\middle|\, \phi=\psi\right\rangle f\right]. \tag{C.20}$$

The first term on the right-hand side can be identified as transport in physical space due to spatial gradients of the PDF and is in close form, while the second term represents transport in scalar space and is unclosed. The term conditioned on ψ can be identified with half the scalar dissipation.

Appendix D

Derivation of the Stochastic Field Equation

The objective of the present appendix is to show the derivation of the stochastic field equation. As with Appendix C, in order to demonstrate the concept in the simplest possible form, we treat the case of a single scalar with a nonlinear source term and constant diffusivity in a constant density flow. The derivation of the stochastic field method was carried out by Valiño (1998) and Sabel'nikov and Soulard (2005) with different methods. The derivation shown here follows the approach of Valiño.

The derivation employs concepts from stochastic calculus, and it is thus necessary to first state some essential results from that. While stochastic calculus is a vast subject, our interest in it here is solely from the point of view of devising a numerical method for solving the PDF transport equation. It should be emphasised that the latter is a deterministic equation, and the need to resort to stochastic methods is necessitated only by its high dimensionality. Similar uses of stochastic methods can be found in many other deterministic problems of numerical analysis, such as the calculation of multidimensional integrals. Only a few essential results from stochastic calculus are needed and these will be reviewed below.

D.1 Essential Results from Stochastic Calculus

The following results from stochastic calculus are stated here in an informal manner and without proof. More information and a rigorous treatment can be found in several books such as Gardiner (1985), Kloeden and Platen (1992) and Stirzaker (2005).

Random variables and stochastic processes were introduced in Appendix C. A stochastic process can have many realisations or *sample paths*. A *stochastic differential equation* is an equation that yields an increment of a sample path of a stochastic process. By simulating several sample paths, the statistics of the process can be collected. It can be shown that the PDF of this stochastic process is equivalent to the one obtained by a deterministic differential equation of a certain form.

This is the key concept that allows the use of SDEs as tools for numerical solution of deterministic differential equations.

The *Wiener process* is the archetypical stochastic process, and its main properties are mentioned first. Afterwards, we state the theorem that provides the equivalence of the stochastic process and differential equation. In the following, we adopt the notation and presentation of Stirzaker (2005).

Wiener Process

The Wiener process (also called standard Wiener process or Brownian motion), $W(t)$ for $t \geq 0$, is the archetypical stochastic process and is defined as follows:

1. $W(0) = 0$.
2. $W(t)$ is continuous in t, where $t \geq 0$.
3. The increments $dW = W(t+s) - W(t)$ are independent and normally distributed.

The *transition density function* of the Wiener process is defined as follows:

$$f(t, y|s, x) = \frac{\partial}{\partial y} P(W(t) \leq y | W(s) = x), \quad s \leq t. \tag{D.1}$$

In other words, the transition density function yields the conditional PDF that the process will assume the value y at time t if it had the value x at time s. The following can now be proven from the fact that the Wiener process has normally distributed increments:

$$f(t, y|s, x) = \frac{1}{\sqrt{2\pi(t-s)}} \exp\left[-\frac{(y-x)^2}{2(t-s)}\right], \quad y \in \mathbb{R}. \tag{D.2}$$

From this, it can be shown that the transition density obeys the following equations:

$$\frac{\partial f}{\partial t} = \frac{1}{2}\frac{\partial^2 f}{\partial y^2}, \tag{D.3}$$

$$\frac{\partial f}{\partial s} = -\frac{1}{2}\frac{\partial^2 f}{\partial x^2}. \tag{D.4}$$

Eqs. D.3 and D.4 are called *forward* and *backward equation*, respectively. It can be seen that these equations have the same form as the diffusion equation. This means that if a number of sample paths of the Wiener process are simulated and the transition density is obtained from their statistics, the latter can be taken as an approximation to the solution of the diffusion equation. The sample paths of a Wiener process can be simulated by generating increments that comply with its properties. This is an example of solving a deterministic equation via a stochastic method.

Stochastic Differential Equations

Stochastic differential equations (SDEs) can be constructed with the aid of the Wiener process. An archetypical SDE is the following:

$$dX = \alpha(t,X) + \sqrt{\beta(t,X)}dW, \tag{D.5}$$

which effectively denotes the following integral equation:

$$X(t) = X(t_0) + \int_{t_0}^{t} \alpha(s,X)ds + \int_{t_0}^{t} \beta(s,X)dW(s). \tag{D.6}$$

The last integral is not a conventional one, since the integration is with respect to the increment of a stochastic process, the Wiener process. It can be interpreted with the aid of Itô's formula, whose explanation is out of the scope of the present appendix.[1] What is important here is that the transition density of the stochastic process corresponding to Eq. D.5 can be shown to be a solution of the following forward equation:

$$\frac{\partial f}{\partial t} + \frac{\partial(\alpha(t,y)f)}{\partial y} = \frac{1}{2}\frac{\partial^2(b(t,y)f)}{\partial y^2}. \tag{D.7}$$

This result is the cornerstone for the following developments, as it indicates that a deterministic equation of the form of Eq. D.7 can be solved by simulating sample paths of a stochastic process whose increments are given by Eq. D.5. In order to solve the PDF transport equation, therefore, we have to bring it to the form of Eq. D.7, so that an equivalent stochastic process can be constructed. This will be described in the next section.

D.2 Derivation of the Stochastic Field Equation for a Single Scalar

The starting point for the following derivation is the PDF transport equation for a single scalar, constant density and diffusivity, equipped with the gradient transport and IEM models (Eq. 5.71), shown again here for reference:

$$\frac{\partial f}{\partial t} + \langle u_i \rangle \frac{\partial f}{\partial x_i} = \frac{\partial}{\partial x_i}\left(\Gamma \frac{\partial f}{\partial x_i}\right) + \frac{1}{2\tau_\phi}\frac{\partial}{\partial \psi}\left[(\psi - \langle \phi \rangle) f\right] - \frac{\partial}{\partial \psi}\left(S(\psi) f\right), \tag{D.8}$$

where $\langle u_i \rangle$ and Γ are regarded as known (i.e. calculated from the RANS equations and turbulence model), while $\langle \phi \rangle = \int_{-\infty}^{\infty} \psi f d\psi$.

As we saw in the previous section, to solve this equation by a stochastic method, we have to construct a stochastic process whose transition density is the same as the one-point scalar PDF. In the stochastic field method, we consider an ensemble

[1] An alternative approach is Stratonovich calculus. For an in-depth exposition of Itô and Stratonovich calculus, see Gardiner (1985) or Kloeden and Platen (1992). For a derivation based on Stratonovich calculus, see Sabel'nikov and Soulard (2005).

of fields, each of which is a realisation of a spatially dependent stochastic process, $\theta(\mathbf{x}, t)$, with sample space z, which is continuous and differentiable in space and continuous but not differentiable in time. The PDF of θ can be expressed as an average of a fine-grained density (cf. Eq. C.7):

$$f(z; \mathbf{x}, t) = \langle \delta(z - \theta(\mathbf{x}, t)) \rangle . \tag{D.9}$$

This stochastic process should obey the same modelled PDF transport equation as the original scalar, that is, Eq. D.8 (note that the arguments will be omitted from now on):

$$\frac{\partial f}{\partial t} + \langle u_i \rangle \frac{\partial f}{\partial x_i} = \frac{\partial}{\partial x_i}\left(\Gamma \frac{\partial f}{\partial x_i}\right) + \frac{1}{2\tau_\phi}\frac{\partial}{\partial z}\left[(z - \langle \theta \rangle) f\right] - \frac{\partial}{\partial z}(S(z) f) , \tag{D.10}$$

where the PDF of $\theta, f(z; \mathbf{x}, t)$, is identical to that of ϕ.

Eq. D.10 must now be brought into the form of Eq. D.7. We proceed by reversing the steps in the derivation of the PDF transport equation, using Identity C.9 as in Eq. C.13:

$$\begin{aligned}\frac{\partial f}{\partial t} + \left\langle \frac{\partial}{\partial z}\left[-\frac{\partial (\langle u_i \rangle \theta)}{\partial x_i}\delta(z - \theta)\right]\right\rangle &= -\left\langle \frac{\partial}{\partial z}\left[\frac{\partial}{\partial x_i}\left(\Gamma \frac{\partial \theta}{\partial x_i}\right)\delta(z - \theta)\right]\right\rangle \\ &\quad + \left\langle \frac{\partial}{\partial z}\left[\frac{1}{2\tau_\phi}(z - \langle \theta \rangle)\,\delta(z - \theta)\right]\right\rangle - \left\langle \frac{\partial}{\partial z}\left[S(\theta)\delta(z - \theta)\right]\right\rangle . \end{aligned} \tag{D.11}$$

We now apply Identity C.8 to arrive at the following:

$$\begin{aligned}\frac{\partial f}{\partial t} - \frac{\partial}{\partial z}\left[\left\langle \frac{\partial (\langle u_i \rangle \theta)}{\partial x_i}\bigg|\, \theta = z\right\rangle f(z)\right] &= -\frac{\partial}{\partial z}\left[\left\langle \frac{\partial}{\partial x_i}\left(\Gamma \frac{\partial \theta}{\partial x_i}\right)\bigg|\, \theta = z\right\rangle f(z)\right] \\ &\quad + \frac{\partial}{\partial z}\left[\left\langle\left(\frac{1}{2\tau_\phi}(z - \langle \theta \rangle)\right)\bigg|\, \theta = z\right\rangle f(z)\right] - \frac{\partial}{\partial z}\left(\langle S(\theta) \,|\, \theta = z\rangle f(z)\right) . \end{aligned} \tag{D.12}$$

The diffusion term can be split into two terms as shown in Section C.4, thus giving rise to a second derivative in composition space:

$$\begin{aligned} -\frac{\partial}{\partial z}\left[\left\langle \frac{\partial}{\partial x_i}\left(\Gamma \frac{\partial \theta}{\partial x_i}\right)\bigg|\, \theta = z\right\rangle f(z)\right] &= -\frac{\partial}{\partial z}\left[\frac{\partial}{\partial x_i}\left(\Gamma \frac{\partial \theta}{\partial x_i}\right)\delta(z - \theta)\right] \\ &= -\left\langle \frac{\partial}{\partial x_i}\left[\Gamma \frac{\partial \theta}{\partial x_i}\frac{\partial}{\partial z}\delta(z - \theta)\right]\right\rangle + \left\langle \Gamma \frac{\partial \theta}{\partial x_i}\frac{\partial}{\partial x_i}\left(\frac{\partial}{\partial z}\delta(z - \theta)\right)\right\rangle \\ &= -\frac{\partial}{\partial z}\left(\frac{\partial \Gamma}{\partial x_i}\left\langle \frac{\partial \theta}{\partial x_i}\bigg|\, \theta = z\right\rangle f(z)\right) - \frac{\partial}{\partial z}\left(\Gamma \frac{\partial}{\partial x_i}\left\langle \frac{\partial \theta}{\partial x_i}\bigg|\, \theta = z\right\rangle f(z)\right) \end{aligned}$$

$$
-\left\langle \Gamma \frac{\partial \theta}{\partial x_i} \frac{\partial}{\partial z}\left[\frac{\partial}{\partial x_i} \delta(z-\theta)\right]\right\rangle
$$
$$
= -\frac{\partial}{\partial z}\left[\frac{\partial}{\partial x_i}\left(\Gamma\left\langle \frac{\partial \theta}{\partial x_i}\middle|\, \theta = z\right\rangle\right) f(z)\right] + \frac{\partial^2}{\partial z^2}\left[\left\langle\left(\Gamma \frac{\partial \theta}{\partial x_i}\frac{\partial \theta}{\partial x_i}\right)\middle|\, \theta = z\right\rangle f(z)\right]. \tag{D.13}
$$

Collecting all first-order derivatives in z, we now have:

$$
\frac{\partial f}{\partial t} + \frac{\partial}{\partial z}\left[-\left\langle \frac{\partial(\langle u_i\rangle \theta)}{\partial x_i}\middle|\, \theta = z\right\rangle f(z) + \frac{\partial}{\partial x_i}\left(\Gamma\left\langle \frac{\partial \theta}{\partial x_i}\middle|\, \theta = z\right\rangle\right) f(z)\right]
$$
$$
+ \frac{\partial}{\partial z}\left[-\left\langle\left(\frac{1}{2\tau_\phi}(z - \langle\theta\rangle)\right)\middle|\, \theta = z\right\rangle f(z) + \langle S(\theta)|\, \theta = z\rangle f(z)\right]
$$
$$
= \frac{\partial^2}{\partial z^2}\left[\left\langle\left(\Gamma \frac{\partial \theta}{\partial x_i}\frac{\partial \theta}{\partial x_i}\right)\middle|\, \theta = z\right\rangle f(z)\right]. \tag{D.14}
$$

The equation is now in the form of Eq. D.7 in composition space. According to the discussion in Section D.1, its solution is equivalent to the transition density of the stochastic process defined by the following SDE in the form of Eq. D.5:

$$
d\theta = -\left\langle \frac{\partial(\langle u_i\rangle \theta)}{\partial x_i}\middle|\, \theta = z\right\rangle dt + \frac{\partial}{\partial x_i}\left(\Gamma\left\langle \frac{\partial \theta}{\partial x_i}\middle|\, \theta = z\right\rangle\right) dt
$$
$$
-\left\langle\left(\frac{1}{2\tau_\phi}(z - \langle\theta\rangle)\right)\middle|\, \theta = z\right\rangle dt + \langle S(\theta)|\, \theta = z\rangle\, dt
$$
$$
+\left[2\left\langle\left(\Gamma \frac{\partial \theta}{\partial x_i}\frac{\partial \theta}{\partial x_i}\right)\middle|\, \theta = z\right\rangle\right]^{1/2} dW. \tag{D.15}
$$

At this point, we may replace the conditional expectations of the fields' derivatives with the actual derivatives, since the SDE now refers to an individual realisation of the stochastic process:

$$
d\theta = -\frac{\partial(\langle u_i\rangle \theta)}{\partial x_i} dt + \frac{\partial}{\partial x_i}\left(\Gamma \frac{\partial \theta}{\partial x_i}\right) dt - \left[\frac{1}{2\tau_\phi}(z - \langle\theta\rangle)\right] dt
$$
$$
+ S(\theta) dt + (2\Gamma)^{1/2} \frac{\partial \theta}{\partial x_i} dW. \tag{D.16}
$$

References

Abberger, T. 2007. Population balance modelling of granulation. Pages 1109–1186 of: Salman, A. D., Hounslow, M. J., and Seville, J. P. K. (eds), *Handbook of Powder Technology*, vol. 11. Elsevier.

Aiyer, A. K., Yang, D., Chamecki, M., and Meneveau, C. 2019. A population balance model for large eddy simulation of polydisperse droplet evolution. *Journal of Fluid Mechanics*, **878**, 700–739.

Akhtar, M. K., Xiong, Y., and Pratsinis, S. E. 1991. Vapor synthesis of titania powder by titanium tetrachloride oxidation. *AIChE Journal*, **37**(10), 1561–1570.

Akridis, P., and Rigopoulos, S. 2017. Modelling of soot formation in laminar diffusion flames using a comprehensive CFD-PBE model with detailed gas-phase chemistry. *Combustion Theory and Modelling*, **21**(1), 35–48.

Amsden, A., O'Rourke, P. J., and Butler, T. D. 1989. *KIVA-II: A computer program for chemically reactive flows with sprays*. Tech. rept. LA-11560-MS. Los Alamos National Laboratory.

Angerhöfer, A. 1994. Untersuchungen zur Kinetik der Fällungskristallisation von Bariumsulfat. Ph.D. thesis, Technische Univetsität München.

Appel, J., Bockhorn, H., and Frenklach, M. 2000. Kinetic modeling of soot formation with detailed chemistry and physics: laminar premixed flames of C_2 hydrocarbons. *Combustion and Flame*, **121**(1–2), 122–136.

Attili, A., and Bisetti, F. 2013. Application of a robust and efficient Lagrangian particle scheme to soot transport in turbulent flames. *Computers and Fluids*, **84**, 164–175.

Austin, L., Shoji, K., Bhatia, V., Jindal, V., Savage, K., and Klimpel, R. 1976. Some results on the description of size reduction as a rate process in various mills. *Industrial and Engineering Chemistry Process Design and Development*, **15**(1), 187–196.

Balachandar, S., and Eaton, J. K. 2010. Turbulent dispersed multiphase flow. *Annual Review of Fluid Mechanics*, **42**, 111–133.

Bałdyga, J. 2016. Mixing and fluid dynamics effects in particle precipitation processes. *KONA Powder and Particle Journal*, **2016**(33), 127–149.

Bałdyga, J., and Bourne, J. R. 1984. A fluid mechanical approach to turbulent mixing and chemical reaction – Part II: micromixing in the light of turbulence theory. *Chemical Engineering Communications*, **28**(4–6), 243–258.

Bałdyga, J., and Bourne, J. R. 1989. Simplification of micromixing calculations. I. Derivation and application of new model. *The Chemical Engineering Journal*, **42**(2), 83–92.

Bałdyga, J., and Bourne, J. R. 1999. *Turbulent Mixing and Chemical Reactions*. John Wiley & Sons.

Bałdyga, J., and Orciuch, W. 1997. Closure problem for precipitation. *Chemical Engineering Research and Design*, **75**(2), 160–170.

Bałdyga, J., and Orciuch, W. 2001. Barium sulphate precipitation in a pipe – an experimental study and CFD modelling. *Chemical Engineering Science*, **56**(7), 2435–2444.

Bałdyga, J., Jasińska, M., and Orciuch, W. 2003. Barium sulphate agglomeration in a pipe – an experimental study and CFD modeling. *Chemical Engineering and Technology*, **26**(3), 334–340.

Bałdyga, J., Makowski, Ł., and Orciuch, W. 2007. Double-feed semibatch precipitation: effects of mixing. *Chemical Engineering Research and Design*, **85**(A5), 745–752.

Ball, R. C., and Jullien, R. 1984. Finite size effects in cluster-cluster aggregation. *Journal de Physique Lettres*, **45**(21), 1031–1035.

Balthasar, M., and Frenklach, M. 2005a. Detailed kinetic modeling of soot aggregate formation in laminar premixed flames. *Combustion and Flame*, **140**(1–2), 130–145.

Balthasar, M., and Frenklach, M. 2005b. Monte-Carlo simulation of soot particle coagulation and aggregation: the effect of a realistic size distribution. *Proceedings of the Combustion Institute*, **30**(1), 1467–1475.

Balthasar, M., and Kraft, M. 2003. A stochastic approach to calculate the particle size distribution function of soot particles in laminar premixed flames. *Combustion and Flame*, **133**(3), 289–298.

Balthasar, M., Mauss, F., Knobel, A., and Kraft, M. 2002. Detailed modeling of soot formation in a partially stirred plug flow reactor. *Combustion and Flame*, **128**(4), 395–409.

Banasiak, J., Lamb, W., and Laurençot, P. 2020a. *Analytic Methods for Coagulation-Fragmentation Models*. Vol. I. CRC Press.

Banasiak, J., Lamb, W., and Laurençot, P. 2020b. *Analytic Methods for Coagulation-Fragmentation Models*. Vol. II. CRC Press.

Barrett, J. C., and Jheeta, J. S. 1996. Improving the accuracy of the moments method for solving the aerosol general dynamic equation. *Journal of Aerosol Science*, **27**(8), 1135–1142.

Barrett, J. C., and Webb, N. A. 1998. A comparison of some approximate methods for solving the aerosol general dynamic equation. *Journal of Aerosol Science*, **29**(1–2), 31–39.

Batterham, R. J., Hall, J. S., and Barton, G. 1981. Pelletizing kinetics and simulation of full scale balling circuits. Pages A136–A150 of: *3rd International Symposium on Agglomeration*, vol. 1.

Beckett, F. M., Witham, C. S., Leadbetter, S. J., Crocker, R., Webster, H. N., Hort, M. C., Jones, A. R., Devenish, B. J., and Thomson, D. J. 2020. Atmospheric dispersion modelling at the London VAAC: a review of developments since the 2010 Eyjafjallajökull volcano ash cloud. *Atmosphere*, **11**(4), 352.

Bemrose, C. R., and Bridgwater, J. 1987. A review of attrition and attrition test methods. *Powder Technology*, **49**(2), 97–126.

Bergthorson, J. M. 2018. Recyclable metal fuels for clean and compact zero-carbon power. *Progress in Energy and Combustion Science*, **68**, 169–196.

Bernard, P. S., and Wallace, J. M. 2002. *Turbulent Flow: Analysis, Measurement, and Prediction*. John Wiley & Sons.

Berry, E. X. 1967. Cloud droplet growth by collection. *Journal of the Atmospheric Sciences*, **24**, 688–701.

Berry, E. X. 1969. A mathematical framework for cloud models. *Journal of the Atmospheric Sciences*, **26**, 109–111.

Bini, M., and Jones, W. P. 2007. Particle acceleration in turbulent flows: a class of nonlinear stochastic models for intermittency. *Physics of Fluids*, **19**(3), 035104.

Bini, M., and Jones, W. P. 2008. Large-eddy simulation of particle-laden turbulent flows. *Journal of Fluid Mechanics*, **614**, 207–252.

Bird, R. B., Stewart, W. E., and Lightfoot, E. N. 1960. *Transport Phenomena*. John Wiley & Sons.

Blanquart, G., and Pitsch, H. 2009. A joint volume-surface-hydrogen multi-variate model for soot formation. Pages 437–463 of: Bockhorn, H., D'Anna, A., Sarofim, A., and Wang, H. (eds), *Combustion Generated Fine Carbonaceous Particles*. KIT Scientific Publishing.

Blanquart, G., Pepiot-Desjardins, P., and Pitsch, H. 2009. Chemical mechanism for high temperature combustion of engine relevant fuels with emphasis on soot precursors. *Combustion and Flame*, **156**(3), 588–607.

Bleck, R. 1970. A fast, approximative method for integrating the stochastic coalescence equation. *Journal of Geophysical Research*, **75**(27), 5165–5171.

Bonadonna, C., Genco, R., Gouhier, M., Pistolesi, M., Cioni, R., Alfano, F., Hoskuldsson, A., and Ripepe, M. 2011. Tephra sedimentation during the 2010 Eyjafjallajökull eruption (Iceland) from deposit, radar, and satellite observations. *Journal of Geophysical Research: Solid Earth*, **116**(12), B12202.

Bouaniche, A., Vervisch, L., and Domingo, P. 2019. A hybrid stochastic/fixed-sectional method for solving the population balance equation. *Chemical Engineering Science*, **209**, 115198.

Bouaniche, A., Yon, J., Domingo, P., and Vervisch, L. 2020. Analysis of the soot particle size distribution in a laminar premixed flame: a hybrid stochastic/fixed-sectional approach. *Flow, Turbulence and Combustion*, **104**(2–3), 753–775.

Bourouiba, L. 2020. Turbulent gas clouds and respiratory pathogen emissions: potential implications for reducing transmission of COVID-19. *JAMA – Journal of the American Medical Association*, **323**(18), 1837–1838.

Bramley, A. S., Hounslow, M. J., and Ryall, R. L. 1996. Aggregation during precipitation from solution: a method for extracting rates from experimental data. *Journal of Colloid and Interface Science*, **183**(1), 155–165.

Brasil, A. M., Farias, T. L., and Carvalho, M. G. 1999. A recipe for image characterization of fractal-like aggregates. *Journal of Aerosol Science*, **30**(10), 1379–1389.

Brock, J. R. 1972. Condensational growth of atmospheric aerosols. *Journal of Colloid and Interface Science*, **39**(1), 32–36.

Brock, J. R., and Hidy, G. M. 1965. Collision-rate theory and the coagulation of free-molecule aerosols. *Journal of Applied Physics*, **36**(6), 1857–1862.

Brookes, S. J., and Moss, J. B. 1999. Predictions of soot and thermal radiation properties in confined turbulent jet diffusion flames. *Combustion and Flame*, **116**(4), 486–503.

Brown, R. J., Bonadonna, C., and Durant, A. J. 2012. A review of volcanic ash aggregation. *Physics and Chemistry of the Earth*, **45–46**, 65–78.

Buaria, D., Clay, M. P., Sreenivasan, K. R., and Yeung, P. K. 2021. Small-scale isotropy and ramp-cliff structures in scalar turbulence. *Physical Review Letters*, **126**(3), 034504.

Buesser, B., and Gröhn, A. J. 2012. Multiscale aspects of modeling gas-phase nanoparticle synthesis. *Chemical Engineering and Technology*, **35**(7), 1133–1143.

Buesser, B., and Pratsinis, S. E. 2012. Design of nanomaterial synthesis by aerosol processes. *Annual Review of Chemical and Biomolecular Engineering*, **3**, 103–127.

Calderón, S. M., Tonttila, J., Buchholz, A., Joutsensaari, J., Komppula, M., Leskinen, A., Hao, L., Moisseev, D., Pullinen, I., Tiitta, P., Xu, J., Virtanen, A., Kokkola, H., and Romakkaniemi, S. 2022. Aerosol-stratocumulus interactions: towards a better process understanding using closures between observations and large eddy simulations. *Atmospheric Chemistry and Physics*, **22**(18), 12417–12441.

Camenzind, A., Schulz, H., Teleki, A., Beaucage, G., Narayanan, T., and Pratsinis, S. E. 2008. Nanostructure evolution: from aggregated to spherical SiO_2 particles made in diffusion flames. *European Journal of Inorganic Chemistry*, **6**, 911–918.

Campos, F. B., and Lage, P. L. C. 2003. A numerical method for solving the transient multidimensional population balance equation using an Euler-Lagrange formulation. *Chemical Engineering Science*, **58**(12), 2725–2744.

Cant, R. S., and Mastorakos, E. 2008. *An Introduction to Turbulent Reacting Flows*. Imperial College Press.

Chatzopoulos, A. K., and Rigopoulos, S. 2013. A chemistry tabulation approach via rate-controlled constrained equilibrium (RCCE) and artificial neural networks (ANNs), with application to turbulent non-premixed $CH_4/H_2/N_2$ flames. *Proceedings of the Combustion Institute*, **34**(1), 1465–1473.

Chen, D., and Luo, K. H. 2020. Reactive sites on the surface of polycyclic aromatic hydrocarbon clusters: a numerical study. *Combustion and Flame*, **211**, 362–373.

Chen, D., Zainuddin, Z., Yapp, E., Akroyd, J., Mosbach, S., and Kraft, M. 2013. A fully coupled simulation of PAH and soot growth with a population balance model. *Proceedings of the Combustion Institute*, **34**(1), 1827–1835.

Chen, J.-Y. 1997. Stochastic modeling of partially stirred reactors. *Combustion Science and Technology*, **122**(1–6), 63–94.

Cichocki, B. (ed). 2017. *Marian Smoluchowski: Selected Scientific Works*. Wydawnictwa Uniwersytetu Warszawskiego.

Cifuentes, L., Sellmann, J., Wlokas, I., and Kempf, A. 2020. Direct numerical simulations of nanoparticle formation in premixed and non-premixed flame-vortex interactions. *Physics of Fluids*, **32**(9), 093605.

Clift, R., Grace, J. R., and Weber, M. E. 1978. *Bubbles, Drops, and Particles*. Academic Press.

Coelho, P. J. 2007. Numerical simulation of the interaction between turbulence and radiation in reactive flows. *Progress in Energy and Combustion Science*, **33**(4), 311–383.

Coulaloglou, C. A., and Tavlarides, L. L. 1977. Description of interaction processes in agitated liquid-liquid dispersions. *Chemical Engineering Science*, **32**(11), 1289–1297.

Crank, J. 1975. *The Mathematics of Diffusion*. 2nd edn. Oxford University Press.

Crowe, C. T., Sommerfeld, M., and Tsuji, Y. 1998. *Multiphase Flows with Droplets and Particles*. CRC Press.

Curl, R. L. 1963. Dispersed phase mixing: I. Theory and effects in simple reactors. *AIChE Journal*, **9**(2), 175–181.

D'Anna, A. 2008. Detailed kinetic modeling of particulate formation in rich premixed flames of ethylene. *Energy and Fuels*, **22**(3), 1610–1619.

D'Anna, A. 2009. Combustion-formed nanoparticles. *Proceedings of the Combustion Institute*, **32**(1), 593–613.

D'Anna, A., and Kent, J. H. 2003. Aromatic formation pathways in non-premixed methane flames. *Combustion and Flame*, **132**(4), 715–722.

D'Anna, A., and Kent, J. H. 2008. A model of particulate and species formation applied to laminar, nonpremixed flames for three aliphatic-hydrocarbon fuels. *Combustion and Flame*, **152**(4), 573–587.

D'Anna, A., Sirignano, M., and Kent, J. 2010. A model of particle nucleation in premixed ethylene flames. *Combustion and Flame*, **157**(11), 2106–2115.

Davidson, P. A. 2004. *Turbulence: An Introduction for Scientists and Engineers*. Oxford University Press.

Davies, S. C., King, J. R., and Wattis, J. A. D. 1999. Self-similar behaviour in the coagulation equations. *Journal of Engineering Mathematics*, **36**(1–2), 57–88.

De La Mora, J. F., and Rosner, D. E. 1982. Effects of inertia on the diffusional deposition of small particles to spheres and cylinders at low Reynolds numbers. *Journal of Fluid Mechanics*, **125**, 379–395.

De Oliveira, P. M., Mesquita, L. C. C., Gkantonas, S., Giusti, A., and Mastorakos, E. 2021. Evolution of spray and aerosol from respiratory releases: theoretical estimates for insight on viral transmission. *Proceedings of the Royal Society A: Mathematical, Physical and Engineering Sciences*, **477**, 20200584.

Derksen, J. J. 2012. Direct simulations of mixing of liquids with density and viscosity differences. *Industrial & Engineering Chemistry Research*, **51**(19), 6948–6957.

Di Veroli, G., and Rigopoulos, S. 2009. A study of turbulence–chemistry interaction in reactive precipitation via a Population Balance-transported PDF method. In: Hanjalić, K, Nagano, Y, and Jakirlić, S (eds), *Turbulence, Heat and Mass Transfer 6. Proceedings of the Sixth International Symposium on Turbulence, Heat and Mass Transfer, Rome, Italy*. Begell House, Inc.

Di Veroli, G., and Rigopoulos, S. 2010. Modeling of turbulent precipitation: a transported population balance-PDF method. *AIChE Journal*, **56**(4), 878–892.

Di Veroli, G. Y. 2009. Stochastic modelling of particle formation in turbulent flows via the transported population balance – PDF method. Ph.D. thesis, The University of Manchester.

Di Veroli, G. Y., and Rigopoulos, S. 2011. Modeling of aerosol formation in a turbulent jet with the transported population balance equation-probability density function approach. *Physics of Fluids*, **23**(4), 043305.

Diemer, R. B., Jr. 2023. Simplifying population balance models to promote broader use in industry. *Chemical Engineering Research and Design*, **190**, 91–105.

Diemer, R. B., and Olson, J. H. 2002. A moment methodology for coagulation and breakage problems: Part 3 – generalized daughter distribution functions. *Chemical Engineering Science*, **57**(19), 4187–4198.

Ding, T., Readshaw, T., Rigopoulos, S., and Jones, W. P. 2021. Machine learning tabulation of thermochemistry in turbulent combustion: an approach based on hybrid flamelet/random data and multiple multilayer perceptrons. *Combustion and Flame*, **231**, 111493.

Ding, T., Rigopoulos, S., and Jones, W. P. 2022. Machine learning tabulation of thermochemistry of fuel blends. *Applications in Energy and Combustion Science*, **12**, 100086.

Dobbins, R. A., and Mulholland, G. W. 1984. Interpretation of optical measurements of flame generated particles. *Combustion Science and Technology*, **40**(5–6), 175–191.

Donde, P., Raman, V., Mueller, M. E., and Pitsch, H. 2013. LES/PDF based modeling of soot-turbulence interactions in turbulent flames. *Proceedings of the Combustion Institute*, **34**(1), 1183–1192.

Donzis, D. A., and Yeung, P. K. 2010. Resolution effects and scaling in numerical simulations of passive scalar mixing in turbulence. *Physica D: Nonlinear Phenomena*, **239**(14), 1278–1287.

Donzis, D. A., Sreenivasan, K. R., and Yeung, P. K. 2010. The Batchelor spectrum for mixing of passive scalars in isotropic turbulence: submitted for the special issue dedicated to S. B. Pope. *Flow, Turbulence and Combustion*, **85**(3–4), 549–566.

Donzis, D. A., Aditya, K., Sreenivasan, K. R., and Yeung, P. K. 2014. The turbulent Schmidt number. *Journal of Fluids Engineering, Transactions of the ASME*, **136**(6), 060912.

Dopazo, C. 1994. Recent developments in pdf methods. Pages 375–474 of: Libby, P. A., and Williams, F. A. (eds), *Turbulent Reacting Flows*. Academic Press London.

Dopazo, C., and O'Brien, E. E. 1974. An approach to the autoignition of a turbulent mixture. *Acta Astronautica*, **1**(9–10), 1239–1266.

Drake, R. 1972. A general mathematical survey of the coagulation equation. Pages 201–376 of: Hidy, G. M., and Brock, J. R. (eds), *Topics in Current Aerosol Research (Part 2)*. Pergamon Press.

Drew, D. A., and Passman, S. L. 1998. *Theory of Multicomponent Fluids, Applied Mathematical Sciences, 135*. Springer.

Drossinos, Y., and Housiadas, C. 2017. Aerosols. Pages 1251–1320 of: Michaelides, E. E., Crowe, C. T., and Schwarzkopf, J. D. (eds), *Multiphase Flow Handbook*, 2nd edn. CRC Press.

Drossinos, Y., and Stilianakis, N. I. 2020. What aerosol physics tells us about airborne pathogen transmission. *Aerosol Science and Technology*, **54**(6), 639–643.

Drossinos, Y., Reid, J. P., Hugentobler, W., and Stilianakis, N. I. 2022. Challenges of integrating aerosol dynamics into SARS-CoV-2 transmission models. *Aerosol Science and Technology*, **56**(9), 777–784.

Druzhinin, O. A. 1995. On the two-way interaction in two-dimensional particle-laden flows: the accumulation of particles and flow modification. *Journal of Fluid Mechanics*, **297**, 49–76.

Duboc, B., Ribert, G., and Domingo, P. 2019. Hybrid transported-tabulated chemistry for partially premixed combustion. *Computers and Fluids*, **179**, 206–227.

Dukowicz, J. K. 1980. A particle-fluid numerical model for liquid sprays. *Journal of Computational Physics*, **35**(2), 229–253.

Dworkin, S. B., Zhang, Q., Thomson, M. J., Slavinskaya, N. A., and Riedel, U. 2011. Application of an enhanced PAH growth model to soot formation in a laminar coflow ethylene/air diffusion flame. *Combustion and Flame*, **158**(9), 1682–1695.

Eibeck, A., and Wagner, W. 2000a. Approximative solution of the coagulation-fragmentation equation by stochastic particle systems. *Stochastic Analysis and Applications*, **18**(6), 921–948.

Eibeck, A., and Wagner, W. 2000b. An efficient stochastic algorithm for studying coagulation dynamics and gelation phenomena. *SIAM Journal on Scientific Computing*, **22**(3), 802–821.

Eibeck, A., and Wagner, W. 2001. Stochastic particle approximations for Smoluchoski's coagulation equation. *Annals of Applied Probability*, **11**(4), 1137–1165.

Eibeck, A., and Wagner, W. 2003. Stochastic interacting particle systems and nonlinear kinetic equations. *Annals of Applied Probability*, **13**(3), 845–889.

Einstein, A. 1926. *Investigations on the Theory of the Brownian Movement, Edited by R. Fürth*. Dover Publications, Inc. (1956 republication).

Elimelech, M., Gregory, J., Jia, X., and Williams, R. 1995. *Particle Deposition and Aggregation: Measurement, Modelling and Simulation*. Butterworth-Heinemann.

Enukashvili, I. M. 1964. On the solution of the kinetic coagulation equation. *Izv. Geophys. Ser.*, **10**, 944–948.

Epstein, P. S. 1924. On the resistance experienced by spheres in their motion through gases. *Physical Review*, **23**(6), 710–733.

Ernst, F. O., and Pratsinis, S. E. 2006. Self-preservation and gelation during turbulence-induced coagulation. *Journal of Aerosol Science*, **37**(2), 123–142.

Falk, L., and Schaer, E. 2001. A PDF modelling of precipitation reactors. *Chemical Engineering Science*, **56**(7), 2445–2457.

Favre, A. 1969. Statistical equations of turbulent gases. Pages 231–266 of: *Problems of Hydrodynamics and Continuum Mechanics*. SIAM, Philadelphia.

Feroughi, O. M., Deng, L., Kluge, S., Dreier, T., Wiggers, H., Wlokas, I., and Schulz, C. 2017. Experimental and numerical study of a HMDSO-seeded premixed laminar low-pressure flame for SiO_2 nanoparticle synthesis. *Proceedings of the Combustion Institute*, **36**(1), 1045–1053.

Ferry, J., and Balachandar, S. 2001. A fast Eulerian method for disperse two-phase flow. *International Journal of Multiphase Flow*, **27**(7), 1199–1226.

Ferziger, J. H., Perić, M., and Street, R. L. 2020. *Computational Methods for Fluid Dynamics*. 4th edn. Springer.

Filbet, F., and Laurençot, P. 2004. Numerical simulation of the Smoluchowski coagulation equation. *SIAM Journal on Scientific Computing*, **25**(6), 2004–2028.

Finke, J., and Sewerin, F. 2023. Combining a population balance approach with detailed chemistry to model the condensation of oxide smoke during aluminum combustion in spatially homogeneous reactors. *Combustion and Flame*, **248**, 112510.

Finlayson, B. A. 2014. *The Method of Weighted Residuals and Variational Principles*. SIAM, republication of the work first published by Academic Press in 1972.

Fletcher, T. H., Ma, J., Rigby, J. R., Brown, A. L., and Webb, B. W. 1997. Soot in coal combustion systems. *Progress in Energy and Combustion Science*, **23**(3), 283–301.

Ford, I. J. 2004. Statistical mechanics of nucleation: a review. *Proceedings of the Institution of Mechanical Engineers, Part C: Journal of Mechanical Engineering Science*, **218**(8), 883–899.

Fox, R. O. 1998. On the relationship between Lagrangian micromixing models and computational fluid dynamics. *Chemical Engineering and Processing: Process Intensification*, **37**(6), 521–535.

Fox, R. O. 2003. *Computational Models for Turbulent Reacting Flows*. Cambridge University Press.

Fox, R. O. 2007. Introduction and fundamentals of modeling approaches for polydisperse multiphase flows. Pages 1–40 of: Marchisio, D. L., and Fox, R. O. (eds), *Multiphase Reacting Flows: Modelling and Simulation*. SpringerWienNewYork.

Franke, L. L. C., Chatzopoulos, A. K., and Rigopoulos, S. 2017. Tabulation of combustion chemistry via Artificial Neural Networks (ANNs): methodology and application to LES-PDF simulation of Sydney flame L. *Combustion and Flame*, **185**, 245–260.

Fredrickson, A. G., and Tsuchiya, H. M. 1963. Continuous propagation of microorganisms. *AIChE Journal*, **9**(4), 459–468.

Fredrickson, A. G., Ramkrishna, D., and Tsuchiya, H. M. 1967. Statistics and dynamics of procaryotic cell populations. *Mathematical Biosciences*, **1**(3), 327–374.

Frenkel, J. 1945. Viscous flow of crystalline bodies under the action of surface tension. *Journal of Physics*, **9**, 385.

Frenklach, M. 1985. Computer modeling of infinite reaction sequences: a chemical lumping. *Chemical Engineering Science*, **40**(10), 1843–1849.

Frenklach, M. 2002a. Method of moments with interpolative closure. *Chemical Engineering Science*, **57**(12), 2229–2239.

Frenklach, M. 2002b. Reaction mechanism of soot formation in flames. *Physical Chemistry Chemical Physics*, **4**(11), 2028–2037.

Frenklach, M., and Harris, S. J. 1987. Aerosol dynamics modeling using the method of moments. *Journal of Colloid and Interface Science*, **118**(1), 252–261.

Frenklach, M., and Wang, H. 1991. Detailed modeling of soot particle nucleation and growth. *Symposium (International) on Combustion*, **23**(1), 1559–1566.

Frenklach, M., and Wang, H. 1994. Detailed mechanism and modeling of soot particle formation. Pages 165–192 of: Bockhorn, H. (ed), *Soot Formation in Combustion*. Springer Series in Chemical Physics 59. Springer-Verlag.

Frenklach, M. and Wang, H. and Goldenberg, M. and Smith, G. P. and Golden, D. M., and Bowman, C. T., and Hanson, R. K., and Gardiner, W. C., and Lissianski, V. 1995. *GRI-MECH: An optimized detailed chemical reaction mechanism for methane combustion*. Gas Research Institute Topical Report No. GRI-95/0058.

Friedlander, S. K. 1960. On the particle size spectrum of a condensing vapor. *Physics of Fluids*, **3**(5), 693–696.

Friedlander, S. K. 2000. *Smoke, Dust, and Haze: Fundamentals of Aerosol Dynamics*. 2nd edn. Oxford University Press.

Friedlander, S. K., and Wang, C. S. 1966. The self-preserving particle size distribution for coagulation by Brownian motion. *Journal of Colloid and Interface Science*, **22**(2), 126–132.

Fröhlich, J., and Rodi, W. 2002. Introduction to large eddy simulation of turbulent flows. Pages 267–298 of: Launder, B. E., and Sandham, N. D. (eds), *Closure Strategies for Turbulent and Transitional Flows*. Cambridge University Press.

Froment, G. F., Bischoff, K. B., and De Wilde, J. 2011. *Chemical Reactor Analysis and Design*. 3rd edn. John Wiley & Sons.

Fuchs, N. A. 1964. *The Mechanics of Aerosols*. Pergamon, New York.

Fuchs, N. A., and Sutugin, A. G. 1971. High-dispersed aerosols. Pages 1–60 of: Hidy, G. M., and Brock, J. R. (eds), *Topics in Current Aerosol Research*. Pergamon Press.

Gany, A., Manheimer-Timnat, Y., and Wolfshtein, M. 1976. Two-phase flow effects on hybrid combustion. *Acta Astronautica*, **3**(3–4), 241–263.

Gao, F., and O'Brien, E. E. 1993. A large-eddy simulation scheme for turbulent reacting flows. *Physics of Fluids A*, **5**(6), 1282–1284.

Garcia Gonzalez, C. E. 2018. Characterization of nanoparticles generated in reacting flows. Ph.D. thesis, Imperial College London.

Gardiner, C. W. 1985. *Handbook of Stochastic Methods for Physics, Chemistry and the Natural Sciences*. Springer.

Garmory, A., and Mastorakos, E. 2008. Aerosol nucleation and growth in a turbulent jet using the Stochastic Fields method. *Chemical Engineering Science*, **63**(16), 4078–4089.

Garrick, S. C., and Khakpour, M. 2004. The effects of differential diffusion on nanoparticle coagulation in temporal mixing layers. *Aerosol Science and Technology*, **38**(8), 851–860.

Gavi, E., Marchisio, D. L., and Barresi, A. A. 2007a. CFD modelling and scale-up of Confined Impinging Jet Reactors. *Chemical Engineering Science*, **62**(8), 2228–2241.

Gavi, E., Rivautella, L., Marchisio, D. L., Vanni, M., Barresi, A. A., and Baldi, G. 2007b. CFD modelling of nano-particle precipitation in confined impinging jet reactors. *Chemical Engineering Research and Design*, **85**(5 A), 735–744.

Gelbard, F. 1990. Modeling multicomponent aerosol particle growth by vapor condensation. *Aerosol Science and Technology*, **12**(2), 399–412.

Gelbard, F., and Seinfeld, J. H. 1978a. Numerical solution of the dynamic equation for particulate systems. *Journal of Computational Physics*, **28**(3), 357–375.

Gelbard, F. M., and Seinfeld, J. H. 1978b. Coagulation and growth of a multicomponent aerosol. *Journal of Colloid And Interface Science*, **63**(3), 472–479.

Gelbard, F., and Seinfeld, J. H. 1979. The general dynamic equation for aerosols. Theory and application to aerosol formation and growth. *Journal of Colloid and Interface Science*, **68**(2), 363–382.

Gelbard, F., and Seinfeld, J. H. 1980. Simulation of multicomponent aerosol dynamics. *Journal of Colloid and Interface Science*, **78**(2), 485–501.

Gelbard, F., Tambour, Y., and Seinfeld, J. H. 1980. Sectional representations for simulating aerosol dynamics. *Journal of Colloid and Interface Science*, **76**(2), 541–556.

Ghadiri, M., and Zhang, Z. 2002. Impact attrition of particulate solids. Part 1: a theoretical model of chipping. *Chemical Engineering Science*, **57**(17), 3659–3669.

Gillespie, D. T. 1975. An exact method for numerically simulating the stochastic coalescence process in a cloud. *Journal of the Atmospheric Sciences*, **32**(10), 1977–1989.

Glassman, I. 1988. Soot formation in combustion processes. *Symposium (International) on Combustion*, **22**(1), 295–311.

Golovin, A. M. 1963. On the spectrum of coagulating cloud droplets, II. *Izv. Geophys. Ser.*, **9**, 880–884.

Goodson, M., and Kraft, M. 2002. An efficient stochastic algorithm for simulating nanoparticle dynamics. *Journal of Computational Physics*, **183**(1), 210–232.

Gosman, A. D., and Ioannides, E. 1983. Aspects of computer simulation of liquid-fueled combustors. *Journal of Energy*, **7**(6), 482–490.

Goudeli, E., Eggersdorfer, M. L., and Pratsinis, S. E. 2015. Aggregate characteristics accounting for the evolving fractal-like structure during coagulation and sintering. *Journal of Aerosol Science*, **89**, 58–68.

Gradl, J., and Peukert, W. 2009. Simultaneous 3D observation of different kinetic subprocesses for precipitation in a T-mixer. *Chemical Engineering Science*, **64**(4), 709–720.

Greenberg, J. B., Silverman, I., and Tambour, Y. 1993. On the origins of spray sectional conservation equations. *Combustion and Flame*, **93**(1–2), 90–96.

Griffiths, P. R., and De Haseth, J. A. 2007. *Fourier Transform Infrared Spectrometry*. John Wiley & Sons.

Grimmett, G., and Stirzaker, D. 2001. *Probability and Random Processes*. 3rd edn. Oxford University Press.

Grinstein, F. F., Margolin, L. G., and Rider, W. J. 2007. *Implicit Large Eddy Simulation: Computing Turbulent Fluid Dynamics*. Cambridge University Press.

Grosshandler, W. L. 1993. *RADCAL: A narrow-band model for radiation calculations in a combustion environment*. NIST Technical Note 1402. National Institute of Standards and Technology.

Gunawan, R., Fusman, I., and Braatz, R. D. 2004. High resolution algorithms for multidimensional population balance equations. *AIChE Journal*, **50**(11), 2738–2749.

Guo, H., Liu, F., Smallwood, G. J., and Gülder, Ö. L. 2002. The flame preheating effect on numerical modelling of soot formation in a two-dimensional laminar ethylene-air diffusion flame. *Combustion Theory and Modelling*, **6**(2), 173–187.

Guo, H., Liu, F., Smallwood, G. J., and Gülder, Ö. L. 2006. Numerical study on the influence of hydrogen addition on soot formation in a laminar ethylene-air diffusion flame. *Combustion and Flame*, **145**(1–2), 324–338.

Gupta, H. C., and Bracco, F. V. 1978. Numerical computations of two-dimensional unsteady sprays for application to engines. *AIAA Journal*, **16**(10), 1053–1061.

Hall, R. J., Smooke, M. D., and Colket, M. B. 1997. Predictions of soot dynamics in opposed jet diffusion flames. Pages 189–229 of: Dryer, F. L., and Sawyer, R. F. (eds), *Physical and Chemical Aspects of Combustion: A Tribute to Irvin Glassman*. Combustion Science and Technology Book Series, vol. 4. Gordon and Breach.

Hanjalić, K., and Launder, B. 2022. *Modelling Turbulence in Engineering and the Environment: Rational Alternative Routes to Closure*. 2nd edn. Cambridge University Press.

Harris, S. J., and Kennedy, I. M. 1988. The coagulation of soot particles with van der Waals forces. *Combustion Science and Technology*, **59**(4–6), 443–454.

Harris, S. J., and Weiner, A. M. 1983a. Surface growth of soot particles in premixed ethylene/air flames. *Combustion Science and Technology*, **31**(3–4), 155–167.

Harris, S. J., and Weiner, A. M. 1983b. Determination of the rate constant for soot surface growth. *Combustion Science and Technology*, **32**(5–6), 267–275.

Harris, S. J., and Weiner, A. M. 1985. Chemical kinetics of soot particle growth. *Annual Review of Physical Chemistry*, **36**, 31–52.

Haworth, D.C. 2010. Progress in probability density function methods for turbulent reacting flows. *Progress in Energy and Combustion Science*, **36**(2), 168–259.

Haynes, B. S., and Wagner, H. Gg. 1981. Soot formation. *Progress in Energy and Combustion Science*, **7**(4), 229–273.

Heath, M. T. 2002. *Scientific Computing: An Introductory Survey*. 2nd edn. McGraw-Hill.

Hidy, G. M. 1965. On the theory of the coagulation of noninteracting particles in Brownian motion. *Journal of Colloid Science*, **20**(2), 123–144.

Hidy, G. M., and Brock, J. R. 1970. *The Dynamics of Aerocolloidal Systems*. Pergamon Press.

Hill, P. J., and Ng, K. M. 1995. New discretization procedure for the breakage equation. *AIChE Journal*, **41**(5), 1204–1216.

Hill, P. J., and Ng, K. M. 1996a. New discretization procedure for the agglomeration equation. *AIChE Journal*, **42**(3), 727–741.

Hill, P. J., and Ng, K. M. 1996b. Statistics of multiple particle breakage. *AIChE Journal*, **42**(6), 1600–1611.

Hirsch, C. 2007. *Numerical Computation of Internal and External Flows: Volume 1, Fundamentals of Computational Fluid Dynamics*. 2nd edn. Butterworth-Heinemann.

Hirschfelder, J. O., Curtiss, C. F., and Bird, R. B. 1964. *Molecular Theory of Gases and Liquids*. John Wiley & Sons.

Hjertager, B. H. 2007. Multi-fluid CFD analysis of chemical reactors. Pages 125–179 of: Marchisio, D. L., and Fox, R. O. (eds), *Multiphase Reacting Flows: Modelling and Simulation*. SpringerWienNewYork.

Hjortsø, M. A. 2004. *Population Balances in Biomedical Engineering: Segregation Through the Distribution of Cell States*. McGraw-Hill.

Hounslow, M. J. 1990. Nucleation, growth, and aggregation rates from steady-state experimental data. *AIChE Journal*, **36**(11), 1748–1752.

Hounslow, M. J. 1998. The population balance as a tool for understanding particle rate processes. *KONA Powder and Particle Journal*, **16**(May), 179–193.

Hounslow, M. J., Ryall, R. L., and Marshall, V. R. 1988. A discretized population balance for nucleation, growth, and aggregation. *AIChE Journal*, **34**(11), 1821–1832.

Hulburt, H. M., and Akiyama, T. 1969. Liouville equations for agglomeration and dispersion processes. *Industrial & Engineering Chemistry Fundamentals*, **8**(2), 319–324.

Hulburt, H. M., and Katz, S. 1964. Some problems in particle technology. A statistical mechanical formulation. *Chemical Engineering Science*, **19**(8), 555–574.

ISF Workshop. 2018. 'Accessed on 01/04/2024' www.adelaide.edu.au/cet/isfworkshop/.

Ishii, M., and Hibiki, T. 2005. *Thermo-fluid Dynamics of Two-Phase Flow*. Springer.

Iyer, S. S., Litzinger, T. A., Lee, S.-Y., and Santoro, R. J. 2007. Determination of soot scattering coefficient from extinction and three-angle scattering in a laminar diffusion flame. *Combustion and Flame*, **149**(1–2), 206–216.

Jain, A., and Xuan, Y. 2019. Effects of large aromatic precursors on soot formation in turbulent non-premixed sooting jet flames. *Combustion Theory and Modelling*, **23**(3), 439–466.

Jakobsen, H. A. 2014. *Chemical Reactor Modeling: Multiphase Reactive Flows*. 2nd edn. Springer.

Jakobsen, H. A., Lindborg, H., and Dorao, C. A. 2005. Modeling of bubble column reactors: progress and limitations. *Industrial & Engineering Chemistry Research*, **44**(14), 5107–5151.

Jensen, D. E. 1974. Prediction of soot formation rates: a new approach. *Proceedings of the Royal Society of London A*, **338**, 375–396.

Jeong, J. I., and Choi, M. 2001. A sectional method for the analysis of growth of polydisperse non-spherical particles undergoing coagulation and coalescence. *Journal of Aerosol Science*, **32**(5), 565–582.

Jones, W. P. 2002. The joint scalar probability density function method. Pages 582–625 of: Launder, B. E., and Sandham, N. D. (eds), *Closure Strategies for Turbulent and Transitional Flows*. Cambridge University Press.

Jones, W. P., and Launder, B. E. 1972. The prediction of laminarization with a two-equation model of turbulence. *International Journal of Heat and Mass Transfer*, **15**(2), 301–314.

Jones, W. P., and Navarro-Martinez, S. 2007. Large eddy simulation of autoignition with a subgrid probability density function method. *Combustion and Flame*, **150**(3), 170–187.

Jones, W. P., and Prasad, V. N. 2010. Large eddy simulation of the Sandia flame series (D-F) using the Eulerian stochastic field method. *Combustion and Flame*, **157**(9), 1621–1636.

Jones, W. P., and Sheen, D.-H. 2000. Probability density function method for modelling liquid fuel sprays. *Flow, Turbulence and Combustion*, **63**(1), 379–394.

Jones, W. P., di Mare, F., and Marquis, A. J. 2002. *LES-BOFFIN: User's Guide*.

Kammler, H. K., Pratsinis, S. E., Morrison Jr., P. W., and Hemmerling, B. 2002. Flame temperature measurements during electrically assisted aerosol synthesis of nanoparticles. *Combustion and Flame*, **128**(4), 369–381.

Kaplan, C. R., and Gentry, J. W. 1988. Agglomeration of chain-like combustion aerosols due to Brownian motion. *Aerosol Science and Technology*, **8**(1), 11–28.

Kazakov, A., and Frenklach, M. 1998. Dynamic modeling of soot particle coagulation and aggregation: implementation with the method of moments and application to high-pressure laminar premixed flames. *Combustion and Flame*, **114**(3–4), 484–501.

Kearney, S. P., Guildenbecher, D. R., Winters, C., Farias, P. A., Grasser, T. W., and Hewson, J. C. 2015. *Temperature, oxygen, and soot-volume-fraction measurements in a turbulent* C_2H_4*-fueled jet flame*. Tech. rept. SAND2015-7968. Sandia National Laboratories.

Kee, R. J., Coltrin, M. E., and Glarborg, P. 2003. *Chemically Reacting Flow: Theory and Practice*. 1st edn. Wiley.

Kelesidis, G. A., and Pratsinis, S. E. 2021. Determination of the volume fraction of soot accounting for its composition and morphology. *Proceedings of the Combustion Institute*, **38**(1), 1189–1196.

Kelesidis, G. A., Goudeli, E., and Pratsinis, S. E. 2017. Flame synthesis of functional nanostructured materials and devices: surface growth and aggregation. *Proceedings of the Combustion Institute*, **36**(1), 29–50.

Kendall, D. G. 1950. An artificial realization of a simple 'birth-and-death' process. *Journal of the Royal Statistical Society, Series B: (Methodological)*, **12**(1), 116–119.

Kennedy, I. M. 1990. Enhancement of silica aerosol coagulation by Van der Waals forces. *Aerosol Science and Technology*, **12**(4), 869–875.

Kennedy, I. M. 1997. Models of soot formation and oxidation. *Progress in Energy and Combustion Science*, **23**(2), 95–132.

Kennedy, I. M., Yam, C., Rapp, D. C., and Santoro, R. J. 1996. Modeling and measurements of soot and species in a laminar diffusion flame. *Combustion and Flame*, **107**(4), 368–382.

Kholghy, M. R., Veshkini, A., and Thomson, M. J. 2016. The core-shell internal nanostructure of soot – a criterion to model soot maturity. *Carbon*, **100**, 508–536.

Khosousi, A., and Dworkin, S. B. 2015. Soot surface reactivity during surface growth and oxidation in laminar diffusion flames. *Combustion and Flame*, **162**(12), 4523–4532.

Kim, Y. P., and Seinfeld, J. H. 1990. Simulation of multicomponent aerosol condensation by the moving sectional method. *Journal of Colloid and Interface Science*, **135**(1), 185–199.

Kiparissides, C. 2006. Challenges in particulate polymerization reactor modeling and optimization: a population balance perspective. *Journal of Process Control*, **16**(3), 205–224.

Kittelson, D. B. 1998. Engines and nanoparticles: a review. *Journal of Aerosol Science*, **29**(5–6), 575–588.

Kloeden, P. E., and Platen, E. 1992. *Numerical Solution of Stochastic Differential Equations*. Springer.

Kobata, A., Kusakabe, K., and Morooka, S. 1991. Growth and transformation of TiO_2 crystallites in aerosol reactor. *AIChE Journal*, **37**(3), 347–359.

Koch, W., and Friedlander, S. K. 1990. The effect of particle coalescence on the surface area of a coagulating aerosol. *Journal of Colloid and Interface Science*, **140**(2), 419–427.

Kodas, T. T., and Hampden-Smith, M. J. 1999. *Aerosol Processing of Materials*. Wiley-VCH.

Kollmann, W. 1990. The pdf approach to turbulent flow. *Theoretical and Computational Fluid Dynamics*, **1**(5), 249–285.

Kollmann, W. 1995. Probability Density Function (PDF) closures for scalar transport in turbulent flows. Pages 621–654 of: Habashi, W. G., and Hafez, M. M. (eds), *Computational Fluid Dynamics Techniques*. Gordon and Breach Publishers.

Kollmann, W., Kennedy, I. M., Metternich, M., and Chen, J.-Y. 1994. Application of a soot model to a turbulent ethylene di usion flame. Pages 503–526 of: Bockhorn, H. (ed), *Soot Formation in Combustion*. Springer Series in Chemical Physics 59. Springer-Verlag.

Koren, B. 1993. A robust upwind discretization method for advection, diffusion and source terms. Pages 117–138 of: Vreugdenhil, C. B., and Koren, B. (eds), *Numerical Methods for Advection-Diffusion Problems*. Notes on Numerical Fluid Mechanics and Multidisciplinary Design, vol. 45. Vieweg Verlag.

Kostoglou, M. 2007a. Extended cell average technique for the solution of coagulation equation. *Journal of Colloid and Interface Science*, **306**(1), 72–81.

Kostoglou, M. 2007b. The linear breakage equation: from fundamental issues to numerical solution techniques. Pages 793–835 of: Williams, J. C., Allen, T., Salman, A. D., Ghadiri, M., and Hounslow, M. J. (eds), *Handbook of Powder Technology, Volume 12: Particle Breakage*. Elsevier.

Kostoglou, M., and Karabelas, A. J. 1994. Evaluation of zero order methods for simulating particle coagulation. *Journal of Colloid and Interface Science*, **163**(2), 420–431.

Kostoglou, M., and Karabelas, A. J. 1995. Evaluation of numerical methods for simulating an evolving particle size distribution in growth processes. *Chemical Engineering Communications*, **136**(1), 177–199.

Kostoglou, M., and Karabelas, A. J. 2001. On the breakage problem with a homogeneous erosion type kernel. *Journal of Physics A: Mathematical and General*, **34**(8), 1725–1740.

Kostoglou, M., and Karabelas, A. J. 2004. Optimal low order methods of moments for solving the fragmentation equation. *Powder Technology*, **143–144**, 280–290.

Kostoglou, M., and Karabelas, A. J. 2005. Toward a unified framework for the derivation of breakage functions based on the statistical theory of turbulence. *Chemical Engineering Science*, **60**(23), 6584–6595.

Kostoglou, M., and Konstandopoulos, A. G. 2001. Evolution of aggregate size and fractal dimension during Brownian coagulation. *Journal of Aerosol Science*, **32**(12), 1399–1420.

Kostoglou, M., Dovas, S., and Karabelas, A. J. 1997. On the steady-state size distribution of dispersions in breakage processes. *Chemical Engineering Science*, **52**(8), 1285–1299.

Kostoglou, M., Konstandopoulos, A. G., and Friedlander, S. K. 2006. Bivariate population dynamics simulation of fractal aerosol aggregate coagulation and restructuring. *Journal of Aerosol Science*, **37**(9), 1102–1115.

Kotalczyk, G., and Kruis, F. E. 2020. Compartmental population balances by means of Monte Carlo methods. Page 519–548 of: Heinrich, S. (ed), *Dynamic Flowsheet Simulation of Solids Processes*. Springer Nature Switzerland AG.

Köylü, Ü. Ö., McEnally, C. S., Rosner, D. E., and Pfefferle, L. D. 1997. Simultaneous measurements of soot volume fraction and particle size / microstructure in flames using a thermophoretic sampling technique. *Combustion and Flame*, **110**(4), 494–507.

Kreyling, W. G., Semmler-Behnke, M., and Möller, W. 2006. Ultrafine particle – lung interactions: does size matter? *Journal of Aerosol Medicine*, **19**(1), 74–83.

Kronenburg, A., Bilger, R. W., and Kent, J. H. 2000. Modeling soot formation in turbulent methane-air jet diffusion flames. *Combustion and Flame*, **121**(1–2), 24–40.

Kruis, F. E., and Kusters, K. A. 1997. The collision rate of particles in turbulent flow. *Chemical Engineering Communications*, **158**, 201–230.

Kruis, F. E., Kusters, K. A., and Pratsinis, S. E. 1993. A simple model for the evolution of the characteristics of aggregate particles undergoing coagulation and sintering. *Aerosol Science and Technology*, **19**(4), 514–526.

Kruis, F. E., Fissan, H., and Peled, A. 1998. Synthesis of nanoparticles in the gas phase for electronic, optical and magnetic applications – a review. *Journal of Aerosol Science*, **29**(5–6), 511–535.

Kruis, F. E., Maisels, A., and Fissan, H. 2000. Direct simulation Monte Carlo method for particle coagulation and aggregation. *AIChE Journal*, **46**(9), 1735–1742.

Kucher, M., Babic, D., and Kind, M. 2006. Precipitation of barium sulfate: experimental investigation about the influence of supersaturation and free lattice ion ratio on particle formation. *Chemical Engineering and Processing: Process Intensification*, **45**(10), 900–907.

Kumar, J., Peglow, M., Warnecke, G., Heinrich, S., and Mörl, L. 2006. Improved accuracy and convergence of discretized population balance for aggregation: the cell average technique. *Chemical Engineering Science*, **61**(10), 3327–3342.

Kumar, S., and Ramkrishna, D. 1996a. On the solution of population balance equations by discretization – I. A fixed pivot technique. *Chemical Engineering Science*, **51**(8), 1311–1332.

Kumar, S., and Ramkrishna, D. 1996b. On the solution of population balance equations by discretization – II. A moving pivot technique. *Chemical Engineering Science*, **51**(8), 1333–1342.

Kumar, S., and Ramkrishna, D. 1997. On the solution of population balance equations by discretization – III. Nucleation, growth and aggregation of particles. *Chemical Engineering Science*, **52**(24), 4659–4679.

Lage, P. L. C. 2002. Comments on the 'An analytical solution to the population balance equation with coalescence and breakage – the special case with constant number of particles' by D.P. Patil and J.R.G. Andrews [*Chemical Engineering Science* **53**(3) 599–601]. *Chemical Engineering Science*, **57**(19), 4253–4254.

Lai, F. S., Friedlander, S. K., Pich, J., and Hidy, G. M. 1972. The self-preserving particle size distribution for Brownian coagulation in the free-molecule regime. *Journal of Colloid and Interface Science*, **39**(2), 395–405.

Landgrebe, J. D., and Pratsinis, S. E. 1989. Gas-phase manufacture of particulates: interplay of chemical reaction and aerosol coagulation in the free-molecular regime. *Industrial & Engineering Chemistry Research*, **28**(10), 1474–1481.

Landgrebe, J. D., and Pratsinis, S. E. 1990. A discrete-sectional model for particulate production by gas-phase chemical reaction and aerosol coagulation in the free-molecular regime. *Journal of Colloid and Interface Science*, **139**(1), 63–86.

Laurent, F., and Massot, M. 2001. Multi-fluid modelling of laminar polydisperse spray flames: origin, assumptions and comparison of sectional and sampling methods. *Combustion Theory and Modelling*, **5**(4), 537–572.

Lee, G., Meyer, X. M., Biscans, B., Le Lann, J. M., and Yoon, E. S. 1999. Adaptive finite difference method for the simulation of batch crystallization. *Computers and Chemical Engineering*, **23**(SUPPL. 1), S363–S366.

Lee, G., Yoon, E. S., Lim, Y.-I., Le Lann, J. M., Meyer, X.-M., and Joulia, X. 2001. Adaptive mesh method for the simulation of crystallization processes including agglomeration and breakage: the potassium sulfate system. *Industrial and Engineering Chemistry Research*, **40**(26), 6228–6235.

Lee, J. 2011. Estimation of emission properties for silica particles using thermal radiation spectroscopy. *Applied Optics*, **50**(22), 4262–4267.

Lee, K. B., Thring, M. W., and Beér, J. M. 1962. On the rate of combustion of soot in a laminar soot flame. *Combustion and Flame*, **6**(C), 137–145.

Lee, K. W. 1983. Change of particle size distribution during Brownian coagulation. *Journal of Colloid and Interface Science*, **92**(2), 315–325.

Lee, K. W., and Chen, H. 1984. Coagulation rate of polydisperse particles. *Aerosol Science and Technology*, **3**(3), 327–334.

Lee, K. W., Chen, H., and Gieseke, J. A. 1984. Log-normally preserving size distribution for Brownian coagulation in the free-molecule regime. *Aerosol Science and Technology*, **3**(1), 53–62.

Lee, M. H. 2000. On the validity of the coagulation equation and the nature of runaway growth. *Icarus*, **143**(1), 74–86.

Leer, M., Pettit, M. W. A., Lipkowicz, J. T., Domingo, P., Vervisch, L., and Kempf, A. M. 2022. A conservative Eulerian-Lagrangian decomposition principle for the solution of multi-scale flow problems at high Schmidt or Prandtl numbers. *Journal of Computational Physics*, **464**, 111216.

Lemons, D. S., and Gythiel, A. 1997. Paul Langevin's 1908 paper 'on the theory of Brownian motion' ['Sur la théorie du mouvement brownien,' C. R. Acad. Sci. (Paris) 146, 530–533 (1908)]. *American Journal of Physics*, **65**(11), 1079–1080.

Leschziner, M. 2016. *Statistical Turbulence Modelling for Fluid Dynamics – Demystified: An Introductory Text for Graduate Engineering Students*. Imperial College Press.

Leung, K. M., Lindstedt, R. P., and Jones, W. P. 1991. A simplified reaction mechanism for soot formation in nonpremixed flames. *Combustion and Flame*, **87**(3–4), 289–305.

Leveque, R. J. 2002. *Finite Volume Methods for Hyperbolic Problems*. Cambridge University Press.

Lewis, A., Seckler, M., Kramer, H., and van Rosmalen, G. 2015. *Industrial Crystallization: Fundamentals and Applications*. Cambridge University Press.

Liao, Y., and Lucas, D. 2009. A literature review of theoretical models for drop and bubble breakup in turbulent dispersions. *Chemical Engineering Science*, **64**(15), 3389–3406.

Libby, P. A., and Williams, F. A. 1980. Fundamental aspects. Pages 1–43 of: Libby, P. A., and Williams, F. A. (eds), *Topics in Applied Physics, Volume 44: Turbulent Reacting Flows*. Springer-Verlag.

Lifshitz, I. M., and Slyozov, V. V. 1961. The kinetics of precipitation from supersaturated solid solutions. *Journal of Physics and Chemistry of Solids*, **19**(1–2), 35–50.

Lindstedt, R. P., and Louloudi, S. A. 2005. Joint-scalar transported PDF modeling of soot formation and oxidation. *Proceedings of the Combustion Institute*, **30**(1), 775–782.

Link, O., Snelling, D. R., Thomson, K. A., and Smallwood, G. J. 2011. Development of absolute intensity multi-angle light scattering for the determination of polydisperse soot aggregate properties. *Proceedings of the Combustion Institute*, **33**(1), 847–854.

Litster, J. 2016. *Design and Processing of Particulate Products*. Cambridge University Press.

Litster, J., and Ennis, B. 2004. *The Science and Engineering of Granulation Processes*. Springer-Science+Business Media, B.V.

Litster, J. D., Smit, D. J., and Hounslow, M. J. 1995. Adjustable discretized population balance for growth and aggregation. *AIChE Journal*, **41**(3), 591–603.

Liu, A., and Rigopoulos, S. 2019. A conservative method for numerical solution of the population balance equation, and application to soot formation. *Combustion and Flame*, **205**, 506–521.

Liu, A., Garcia, C. E., Sewerin, F., Williams, B. A. O., and Rigopoulos, S. 2020. Population balance modelling and laser diagnostic validation of soot particle evolution in laminar ethylene diffusion flames. *Combustion and Flame*, **221**, 384–400.

Liu, F., Guo, H., Smallwood, G. J., and Gülder, Ö. L. 2003. Numerical modelling of soot formation and oxidation in laminar coflow non-smoking and smoking ethylene diffusion flames. *Combustion Theory and Modelling*, **7**(2), 301–315.

Lombart, M., and Laibe, G. 2021. Grain growth for astrophysics with discontinuous Galerkin schemes. *Monthly Notices of the Royal Astronomical Society*, **501**(3), 4298–4316.

Lucchesi, M., Abdelgadir, A., Attili, A., and Bisetti, F. 2017. Simulation and analysis of the soot particle size distribution in a turbulent nonpremixed flame. *Combustion and Flame*, **178**, 35–45.

Lundgren, T. S. 1967. Distribution functions in the statistical theory of turbulence. *Physics of Fluids*, **10**(5), 969–975.

Ma, D. L., Tafti, D. K., and Braatz, R. D. 2002. High-resolution simulation of multidimensional crystal growth. *Industrial & Engineering Chemistry Research*, **41**(25), 6217–6223.

Mahoney, A. W., and Ramkrishna, D. 2002. Efficient solution of population balance equations with discontinuities by finite elements. *Chemical Engineering Science*, **57**(7), 1107–1119.

Mansurov, Z. A. 2005. Soot formation in combustion processes (review). *Combustion, Explosion and Shock Waves*, **41**(6), 727–744.

Mantzaris, N. V., Daoutidis, P., and Srienc, F. 2001a. Numerical solution of multivariable cell population balance models: I. Finite difference methods. *Computers and Chemical Engineering*, **25**(11–12), 1411–1440.

Mantzaris, N. V., Daoutidis, P., and Srienc, F. 2001b. Numerical solution of multi-variable cell population balance models. II. Spectral methods. *Computers and Chemical Engineering*, **25**(11–12), 1441–1462.

Mantzaris, N. V., Daoutidis, P., and Srienc, F. 2001c. Numerical solution of multi-variable cell population balance models. III. Finite element methods. *Computers and Chemical Engineering*, **25**(11–12), 1463–1481.

Marchal, C. 2008. Soot formation and oxidation modelling in an automotive engine. Ph.D. thesis, Université d'Orléans.

Marchal, P., David, R., Klein, J. P., and Villermaux, J. 1988. Crystallization and precipitation engineering – I. An efficient method for solving population balance in crystallization with agglomeration. *Chemical Engineering Science*, **43**(1), 59–67.

Marchisio, D. L., and Fox, R. O. 2005. Solution of population balance equations using the direct quadrature method of moments. *Journal of Aerosol Science*, **36**(1), 43–73.

Marchisio, D. L., and Fox, R. O. 2013. *Computational Models for Polydisperse Particulate and Multiphase Systems*. Cambridge University Press.

Marchisio, D. L., Fox, R. O., Barresi, A. A., and Baldi, G. 2001a. On the comparison between presumed and full PDF methods for turbulent precipitation. *Industrial & Engineering Chemistry Research*, **40**(23), 5132–5139.

Marchisio, D. L., Fox, R. O., Barresi, A. A., Garbero, M., and Baldi, G. 2001b. On the simulation of turbulent precipitation in a tubular reactor via computational fluid dynamics (CFD). *Chemical Engineering Research and Design*, **79**(8), 998–1004.

Marchisio, D. L., Barresi, A. A., and Fox, R. O. 2001c. Simulation of turbulent precipitation in a semi-batch Taylor-Couette reactor using CFD. *AIChE Journal*, **47**(3), 664–676.

Marchisio, D. L., Vigil, R. D., and Fox, R. O. 2003. Quadrature method of moments for aggregation-breakage processes. *Journal of Colloid and Interface Science*, **258**(2), 322–334.

Marchisio, D. L., Rivautella, L., and Barresi, A. A. 2006. Design and scale-up of chemical reactors for nanoparticle precipitation. *AIChE Journal*, **52**(5), 1877–1887.

Marinov, N. M., Pitz, W. J., Westbrook, C. K., Castaldi, M. J., and Senkan, S. M. 1996. Modeling of aromatic and polycyclic aromatic hydrocarbon formation in premixed methane and ethane flames. *Combustion Science and Technology*, **116–117**(1–6), 211–287.

Martynov, G. A., and Bakanov, S. P. 1961. Solution of the kinetic equation for coagulation. Pages 182–190 of: Derjaguin, B. (ed), *Surface Forces*.

Mashayek, F., and Pandya, R. V. R. 2003. Analytical description of particle/droplet-laden turbulent flows. *Progress in Energy and Combustion Science*, **29**(4), 329–378.

Massot, M. 2007. Eulerian multi-fluid models for polydisperse evaporating sprays. Pages 79–123 of: Marchisio, D. L., and Fox, R. O. (eds), *Multiphase Reacting Flows: Modelling and Simulation*. SpringerWienNewYork.

Mathieu, J., and Scott, J. 2000. *An Introduction to Turbulent Flow*. Cambridge University Press.

Matsoukas, T., and Friedlander, S. K. 1991. Dynamics of aerosol agglomerate formation. *Journal of Colloid And Interface Science*, **146**(2), 495–506.

Maxey, M. R. 1987. The gravitational settling of aerosol particles in homogeneous turbulence and random flow fields. *Journal of Fluid Mechanics*, **174**, 441–465.

Maxey, M. R., and Riley, J. J. 1983. Equation of motion for a small rigid sphere in a nonuniform flow. *Physics of Fluids*, **26**(4), 883–889.

McDermott, R., and Pope, S. B. 2007. A particle formulation for treating differential diffusion in filtered density function methods. *Journal of Computational Physics*, **226**(1), 947–993.

McEnally, C. S., Pfefferle, L. D., Atakan, B., and Kohse-Höinghaus, K. 2006. Studies of aromatic hydrocarbon formation mechanisms in flames: progress towards closing the fuel gap. *Progress in Energy and Combustion Science*, **32**(3), 247–294.

McGrady, E. D., and Ziff, R. M. 1987. 'Shattering' transition in fragmentation. *Physical Review Letters*, **58**(9), 892–895.

McGrady, E. D., and Ziff, R. M. 1988. Analytical solutions to fragmentation equations with flow. *AIChE Journal*, **34**(12), 2073–2076.

McGraw, R. 1997. Description of aerosol dynamics by the quadrature method of moments. *Aerosol Science and Technology*, **27**(2), 255–265.

McLeod, J. B. 1962. On an infinite set of non-linear differential equations. *Quarterly Journal of Mathematics*, **13**(1), 119–128.

Megaridis, C. M., and Dobbins, R. A. 1989. Comparison of soot growth and oxidation in smoking and non-smoking ethylene diffusion flames. *Combustion Science and Technology*, **66**(1–3), 1–16.

Meierhofer, F., and Fritsching, U. 2021. Synthesis of metal oxide nanoparticles in flame sprays: review on process technology, modeling, and diagnostics. *Energy and Fuels*, **35**(7), 5495–5537.

Melzak, Z. A. 1957a. A scalar transport equation. *Transactions of the American Mathematical Society*, **85**(2), 547–560.

Melzak, Z. A. 1957b. A scalar transport equation, II. *Michigan Mathematical Journal*, **4**(3), 193–206.

Menz, W. J., Akroyd, J., and Kraft, M. 2014. Stochastic solution of population balance equations for reactor networks. *Journal of Computational Physics*, **256**, 615–629.

Mersmann, A. (ed). 2001. *Crystallization Technology Handbook*. 2nd edn. Marcel Dekker.

Metzger, L., and Kind, M. 2017. The influence of mixing on fast precipitation processes – a coupled 3D CFD-PBE approach using the direct quadrature method of moments (DQMOM). *Chemical Engineering Science*, **169**, 284–298.

Michaelides, E. E. 2003. Hydrodynamic force and heat/mass transfer from particles, bubbles, and drops – the Freeman Scholar Lecture. *Journal of Fluids Engineering, Transactions of the ASME*, **125**(2), 209–238.

Michaelides, E. E., and Feng, Z.-G. 2017. Fundamentals of multiphase flow. Pages 1–78 of: Michaelides, E. E., Crowe, C. T., and Schwarzkopf, J. D. (eds), *Multiphase Flow Handbook*. 2nd edn. CRC Press.

Michelsen, H. A., Colket, M. B., Bengtsson, P.-E., D'Anna, A., Desgroux, P., Haynes, B. S., Miller, J. H., Nathan, G. J., Pitsch, H., and Wang, H. 2020. A review of terminology used to describe soot formation and evolution under combustion and pyrolytic conditions. *ACS Nano*, **14**(10), 12470–12490.

Mitchell, P., and Frenklach, M. 1998. Monte Carlo simulation of soot aggregation with simultaneous surface growth – why primary particles appear spherical. *Symposium (International) on Combustion*, **27**(1), 1507–1514.

Mitchell, P., and Frenklach, M. 2003. Particle aggregation with simultaneous surface growth. *Physical Review E – Statistical Physics, Plasmas, Fluids, and Related Interdisciplinary Topics*, **67**(6), 061407.

Modest, M. F., and Mazumder, S. 2022. *Radiative Heat Transfer*. 4th edn. Academic Press.

Moin, P., and Mahesh, K. 1998. Direct numerical simulation: a tool in turbulence research. *Annual Review of Fluid Mechanics*, **30**, 539–578.

Monnin, C. 1999. A thermodynamic model for the solubility of barite and celestite in electrolyte solutions and seawater to 200°C and to 1 kbar. *Chemical Geology*, **153**(1–4), 187–209.

Morrison Jr., P. W., Raghavan, R., Timpone, A. J., Artelt, C. P., and Pratsinis, S. E. 1997. In situ Fourier transform infrared characterization of the effect of electrical fields on the flame synthesis of TiO_2 particles. *Chemistry of Materials*, **9**(12), 2702–2708.

Mueller, M. E., Blanquart, G., and Pitsch, H. 2009a. Hybrid method of moments for modeling soot formation and growth. *Combustion and Flame*, **156**(6), 1143–1155.

Mueller, M. E., Blanquart, G., and Pitsch, H. 2009b. A joint volume-surface model of soot aggregation with the method of moments. *Proceedings of the Combustion Institute*, **32**(1), 785–792.

Mühlenweg, H., Gutsch, A., Schild, A., and Pratsinis, S. E. 2002. Process simulation of gas-to-particle-synthesis via population balances: investigation of three models. *Chemical Engineering Science*, **57**(12), 2305–2322.

Müller, H. 1928. Zur allgemeinen Theorie ser raschen Koagulation. *Kolloidchemische Beihefte*, **27**, 223–250.

Mullin, J. W. 2001. *Crystallization*. 4th edn. Butterworth-Heinemann.

Mustata, R., Valiño, L., Jiménez, C., Jones, W. P., and Bondi, S. 2006. A probability density function Eulerian Monte Carlo field method for large eddy simulations: application to a turbulent piloted methane/air diffusion flame (Sandia D). *Combustion and Flame*, **145**(1–2), 88–104.

Myerson, A. S., Erdemir, D., and Lee, A. Y. (eds). 2019. *Handbook of Industrial Crystallization*. 3rd edn. Cambridge University Press.

Nagle, J., and Strickland-Constable, R. F. 1962. Oxidation of carbon between 1000-2000°C. Pages 154–164 of: *Proceedings of the Fifth Carbon Conference*, vol. 1.

Narsimhan, G., Gupta, J. P., and Ramkrishna, D. 1979. A model for transitional breakage probability of droplets in agitated lean liquid-liquid dispersions. *Chemical Engineering Science*, **34**(2), 257–265.

Neoh, K. G., Howard, J. B., and Sarofim, A. F. 1985. Effect of oxidation on the physical structure of soot. *Symposium (International) on Combustion*, **20**(1), 951–957.

Neuber, G., Kronenburg, A., Stein, O. T., and Cleary, M. J. 2017. MMC-LES modelling of droplet nucleation and growth in turbulent jets. *Chemical Engineering Science*, **167**, 204–218.

Nicmanis, M., and Hounslow, M. J. 1998. Finite-element methods for steady-state population balance equations. *AIChE Journal*, **44**(10), 2258–2272.

O'Brien, E. E. 1980. The probability density function (pdf) approach to reacting turbulent flows. Pages 185–218 of: Libby, P. A., and Williams, F. A. (eds), *Topics in Applied Physics, Volume 44: Turbulent Reacting Flows*. Springer-Verlag.

Olivieri, D. A., Fairweather, M., and Falle, S. A. E. G. 2009. An adaptive mesh refinement method for solution of the transported PDF equation. *International Journal for Numerical Methods in Engineering*, **79**(12), 1536–1556.

Oseen, C. W. 1910. Über die Stokes'sche Formel und über eine verwandte Aufgabe in der Hydrodynamik. *Arkiv för Matematik, Astronomi och Fysik*, **6**(29).

O'Sullivan, D., and Rigopoulos, S. 2022. A conservative finite volume method for the population balance equation with aggregation, fragmentation, nucleation and growth. *Chemical Engineering Science*, **263**, 117925.

Papoulis, A. 1991. *Probability, Random Variables, and Stochastic Processes*. 3rd edn. McGraw-Hill, Inc.

Park, S. H., and Rogak, S. N. 2003. A one-dimensional model for coagulation, sintering, and surface growth of aerosol agglomerates. *Aerosol Science and Technology*, **37**(12), 947–960.

Park, S. H., and Rogak, S. N. 2004. A novel fixed-sectional model for the formation and growth of aerosol agglomerates. *Journal of Aerosol Science*, **35**(11), 1385–1404.

Park, S. H., Rogak, S. N., Bushe, W. K., Wen, J. Z., and Thomson, M. J. 2005. An aerosol model to predict size and structure of soot particles. *Combustion Theory and Modelling*, **9**(3), 499–513.

Partheniades, E. 2009. *Cohesive Sediments in Open Channels*. Elsevier Inc.

Patil, D. P., and Andrews, J. R. G. 1998. An analytical solution to continuous population balance model describing floc coalescence and breakage – a special case. *Chemical Engineering Science*, **53**(3), 599–601.

Patruno, L. E., Dorao, C. A., Svendsen, H. F., and Jakobsen, H. A. 2009. Analysis of breakage kernels for population balance modelling. *Chemical Engineering Science*, **64**(3), 501–508.

Patterson, R. I. A., and Kraft, M. 2007. Models for the aggregate structure of soot particles. *Combustion and Flame*, **151**(1–2), 160–172.

Pavliotis, G. A. 2014. *Stochastic Processes and Applications: Diffusion Processes, the Fokker-Planck and Langevin Equations*. Springer.

Peirano, E., Chibbaro, S., Pozorski, J., and Minier, J.-P. 2006. Mean-field/PDF numerical approach for polydispersed turbulent two-phase flows. *Progress in Energy and Combustion Science*, **32**(3), 315–371.

Pesmazoglou, I., Kempf, A. M., and Navarro-Martinez, S. 2014. Aerosol nucleation in a turbulent jet using Large Eddy Simulations. *Chemical Engineering Science*, **116**, 383–397.

Pesmazoglou, I., Kempf, A. M., and Navarro-Martinez, S. 2017. Large eddy simulation of particle aggregation in turbulent jets. *Journal of Aerosol Science*, **111**, 1–17.

Peters, N. 2000. *Turbulent Combustion*. Cambridge University Press.

Peterson, T. W., Gelbard, F., and Seinfeld, J. H. 1978. Dynamics of source-reinforced, coagulating, and condensing aerosols. *Journal of Colloid And Interface Science*, **63**(3), 426–445.

Pilou, M., Tsangaris, S., Neofytou, P., Housiadas, C., and Drossinos, Y. 2011. Inertial particle deposition in a 90° laminar flow bend: an Eulerian fluid particle approach. *Aerosol Science and Technology*, **45**(11), 1376–1387.

Piomelli, U., and Chasnov, J. R. 1996. Large eddy simulations: theory and applications. Pages 269–336 of: Hallbäck, M., Henningson, D. S., Johansson, A.V., and Alfredsson, P. H. (eds), *ERCOFTAC Series: Turbulence and Transition Modelling*. Kluwer Academic Publishers.

Piton, D., Fox, R. O., and Marcant, B. 2000. Simulation of fine particle formation by precipitation using computational fluid dynamics. *Canadian Journal of Chemical Engineering*, **78**(5), 983–993.

Poinsot, T., and Veynante, D. 2005. *Theoretical and Numerical Combustion*. 2nd edn. Edwards.

Pope, S. B. 1976. The probability approach to the modelling of turbulent reacting flows. *Combustion and Flame*, **27**(C), 299–312.

Pope, S. B. 1981. A Monte Carlo method for the PDF equations of turbulent reactive flow. *Combustion Science and Technology*, **25**(5–6), 159–174.

Pope, S. B. 1985. PDF methods for turbulent reactive flows. *Progress in Energy and Combustion Science*, **11**(2), 119–192.

Pope, S. B. 1991. Mapping closures for turbulent mixing and reaction. *Theoretical and Computational Fluid Dynamics*, **2**(5–6), 255–270.

Pope, S. B. 1997. Computationally efficient implementation of combustion chemistry using *in situ* adaptive tabulation. *Combustion Theory and Modelling*, **1**(1), 41–63.

Pope, S. B. 2000. *Turbulent Flows*. Cambridge University Press.

Prasad, V. N. 2011. Large Eddy simulation of partially premixed turbulent combustion. Ph.D. thesis, Imperial College London.

Pratsinis, S. E. 1988. Simultaneous nucleation, condensation, and coagulation in aerosol reactors. *Journal of Colloid and Interface Science*, **124**(2), 416–427.

Pratsinis, S. E. 1998. Flame aerosol synthesis of ceramic powders. *Progress in Energy and Combustion Science*, **24**(3), 197–219.

Pratsinis, S. E. 2010. Aerosol-based technologies in nanoscale manufacturing: from functional materials to devices through core chemical engineering. *AIChE Journal*, **56**(12), 3028–3035.

Press, W. H., Teukolsky, S. A., Vetterling, W. T., and Flannery, B. P. 1997. *Numerical Recipes in Fortran 77: The Art of Scientific Computing, Vol. 1 of Fortran Numerical Recipes*. 2nd edn. Cambridge University Press.

Pruppacher, H. R., and Klett, J. D. 1996. *Microphysics of Clouds and Precipitation*. 2nd edn. Springer.

Qamar, S., Elsner, M. P., Angelov, I. A., Warnecke, G., and Seidel-Morgenstern, A. 2006. A comparative study of high resolution schemes for solving population balances in crystallization. *Computers & Chemical Engineering*, **30**(6–7), 1119–1131.

Qamar, S., Ashfaq, A., Warnecke, G., Angelov, I., Elsner, M.P., and Seidel-Morgenstern, A. 2007. Adaptive high-resolution schemes for multidimensional population balances in crystallization processes. *Computers & Chemical Engineering*, **31**(10), 1296–1311.

Rahmani, N. H. G., Masliyah, J. H., and Dabros, T. 2003. Characterization of asphaltenes aggregation and fragmentation in a shear field. *AIChE Journal*, **49**(7), 1645–1655.

Raj, A., Celnik, M., Shirley, R., Sander, M., Patterson, R., West, R., and Kraft, M. 2009. A statistical approach to develop a detailed soot growth model using PAH characteristics. *Combustion and Flame*, **156**(4), 896–913.

Rajamani, K., Pate, W. T., and Kinneberg, D. J. 1986. Time-driven and event-driven Monte Carlo simulations of liquid-liquid dispersions: a comparison. *Industrial & Engineering Chemistry Fundamentals*, **25**(4), 746–752.

Ramabhadran, T. E., Peterson, T. W., and Seinfeld, J. H. 1976. Dynamics of aerosol coagulation and condensation. *AIChE Journal*, **22**(5), 840–851.

Raman, V., and Fox, R. O. 2016. Modeling of fine-particle formation in turbulent flames. *Annual Review of Fluid Mechanics*, **48**, 159–190.

Ramkrishna, D. 1971. Solution of population balance equations. *Chemical Engineering Science*, **26**(7), 1134–1136.

Ramkrishna, D. 1981. Analysis of population balance-IV. The precise connection between Monte Carlo simulation and population balances. *Chemical Engineering Science*, **36**(7), 1203–1209.

Ramkrishna, D. 1985. The status of population balances. *Reviews in Chemical Engineering*, **3**(1), 49–95.

Ramkrishna, D. 2000. *Population Balances: Theory and Applications to Particulate Systems in Engineering*. Academic Press.

Ramkrishna, D., and Mahoney, A. W. 2002. Population balance modeling. Promise for the future. *Chemical Engineering Science*, **57**(4), 595–606.

Ramkrishna, D., and Singh, M. R. 2014. Population balance modeling: current status and future prospects. *Annual Review of Chemical and Biomolecular Engineering*, **5**, 123–146.

Randolph, A. D. 1964. A population balance for countable entities. *The Canadian Journal of Chemical Engineering*, **42**(6), 280–281.

Randolph, A. D., and Larson, M. A. 1962. Transient and steady state size distributions in continuous mixed suspension crystallizers. *AIChE Journal*, **8**(5), 639–645.

Randolph, A. D., and Larson, M. A. 1971. *Theory of Particulate Processes: Analysis and Techniques of Continuous Crystallization*. 1st edn. Academic Press.

Randolph, A. D., and Larson, M. A. 1988. *Theory of Particulate Processes: Analysis and Techniques of Continuous Crystallization*. 2nd edn. Academic Press.

Randolph, A. D., and White, E. T. 1977. Modeling size dispersion in the prediction of crystal-size distribution. *Chemical Engineering Science*, **32**(9), 1067–1076.

Ranjan, R., and Menon, S. 2021. Two level simulation of Schmidt number effect on passive scalar transport in wall-bounded turbulent flows. *Physics of Fluids*, **33**(3), 035124.

Rawlings, J. B., and Ray, W. H. 1988. The modeling of batch and continuous emulsion polymerization reactors. Part I: model formulation and sensitivity to parameters. *Polymer Engineering & Science*, **28**(5), 237–256.

Readshaw, T., Ding, T., Rigopoulos, S., and Jones, W. P. 2021. Modeling of turbulent flames with the large eddy simulation-probability density function (LES-PDF) approach, stochastic fields, and artificial neural networks. *Physics of Fluids*, **33**(3), 035154.

Readshaw, T., Jones, W. P., and Rigopoulos, S. 2023. On the incorporation of conservation laws in machine learning tabulation of kinetics for reacting flow simulation. *Physics of Fluids*, **35**(4).

Reddy, J. N. 1993. *An Introduction to the Finite Element Method*. 2nd edn. McGraw-Hill, Inc.

Reeks, M. W. 2021. The development and application of a kinetic theory for modeling dispersed particle flows. *Journal of Fluids Engineering*, **143**(8), 080803.

Reeks, M. W., Simonin, O., and Fede, P. 2017. PDF models for particle transport, mixing and collisions in turbulent gas flows. Pages 144–202 of: Michaelides, E. E., Crowe, C. T., and Schwarzkopf, J. D. (eds), *Multiphase Flow Handbook*. 2nd edn. CRC Press.

Reynolds, G. K., Fu, J. S., Cheong, Y. S., Hounslow, M. J., and Salman, A. D. 2005. Breakage in granulation: a review. *Chemical Engineering Science*, **60**(14), 3969–3992.

Ribert, G., Vervisch, L., Domingo, P., and Niu, Y.-S. 2014. Hybrid transported-tabulated strategy to downsize detailed chemistry for numerical simulation of premixed flames. *Flow, Turbulence and Combustion*, **92**(1–2), 175–200.

Richter, H., and Howard, J. B. 2000. Formation of polycyclic aromatic hydrocarbons and their growth to soot – a review of chemical reaction pathways. *Progress in Energy and Combustion Science*, **26**(4), 565–608.

Richter, H., Granata, S., Green, W. H., and Howard, J. B. 2005. Detailed modeling of PAH and soot formation in a laminar premixed benzene/oxygen/argon low-pressure flame. *Proceedings of the Combustion Institute*, **30**(1), 1397–1404.

Rigopoulos, S. 2007. PDF method for population balance in turbulent reactive flow. *Chemical Engineering Science*, **62**(23), 6865–6878.

Rigopoulos, S. 2010. Population balance modelling of polydispersed particles in reactive flows. *Progress in Energy and Combustion Science*, **36**(4), 412–443.

Rigopoulos, S. 2019. Modelling of soot aerosol dynamics in turbulent flow. *Flow, Turbulence and Combustion*, **103**(3), 565–604.

Rigopoulos, S. 2020. Sooting flames. In: Vervisch, L., and Roekaerts, D. (eds), *Computational Fluid Dynamics of Turbulent Combustion*. 2nd edn. ERCOFTAC.

Rigopoulos, S., and Jones, A.G. 2001. Dynamic modelling of a bubble column for particle formation via a gas-liquid reaction. *Chemical Engineering Science*, **56**(21–22), 6177–6184.

Rigopoulos, S., and Jones, A. 2003a. A hybrid CFD-reaction engineering framework for multiphase reactor modelling: basic concept and application to bubble column reactors. *Chemical Engineering Science*, **58**(14), 3077–3089.

Rigopoulos, S., and Jones, A. 2003b. Modeling of semibatch agglomerative gas-liquid precipitation of $CaCO_3$ in a bubble column reactor. *Industrial & Engineering Chemistry Research*, **42**(25), 6567–6575.

Rigopoulos, S., and Jones, A.G. 2003c. Finite-element scheme for solution of the dynamic population balance equation. *AIChE Journal*, **49**(5), 1127–1139.

Rittler, A., Deng, L., Wlokas, I., and Kempf, A. M. 2017. Large eddy simulations of nanoparticle synthesis from flame spray pyrolysis. *Proceedings of the Combustion Institute*, **36**(1), 1077–1087.

Robinson, M., Stilianakis, N. I., and Drossinos, Y. 2012. Spatial dynamics of airborne infectious diseases. *Journal of Theoretical Biology*, **297**, 116–126.

Rod, V., and Mĭšek, T. 1982. Stochastic modelling of dispersion formation in agitated liquid-liquid systems. *Transactions of the Institution of Chemical Engineers*, **60**(1), 48–53.

Rodrigues, P., Franzelli, B., Vicquelin, R., Gicquel, O., and Darabiha, N. 2018. Coupling an LES approach and a soot sectional model for the study of sooting turbulent non-premixed flames. *Combustion and Flame*, **190**, 477–499.

Rogak, S. N. 1997. Modeling small cluster deposition on the primary particles of aerosol agglomerates. *Aerosol Science and Technology*, **26**(2), 127–140.

Rogallo, R. S., and Moin, P. 1984. Numerical simulation of turbulent flows. *Annual Review of Fluid Mechanics*, **16**, 99–137.

Roussos, A. I., Alexopoulos, A. H., and Kiparissides, C. 2005. Part III: Dynamic evolution of the particle size distribution in batch and continuous particulate processes: a Galerkin on finite elements approach. *Chemical Engineering Science*, **60**(24), 6998–7010.

Sabel'nikov, V., and Soulard, O. 2005. Rapidly decorrelating velocity-field model as a tool for solving one-point Fokker-Planck equations for probability density functions of turbulent reactive scalars. *Physical Review E – Statistical, Nonlinear, and Soft Matter Physics*, **72**(1), 016301.

Saffman, P. G., and Turner, J. S. 1956. On the collision of drops in turbulent clouds. *Journal of Fluid Mechanics*, **1**(1), 16–30.

Sagaut, P. 2006. *Large Eddy Simulation for Incompressible Flows: An Introduction*. 3rd edn. Springer.

Saggese, C., Ferrario, S., Camacho, J., Cuoci, A., Frassoldati, A., Ranzi, E., Wang, H., and Faravelli, T. 2015. Kinetic modeling of particle size distribution of soot in a premixed burner-stabilized stagnation ethylene flame. *Combustion and Flame*, **162**(9), 3356–3369.

Salehi, F., Cleary, M. J., and Masri, A. R. 2017. Population balance equation for turbulent polydispersed inertial droplets and particles. *Journal of Fluid Mechanics*, **831**, 719–742.

Samaras, S. N. 2006. Modelling of microstructure evolution during precipitation processes: a population balance approach of the KWN model. *Modelling and Simulation in Materials Science and Engineering*, **14**(8), 1271–1292.

Santoro, R. J., Semerjian, H. G., and Dobbins, R. A. 1983. Soot particle measurements in diffusion flames. *Combustion and Flame*, **51**(C), 203–218.

Santoro, R. J., Yeh, T. T., Horvath, J. J., and Semerjian, H. G. 1987. The transport and growth of soot particles in laminar diffusion flames. *Combustion Science and Technology*, **53**(2–3), 89–115.

Sathyagal, A. N., Ramkrishna, D., and Narsimhan, G. 1995. Solution of inverse problems in population balances – II. Particle break-up. *Computers and Chemical Engineering*, **19**(4), 437–451.

Schenk, M., Lieb, S., Vieker, H., Beyer, A., Gölzhäuser, A., Wang, H., and Kohse-Höinghaus, K. 2015. Morphology of nascent soot in ethylene flames. *Proceedings of the Combustion Institute*, **35**(2), 1879–1886.

Schlichting, H. 1979. *Boundary-Layer Theory*. 7th edn. McGraw-Hill.

Schubert, H. 1998. Keimbildung bei der Kristallisation schwerlöslicher Feststoffe. Ph.D. thesis, Technische Univetsität München.

Schumann, T. E. W. 1940. Theoretical aspects of the size distribution of fog particles. *Quarterly Journal of the Royal Meteorological Society*, **66**(285), 195–208.

Schwarzer, H.-C., and Peukert, W. 2002. Experimental investigation into the influence of mixing on nanoparticle precipitation. *Chemical Engineering and Technology*, **25**(6), 657–661.

Schwarzer, H.-C., and Peukert, W. 2004. Combined experimental/numerical study on the precipitation of nanoparticles. *AIChE Journal*, **50**(12), 3234–3247.

Schwarzer, H.-C., and Peukert, W. 2005. Prediction of aggregation kinetics based on surface properties of nanoparticles. *Chemical Engineering Science*, **60**(1), 11–25.

Schwarzer, H.-C., Schwertfirm, F., Manhart, M., Schmid, H.-J., and Peukert, W. 2006. Predictive simulation of nanoparticle precipitation based on the population balance equation. *Chemical Engineering Science*, **61**(1), 167–181.

Schwertfirm, F., Gradl, J., Schwarzer, H. C., Peukert, W., and Manhart, M. 2007. The low Reynolds number turbulent flow and mixing in a confined impinging jet reactor. *International Journal of Heat and Fluid Flow*, **28**(6), 1429–1442.

Scott, W. T. 1968. Analytic studies of cloud droplet coalescence I. *Journal of the Atmospheric Sciences*, **25**(1), 54–65.

Seinfeld, J. H., and Pandis, S. N. 2016. *Atmospheric Chemistry and Physics: From Air Pollution to Climate Change*. 3rd edn. Wiley.

Seltz, A., Domingo, P., and Vervisch, L. 2021. Solving the population balance equation for non-inertial particles dynamics using probability density function and neural networks: application to a sooting flame. *Physics of Fluids*, **33**(1), 013311.

Settumba, N., and Garrick, S. C. 2003. Direct numerical simulation of nanoparticle coagulation in a temporal mixing layer via a moment method. *Journal of Aerosol Science*, **34**(2), 149–167.

Settumba, N., and Garrick, S. C. 2004. A comparison of diffusive transport in a moment method for nanoparticle coagulation. *Journal of Aerosol Science*, **35**(1), 93–101.

Sewerin, F. 2022. A methodology for solving the population balance equation based on an embedded reduced order representation. *Chemical Engineering Science*, **252**, 117101.

Sewerin, F. 2023. An efficient implementation of a conservative finite volume scheme with constant and linear reconstructions for solving the coagulation equation. *Chemical Engineering Science*, **280**, 119020.

Sewerin, F., and Rigopoulos, S. 2017a. An explicit adaptive grid approach for the numerical solution of the population balance equation. *Chemical Engineering Science*, **168**, 250–270.

Sewerin, F., and Rigopoulos, S. 2017b. An LES-PBE-PDF approach for modeling particle formation in turbulent reacting flows. *Physics of Fluids*, **29**(10), 105105.

Sewerin, F., and Rigopoulos, S. 2018. An LES-PBE-PDF approach for predicting the soot particle size distribution in turbulent flames. *Combustion and Flame*, **189**, 62–76.

Sewerin, F., and Rigopoulos, S. 2019. Algorithmic aspects of the LES-PBE-PDF method for modeling soot particle size distributions in turbulent flames. *Combustion Science and Technology*, **191**(5–6), 766–796.

Shaddix, C. R., Zhang, J., Schefer, R. W., Doom, J., Oefelein, J. C., Kook, S., Pickett, L. M., and Wang, H. 2010. *Understanding and predicting soot generation in turbulent non-premixed jet flames*. Tech. rept. SAND2010-7178. Sandia National Laboratories.

Shah, B. H., Ramkrishna, D., and Borwanker, J. D. 1977. Simulation of particulate systems using the concept of the interval of quiescence. *AIChE Journal*, **23**(6), 897–904.

Shekar, S., Smith, A. J., Menz, W. J., Sander, M., and Kraft, M. 2012. A multidimensional population balance model to describe the aerosol synthesis of silica nanoparticles. *Journal of Aerosol Science*, **44**, 83–98.

Sheth, R. K., and Pitman, J. 1997. Coagulation and branching process models of gravitational clustering. *Monthly Notices of the Royal Astronomical Society*, **289**(1), 66–82.

Shohat, J. A., and Tamarkin, J. D. 1970. *The Problem of Moments*. 4th edn. American Mathematical Society.

Sirignano, M., Kent, J., and D'Anna, A. 2010. Detailed modeling of size distribution functions and hydrogen content in combustion-formed particles. *Combustion and Flame*, **157**(6), 1211–1219.

Sirignano, M., Kent, J., and D'Anna, A. 2013. Modeling formation and oxidation of soot in nonpremixed flames. *Energy and Fuels*, **27**(4), 2303–2315.

Sirignano, W. A. 2010. *Fluid Dynamics and Transport of Droplets and Sprays*. 2nd edn. Cambridge University Press.

Slavinskaya, N. A., and Frank, P. 2009. A modelling study of aromatic soot precursors formation in laminar methane and ethene flames. *Combustion and Flame*, **156**(9), 1705–1722.

Smagorinsky, J. 1963. General circulation experiments with the primitive equations: I. The basic experiment. *Monthly Weather Review*, **91**(3), 99–164.

Smith, M., and Matsoukas, T. 1998. Constant-number Monte Carlo simulation of population balances. *Chemical Engineering Science*, **53**(9), 1777–1786.

Smoluchowski, M. v. 1917. Versuch einer mathematischen Theorie der Koagulationskinetik kolloider Lösungen. *Zeitschrift für Physikalische Chemie*, **92**, 129–168.

Söhnel, O., and Garside, J. 1992. *Precipitation: Basic Principles and Industrial Applications*. Butterworth-Heinemann.

Solsvik, J., and Jakobsen, H. A. 2015. The foundation of the population balance equation: a review. *Journal of Dispersion Science and Technology*, **36**(4), 510–520.

Spalding, D. B. 1996. Multi-fluid models of turbulent combustion. Pages 59–81 of: May, R. L., and Easton, A. K. (eds), *Computational Techniques and Applications: CTAC 95*. World Scientific Publishing Co.

Spielman, L. A., and Levenspiel, O. 1965. A Monte Carlo treatment for reacting and coalescing dispersed phase systems. *Chemical Engineering Science*, **20**(3), 247–254.

Sporleder, F., Borka, Z., Solsvik, J., and Jakobsen, H. A. 2012. On the population balance equation. *Reviews in Chemical Engineering*, **28**(2–3), 149–169.

Stanmore, B. R., Brilhac, J. F., and Gilot, P. 2001. The oxidation of soot: a review of experiments, mechanisms and models. *Carbon*, **39**(15), 2247–2268.

Stettler, M. E. J., Boies, A. M., Petzold, A., and Barrett, S. R. H. 2013a. Global civil aviation black carbon emissions. *Environmental Science and Technology*, **47**(18), 10397–10404.

Stettler, M. E. J., Swanson, J. J., Barrett, S. R. H., and Boies, A. M. 2013b. Updated correlation between aircraft smoke number and black carbon concentration. *Aerosol Science and Technology*, **47**(11), 1205–1214.

Stieltjes, T. J. 1894. Recherches sur les fractions continues. *Annales de la Faculté des sciences de Toulouse : Mathématiques*, **1e série, 8**(4), J1–J122.

Stieltjes, T. J. 1895. Recherches sur les fractions continues [Suite et fin]. *Annales de la Faculté des sciences de Toulouse : Mathématiques*, **1e série, 9**(1), A5–A47.

Stilianakis, N. I., and Drossinos, Y. 2010. Dynamics of infectious disease transmission by inhalable respiratory droplets. *Journal of the Royal Society Interface*, **7**(50), 1355–1366.

Stirzaker, D. 2005. *Stochastic Processes & Models*. Oxford University Press.

Stokes, G. G. 1851. On the effect of the internal friction of fluids on the motion of pendulums. *Transactions of the Cambridge Philosophical Society, Part II*, **9**, 8–106.

Subramaniam, S. 2001. Statistical modeling of sprays using the droplet distribution function. *Physics of Fluids*, **13**(3), 624–642.

Subramaniam, S., and Pope, S. B. 1998. A mixing model for turbulent reactive flows based on Euclidean minimum spanning trees. *Combustion and Flame*, **115**(4), 487–514.

Suman, A., Casari, N., Fabbri, E., di Mare, L., Montomoli, F., and Pinelli, M. 2019. Generalization of particle impact behavior in gas turbine via non-dimensional grouping. *Progress in Energy and Combustion Science*, **74**, 103–151.

Sun, B., and Rigopoulos, S. 2022. Modelling of soot formation and aggregation in turbulent flows with the LES-PBE-PDF approach and a conservative sectional method. *Combustion and Flame*, **242**, 112152.

Sun, B., Rigopoulos, S., and Liu, A. 2021. Modelling of soot coalescence and aggregation with a two-population balance equation model and a conservative finite volume method. *Combustion and Flame*, **229**, 111382.

Sutton, R. D., Hines, W. S., and Combs, L. P. 1972. Development and application of a comprehensive analysis of liquid-rocket combustion. *AIAA Journal*, **10**(2), 194–203.

Swift, D. L., and Friedlander, S. K. 1964. The coagulation of hydrosols by Brownian motion and laminar shear flow. *Journal of Colloid Science*, **19**(7), 621–647.

Sztucki, M., Narayanan, T., and Beaucage, G. 2007. In situ study of aggregation of soot particles in an acetylene flame by small-angle x-ray scattering. *Journal of Applied Physics*, **101**(11), 114304.

Tang, H. Y., Rigopoulos, S., and Papadakis, G. 2020. A methodology for coupling DNS and discretised population balance for modelling turbulent precipitation. *International Journal of Heat and Fluid Flow*, **86**, 108689.

Tang, H. Y., Rigopoulos, S., and Papadakis, G. 2022. On the interaction of turbulence with nucleation and growth in reaction crystallisation. *Journal of Fluid Mechanics*, **944**, A48.

Tang, H. Y., Rigopoulos, S., and Papadakis, G. 2023. On the effect of turbulent fluctuations on precipitation: a direct numerical simulation – population balance study. *Chemical Engineering Science*, **270**, 118511.

Tavare, N. S. 1985. Crystal growth rate dispersion. *The Canadian Journal of Chemical Engineering*, **63**(3), 436–442.

Tavare, N. S. 1986. Mixing in continuous crystallizers. *AIChE Journal*, **32**(5), 705–732.

Telib, H., Manhart, M., and Iollo, A. 2004. Analysis and low-order modeling of the inhomogeneous transitional flow inside a T-mixer. *Physics of Fluids*, **16**(8), 2717–2731.

Tennekes, H., and Lumley, J. L. 1974. *A First Course in Turbulence*. The MIT Press.

Thomas, D. N., Judd, S. J., and Fawcett, N. 1999. Flocculation modelling: a review. *Water Research*, **33**(7), 1579–1592.

Thompson, P. D. 1968. A transformation of the stochastic equation for droplet coalescence. Pages 115–126 of: *Proceedings of the International Conference on Cloud Physics*.

Tian, K., Thomson, K. A., Liu, F., Snelling, D. R., Smallwood, G. J., and Wang, D. 2006. Determination of the morphology of soot aggregates using the relative optical density method for the analysis of TEM images. *Combustion and Flame*, **144**(4), 782–791.

Tree, D. R., and Svensson, K. I. 2007. Soot processes in compression ignition engines. *Progress in Energy and Combustion Science*, **33**(3), 272–309.

Tsagkaridis, M., Rigopoulos, S., and Papadakis, G. 2022. Analysis of turbulent coagulation in a jet with discretised population balance and DNS. *Journal of Fluid Mechanics*, **937**, A25.

Tsagkaridis, M., Rigopoulos, S., and Papadakis, G. 2023. Modeling of silica synthesis in a laminar flame by coupling an extended population balance model with computational fluid dynamics. *Aerosol Science and Technology*, **57**(4), 296–317.

Tsang, T. H., and Brock, J. R. 1983. Simulation of condensation aerosol growth by condensation and evaporation. *Aerosol Science and Technology*, **2**(3), 311–320.

Tsang, T. H., and Rao, A. 1990. A moving finite element method for the population balance equation. *International Journal for Numerical Methods in Fluids*, **10**(7), 753–769.

Tsantilis, S., and Pratsinis, S. E. 2000. Evolution of primary and aggregate particle-size distributions by coagulation and sintering. *AIChE Journal*, **46**(2), 407–415.

Tsantilis, S., and Pratsinis, S. E. 2004. Soft- and hard-agglomerate aerosols made at high temperatures. *Langmuir*, **20**(14), 5933–5939.

Tsantilis, S., Pratsinis, S. E., and Briesen, H. 2001. Sintering time for silica particle growth. *Aerosol Science and Technology*, **34**(3), 237–246.

Tsuchiya, H. M., Fredrickson, A. G., and Aris, R. 1966. Dynamics of microbial cell populations. *Advances in Chemical Engineering*, **6**(C), 125–206.

Valentas, K. J., Bilous, O., and Amundson, N. R. 1966. Analysis of breakage in dispersed phase systems. *Industrial & Engineering Chemistry Fundamentals*, **5**(2), 271–279.

Valiño, L. 1998. Field Monte Carlo formulation for calculating the probability density function of a single scalar in a turbulent flow. *Flow, Turbulence and Combustion*, **60**(2), 157–172.

Van Peborgh Gooch, J. R., and Hounslow, M. J. 1996. Monte Carlo simulation of size-enlargement mechanisms in crystallization. *AIChE Journal*, **42**(7), 1864–1874.

Vanni, M. 2000. Approximate population balance equations for aggregation-breakage processes. *Journal of Colloid and Interface Science*, **221**(2), 143–160.

Vanni, M., and Baldi, G. 2002. Coagulation efficiency of colloidal particles in shear flow. *Advances in Colloid and Interface Science*, **97**(1–3), 151–177.

Versteeg, H. K., and Malalasekera, W. 2007. *An Introduction to Computational Fluid Dynamics: The Finite Volume Method*. 2nd edn. Pearson Education Limited.

Vervisch, L., Lodato, G., and Domingo, P. 2023. High-order polynomial approximations for solving non-inertial particle size density in flames. *Proceedings of the Combustion Institute*, **39**(4), 5385–5393.

Veshkini, A., Dworkin, S. B., and Thomson, M. J. 2014. A soot particle surface reactivity model applied to a wide range of laminar ethylene/air flames. *Combustion and Flame*, **161**(12), 3191–3200.

Veshkini, A., Dworkin, S. B., and Thomson, M. J. 2016. Understanding soot particle size evolution in laminar ethylene/air diffusion flames using novel soot coalescence models. *Combustion Theory and Modelling*, **20**(4), 707–734.

Veynante, D., and Vervisch, L. 2002. Turbulent combustion modeling. *Progress in Energy and Combustion Science*, **28**(3), 193–266.

Vicum, L., and Mazzotti, M. 2007. Multi-scale modeling of a mixing-precipitation process in a semibatch stirred tank. *Chemical Engineering Science*, **62**(13), 3513–3527.

Vicum, L., Mazzotti, M., and Bałdyga, J. 2003. Applying a thermodynamic model to the non-stoichiometric precipitation of barium sulfate. *Chemical Engineering and Technology*, **26**(3), 325–333.

Vilas Bôas Fávero, C., Maqbool, T., Hoepfner, M., Haji-Akbari, N., and Fogler, H. S. 2017. Revisiting the flocculation kinetics of destabilized asphaltenes. *Advances in Colloid and Interface Science*, **244**, 267–280.

Villermaux, J., and Devillon, J. C. 1972. Représentation de la coalescence et de la redispersion des domaines de ségrégation dans un fluide par un modèle d'interaction phénoménologique. Pages 1–13 of: *Proceedings of the 2nd International Symposium on Chemical Reaction Engineering*, vol. 26. Elsevier New York.

Vogel, L., and Peukert, W. 2003. Breakage behaviour of different materials – construction of a mastercurve for the breakage probability. *Powder Technology*, **129**(1–3), 101–110.

Vovelle, C., Delfau, J-L., and Reuillon, M. 1994. Formation of aromatic hydrocarbons in decane and kerosene flames at reduced pressure. Pages 50–65 of: Bockhorn, H. (ed), *Soot Formation in Combustion*. Springer Series in Chemical Physics 59. Springer-Verlag.

Wang, H. 2011. Formation of nascent soot and other condensed-phase materials in flames. *Proceedings of the Combustion Institute*, **33**(1), 41–67.

Wang, H., and Frenklach, M. 1997. A detailed kinetic modeling study of aromatics formation in laminar premixed acetylene and ethylene flames. *Combustion and Flame*, **110**(1–2), 173–221.

Warshaw, M. 1967. Cloud droplet coalescence: statistical foundations and a one-dimensional sedimentation model. *Journal of the Atmospheric Sciences*, **24**, 278–286.

Wen, J. Z., Thomson, M. J., Park, S. H., Rogak, S. N., and Lightstone, M. F. 2005. Study of soot growth in a plug flow reactor using a moving sectional model. *Proceedings of the Combustion Institute*, **30**(1), 1477–1483.

Wen, J. Z., Thomson, M. J., Lightstone, M. F., Park, S. H., and Rogak, S. N. 2006. An improved moving sectional aerosol model of soot formation in a plug flow reactor. *Combustion Science and Technology*, **178**(5), 921–951.

Westbrook, C. K. 1977. Three dimensional numerical modeling of liquid fuel sprays. *Symposium (International) on Combustion*, **16**(1), 1517–1526.

Whitby, E. R., and McMurry, P. H. 1997. Modal aerosol dynamics modeling. *Aerosol Science and Technology*, **27**(6), 673–688.

White, F. M. 2006. *Viscous Fluid Flow*. 3rd edn. McGraw-Hill.

Williams, F. A. 1958. Spray combustion and atomization. *Physics of Fluids*, **1**(6), 541–545.

Williams, F. A. 1961. Progress in spray-combustion analysis. *Symposium (International) on Combustion*, **8**(1), 50–69.

Williams, J. C., Allen, T., Salman, A. D., Ghadiri, M., and Hounslow, M. J. (eds). 2007. *Handbook of Powder Technology*, Volume 12: Particle Breakage. Elsevier.

Williams, M. M. R. 1986. Some topics in nuclear aerosol dynamics. *Progress in Nuclear Energy*, **17**(1), 1–52.

Williams, M. M. R. 1990. An exact solution of the fragmentation equation. *Aerosol Science and Technology*, **12**(3), 538–546.

Williams, M. M. R., and Loyalka, S. K. 1991. *Aerosol Science – Theory and Practice: With Special Applications to the Nuclear Industry*. Pergamon Press, Oxford.

Wright, D. L., McGraw, R., and Rosner, D. E. 2001. Bivariate extension of the quadrature method of moments for modeling simultaneous coagulation and sintering of particle populations. *Journal of Colloid and Interface Science*, **236**(2), 242–251.

Wright, H., and Ramkrishna, D. 1992. Solutions of inverse problems in population balances – I. Aggregation kinetics. *Computers and Chemical Engineering*, **16**(12), 1019–1038.

Wu, J. J., and Flagan, R. C. 1988. A discrete-sectional solution to the aerosol dynamic equation. *Journal of Colloid and Interface Science*, **123**(2), 339–352.

Wulkow, M. 1996. The simulation of molecular weight distributions in polyreaction kinetics by discrete Galerkin methods. *Macromolecular Theory and Simulations*, **5**(3), 393–416.

Xiong, Y., and Pratsinis, S. E. 1993. Formation of agglomerate particles by coagulation and sintering – Part I. A two-dimensional solution of the population balance equation. *Journal of Aerosol Science*, **24**(3), 283–300.

Xiong, Y., Kamal Akhtar, M., and Pratsinis, S. E. 1993. Formation of agglomerate particles by coagulation and sintering – Part II. The evolution of the morphology of aerosol-made titania, silica and silica-doped titania powders. *Journal of Aerosol Science*, **24**(3), 301–313.

Xuan, Y., and Blanquart, G. 2015. Effects of aromatic chemistry-turbulence interactions on soot formation in a turbulent non-premixed flame. *Proceedings of the Combustion Institute*, **35**(2), 1911–1919.

Yang, D., Chen, B., Socolofsky, S. A., Chamecki, M., and Meneveau, C. 2016. Large-eddy simulation and parameterization of buoyant plume dynamics in stratified flow. *Journal of Fluid Mechanics*, **794**, 798–833.

Yang, S., Lew, J. K., and Mueller, M. E. 2019. Large Eddy Simulation of soot evolution in turbulent reacting flows: presumed subfilter PDF model for soot–turbulence–chemistry interactions. *Combustion and Flame*, **209**, 200–213.

Yashima, S., Kanda, Y., and Sano, S. 1987. Relationships between particle size and fracture energy or impact velocity required to fracture as estimated from single particle crushing. *Powder Technology*, **51**(3), 277–282.

Yeoh, G. H., Cheung, C. P., and Tu, J. 2014. *Multiphase Flow Analysis Using Population Balance Modeling: Bubbles, Drops and Particles*. Elsevier.

Yu, M., Lin, J., and Chan, T. 2008. A new moment method for solving the coagulation equation for particles in Brownian motion. *Aerosol Science and Technology*, **42**(9), 705–713.

Zamuner, B., and Dupoirieux, F. 2000. Numerical simulation of soot formation in a turbulent flame with a Monte-Carlo PDF approach and detailed chemistry. *Combustion Science and Technology*, **158**, 407–438.

Zhang, J., Shaddix, C. R., and Schefer, R. W. 2011. Design of 'model-friendly' turbulent non-premixed jet burners for C_{2+} hydrocarbon fuels. *Review of Scientific Instruments*, **82**(7), 074101.

Zhang, Q., Guo, H., Liu, F., Smallwood, G. J., and Thomson, M. J. 2008. Implementation of an advanced fixed sectional aerosol dynamics model with soot aggregate formation in a laminar methane/air coflow diffusion flame. *Combustion Theory and Modelling*, **12**(4), 621–641.

Zhang, Q., Guo, H., Liu, F., Smallwood, G.J., and Thomson, M.J. 2009a. Modeling of soot aggregate formation and size distribution in a laminar ethylene/air coflow diffusion flame with detailed PAH chemistry and an advanced sectional aerosol dynamics model. *Proceedings of the Combustion Institute*, **32**(1), 761–768.

Zhang, Q., Thomson, M. J., Guo, H., Liu, F., and Smallwood, G. J. 2009b. A numerical study of soot aggregate formation in a laminar coflow diffusion flame. *Combustion and Flame*, **156**(3), 697–705.

Zhang, X., Chen, X., and Wang, J. 2019. A number-based inventory of size-resolved black carbon particle emissions by global civil aviation. *Nature Communications*, **10**(1), 534.

Zhang, Z., and Ghadiri, M. 2002. Impact attrition of particulate solids. Part 2: Experimental work. *Chemical Engineering Science*, **57**(17), 3671–3686.

Ziff, R. M. 1980. Kinetics of polymerization. *Journal of Statistical Physics*, **23**(2), 241–263.

Ziff, R. M. 1991. New solutions to the fragmentation equation. *Journal of Physics A: General Physics*, **24**(12), 2821–2828.

Ziff, R. M., and McGrady, E. D. 1985. The kinetics of cluster fragmentation and depolymerisation. *Journal of Physics A: Mathematical and General*, **18**, 3027–3037.

Ziff, R. M., and McGrady, E. D. 1986. Kinetics of polymer degradation. *Macromolecules*, **19**(10), 2513–2519.

Index

For EU product safety concerns, contact us at Calle de José Abascal, 56–1°, 28003 Madrid, Spain or eugpsr@cambridge.org.

www.ingramcontent.com/pod-product-compliance
Lightning Source LLC
LaVergne TN
LVHW062205200926
848484LV00006B/1323
* 9 7 8 1 3 1 6 5 1 2 5 7 9 *